Life of the Past

THIRD EDITION

N. GARY LANE
Indiana University

Macmillan Publishing Company
New York

Maxwell Macmillan Canada
Toronto

Maxwell Macmillan International
New York   Oxford   Singapore   Sydney

Editor: Robert A. McConnin
Production Editor: Sheryl Glicker Langner
Art Coordinator: Peter A. Robison
Cover Designer: Russ Maselli
Production Buyer: Patricia A. Tonneman

This book was set in Meridien by Monotype Composition Company, Inc. and was printed and bound by Book Press, Inc., a Quebecor America Book Group Company. The cover was printed by Phoenix Color Corp.

Macmillan Publishing Company
866 Third Avenue
New York, New York 10022

Macmillan Publishing Company is part of the
Maxwell Communication Group of Companies.

Maxwell Macmillan Canada, Inc.
1200 Eglinton Avenue East, Suite 200
Don Mills, Ontario M3C 3N1

Library of Congress Cataloging-in-Publication Data
Lane, N. Gary.
    Life of the past / N. Gary Lane.—3rd ed.
        p.   cm.
    Includes index.
    ISBN 0-02-367405-9
    1. Paleontology.   I. Title.
QE711.2.L35   1992
560—dc20   91-18057
                                                                           CIP

Printing: 1 2 3 4 5 6 7 8 9 Year: 2 3 4 5

Preface

Think about the world we live in. We are surrounded by green—green plants. These green plants range from tiny woodland wildflowers to giant trees. They clothe the landscape. The vast majority of them, over 250,000 species, reproduce by flowers and bear seeds enclosed in fruits. The largest animals today, both on land and in the oceans, are all mammals—animals that have hair, are warm-blooded, suckle their young, and have live birth outside of a shelled egg. In the oceans the most abundant forms of life are thousands of species of clams and snails and hordes of bony fishes.

The fact is, the earth has not always been like this. Our modern world of life has only been in existence in its present form for about the last 60 million years, which is slightly more than 1 percent of earth history. Life on earth has changed continuously through time. No species has lived more than a few million years. That is what this book is about—the continuous succession of life that has inhabited the earth. *Life of the Past* concerns the history of life from the earliest records (which are over 3 billion years old) to the present. The evolutionary history of plants, and animals with or without backbones, as well as smaller, mostly one-celled life forms, is covered in the text.

If we go back far enough in time we come to an era when there were no green plants on dry land. Later, when they first appeared they did not have seeds, or flowers, or fruits but instead reproduced by simple spores. We can go back in time until ocean life was dominated by groups that are now rare or extinct. Clams and snails were few and far between. Fishes lacked jaws or teeth and were harmless mud-suckers. Other animals, relatives of squid and octopus, were the dominant predatory animals on earth. The vast majority of all species that have existed on earth are now extinct. Only a few million species are still alive today.

Instead of focusing on a specific group or age of fossils, *Life of the Past* emphasizes the complexity and diversity of life as revealed by fossils and the evolutionary and ecological development of organisms through time. The book introduces a minimum of terminology—only those terms necessary to understand basic geologic and biologic concepts are included.

Several recent developments in paleontology have been added to this third edition. A short section on the early development of paleontology is in the first chapter. Further amplification of cyclic phenomena and their role in creation of the ice ages and in extinction theory is provided. Recent discoveries of many dinosaur nests, eggs with embryos, and other features that relate to the paleobiology and social history of dinosaurs are also added.

The text is arranged in three parts. First, general principles of earth history and paleontology that are prerequisite to an understanding of life of the past are discussed. These principles include geologic time; the relationship of fossils to the rocks in which they occur; the origin of earth and life; the earliest Precambrian fossil record; and principles of evolution, continental drift, and paleobiogeography. These concepts provide the foundation on which the latter two parts are based.

The second segment consists of the record of life in the marine environment. The early history of fossils in marine rocks and the record of marine planktonic life, including primary producers, filter feeders, detritus feeders, and marine predators, are discussed. The role of these major groups in marine communities is emphasized, and most of the invertebrates are briefly described in this section.

The final part of the book concerns the history of life on land, the origin of land organisms, and the history of land plants and animals. Because most students are more interested in plants and vertebrates, these chapters contain a more detailed discussion of specific groups than do sections on ancient marine life.

Every attempt has been made to illustrate fossils with photographs of actual specimens. Dioramas, shaded sketches, and models often seem to have much less of a visual impact on students than do pictures of the "real thing." In a few instances, photographs of models have been used for purposes of clarity, or when pictures of specimens have been difficult to obtain.

Like any other special discipline, whether in the sciences or the humanities, paleontology has its own special vocabulary of technical terms. These words commonly capsulize important concepts and their precise use helps to sharpen our understanding of major ideas. Trying to remember many unfamiliar words taxes the memory and is a barrier to full appreciation of new knowledge. This book tries to alleviate the problem by keeping technical terms to a minimum, substituting common words for uncommon words where appropriate, and providing short lists of important key terms at the end of each chapter. Each key term is defined in the glossary at the end of the book. Some of the key term lists have been expanded in this edition.

A few pertinent references are included at the end of each chapter. These references are all annotated as to their level of difficulty or content. They have been updated with more recent works although older books are included simply because in many cases they are still the best sources of information.

ACKNOWLEDGMENTS

The following individuals read all or part of the manuscript for the first edition: Ronald L. Parsley, Robert L. Anstey, John C. Kraft, and Walter C. Sweet.

I would like to thank John H. Ostrom, Robert E. Sloan, Norman M. Savage, Lawrence H. Balthaser, and Stanley S. Beus, all of whom made numerous suggestions for the improvement of the second edition. I also thank Robert E. Sloan, John H. Ostrom, Norman M. Savage, and Richard H. Miller who carefully read a preliminary draft of the new chapter on primate and human evolution.

For the third edition, I would like to thank the following reviewers who provided valuable comments and suggestions: Lawrence H. Balthaser, California Poly & State University—San Luis Obispo; Allen J. Kihm, Minot State University; Norman Savage, University of Oregon; Daniel B. Blake, University of Illinois; William Miller III, Humboldt State University; Robert J. Foster, San Jose State University; Ronald E. Martin, University of Delaware; and Ronald L. Parsley, Tulane University.

Acknowledgment is due to my former paleontology teachers, including Robert W. Baxter, Harold K. Brooks, Raymond C. Moore, Frank Peabody, Peter P. Vaughn, and Robert W. Wilson. For several years I team-taught an introductory course in paleontology at U.C.L.A. with Clarence A. Hall, Jr. and J. William Schopf; I owe both of them a debt of gratitude for this association.

Those who aided in acquiring photographs include John A. Barron, Joyce R. Blueford, Derek Briggs, David Dilcher, J. Wyatt Durham, Thomas M. Gibson, Donald E. Hattin, Francis M. Hueber, Rebecca Lindsay, D.B. Macurda, Jr., W.G. Melton, Jr., Carl Rexroad, Eugene Richardson, George Ringer, Raymond T. Rye II, J. William Schopf, Takeo Susuki, Mary Wade, E. Reed Wicander, Edward C. Wilson, and Robert J. Zakrzewski.

Those who have helped especially in preparation of the third edition include Alan Horowitz, who read various segments of new text; Lois Heiser, librarian of the Indiana University Geology library; Mary LaReau, who prepared new text; and my wife, Mary, who helped with a variety of revisionary chores, just as she did with the first two editions.

Contents

1 TIME AND FOSSILS   1

Introduction   1
The Meaning of Fossils   2
The Beginnings of Paleontology   2
Geologic Time   6
   How Do We Tell Geologic Time?   7
   How Old Is the Earth?   13

2 THE ORGANIZATION OF LIFE   18

Kingdom Monera   19
Protistan Kingdom   21
Animal Kingdom   22
Plant Kingdom   28
   Algae   29
   Bryophytes and Fungi   31
   Tracheophytes   31

3   **ROCKS AND FOSSILS**   **33**

Fossilization   33
Preservation   35
Environments of Deposition   47
The Time-Space Problem in Paleontology   48
   Stratigraphy   48
   Facies   52

4   **ORIGINS OF THE EARTH, OCEANS,
ATMOSPHERE, AND LIFE**   **57**

Origin of the Earth   57
Origin of the Oceans   60
Origin of the Atmosphere and Life   61

5   **THE PRECAMBRIAN FOSSIL RECORD**   **66**

The Oldest Known Fossils   67
The Gunflint and Bitter Springs Fossils   69
Chemical Indicators   70
Banded Iron Formations   71
Evolution of Metaphytes and Metazoans   71
   The Ediacaran and Tommotian Faunas   72

6   **ORGANIC EVOLUTION AND EXTINCTION**   **77**

The Basis of Evolution   78
The Mechanism of Evolution   80
   Variation   80
   Natural Selection   81
   Isolation   83
Patterns of Evolution   85
   Adaptive Radiation   85
   Megaevolution   87
   Parallel and Convergent Evolution   89
Extinction   93
Cycles in Earth History   93
Evolution versus Creationism   96

7 CONTINENTAL DRIFT AND PLATE TECTONICS 99

Continental Drift 100
Sea-Floor Spreading 105
Plate Tectonics 109
The Effect of Continental Drift on Life 111

8 PALEOBIOGEOGRAPHY 115

Faunal and Floral Provinces 117
Migration, Dispersal, and Barriers 121
Refuges and Living Fossils 123
Thermometers of the Past 129
Older Paleoclimates 131

**9 ORIGIN AND EVOLUTION OF
MARINE COMMUNITIES** 133

The Fossil Record 133
  Precambrian Life 133
Base of the Cambrian 134
  Cambrian Life 136
  The Burgess Shale 139
  The Acquisition of Preservable Skeletons 139
The Structure of Marine Communities 142
  The Relationship of Organisms to the
    Sediment-Water Interface and the
    Water Air Interface 142
  The Trophic or Feeding Levels of Different
    Organisms 144
  Feeding Strategies of Heterotrophic
    Organisms 144
Changes in Dominance of Marine Communities 145

**10 THE FOSSIL RECORD OF PLANKTON
AND NEKTON** 150

Marine Phytoplankton 151
  Acritarchs 151
  Diatoms and Coccoliths 154

Marine Zooplankton   156
  Radiolarians and Foraminifera   156
  Graptolites   159

Marine Nekton   163
  Jellyfishes   163
  Conodonts   165

**11   HISTORY OF THE FILTER FEEDERS
AND DETRITUS FEEDERS   168**

Dominance among Small-Particle Feeders   169
  Brachiopods   170
  Bryozoans   174
  Stalked Echinoderms   176
  Sponges   179

Post-Paleozoic Filter Feeders   179
  Bivalves   180

Detritus Feeders   183
  Gastropods   183
  Other Mollusks   183
  Echinoids   185

12   MARINE PREDATORS   188

Corals   188
Cephalopods   193
Fishes   198
Later Predators   202
  Marine Reptiles   202
  Marine Mammals   203

**13   ORIGIN AND EARLY EVOLUTION OF
TERRESTRIAL COMMUNITIES   205**

Origin and Early Evolution of Land Plants   206
  The Transition from Water to Land   206
  The Fossil Record   208
  Changes in the Reproductive Cycle   210

Origin and Early Evolution of Animals on Land   211
   Terrestrial Invertebrates   211
   Terrestrial Vertebrates   212

14   **TERRESTRIAL PRIMARY PRODUCERS:
     THE LAND PLANTS**   218

Evolution of Leaves   219
Evolution of Vascular Tissue   221
Evolution of Reproductive Structures   221
   Silurian and Devonian Plants   224
   Coal-Swamp Plants   228
   Permian Plants   231

Plant Communities of the Mesozoic   232
   Evolution of Flowers   235

Plant Communities of the Cenozoic   238

15   **EARLY CONSUMERS ON LAND: REPTILES**   247

Diagnostic Features   247
Late Paleozoic Reptiles   249
Mesozoic Reptiles   253
   The Mammal-like Reptiles of the Triassic   253
   The Ruling Reptiles   254
   Warm-blooded Dinosaurs   262
   The Social Relationships of Dinosaurs   262
   How to Measure a Dinosaur   264
   The Extinction of Dinosaurs   266

The Fossil Record of Birds   268

16   **ADVANCED CONSUMERS ON LAND:
     THE MAMMALS**   273

The Earliest Mammals   273
   Marsupials and Placental Mammals   275

Cenozoic Mammals   275
   The Archaic Mammal Fauna   276
   Modern Mammal Fauna   281

Some Aspects of Special Interest   284
   Centers of Evolution   285
   South American Isolation   289
   Mammalian Extinctions   291
Comparisons and Contrasts   292

17   PRIMATE AND HUMAN EVOLUTION   295
The Primates   298
The Fossil Record of Primates   300
The Human Condition   301

18   CONCLUSIONS   304

GLOSSARY   307

INDEX   322

1
Time and Fossils

INTRODUCTION

This book is about **fossils**, which are the preserved remains of past life on earth, or indirect traces of such life. **Paleontology** is the study of such fossils. The fossils we will study are preserved in rocks that are of different ages—some quite young, others very ancient. The enclosing rocks provide a framework in time within which all fossils occur. We can think about fossils as occurring within a four-dimensional framework—three-dimensional space occupied by rocks on and under the surface of the earth, and one-dimensional time. Fossils thus occupy a four-dimensional space time continuum. In order to understand fossils one must also understand time, the subject of this chapter.

What is time? That is a difficult question to answer, and one that has occupied the minds of philosophers for centuries. A dictionary definition is as follows:

> That character and relation of all events and things with respect to which they are distinguished as simultaneous or successive, and as becoming, enduring, and passing away; usually conceived as a dimension of reality, distinguished from the spatial by the fact that the order of temporal succession is irreversible.*

We can judge from this definition that time is real, that it is filled or can be recognized by a series of events, and that it is irreversible. You and I can return to

---

* Webster's *New International Dictionary*, 3rd ed., s.v. "time."

a specific point in space on earth, but we can never return to a past moment of time.

Life on this planet is a multitudinous series of events that occur in the time dimension. Fossils are the record, preserved in rocks, of past events.

THE MEANING OF FOSSILS

The word fossil comes from the Latin *fossilium*, which means "dug up from beneath the surface of the ground." For example, a mole is spoken of as having a fossorial way of life. As originally used by medieval writers, a fossil was any stone, ore, mineral, or gem that came from an underground source. Some of the earliest books on mineralogy are called books of fossils. This broad meaning gradually was restricted in the eighteenth century to objects in rocks that are parts of once living organisms—bones, shells, leaves, wood, and so on.

For many years there was heated debate about the significance of fossils. Some believed that all fossils resulted from a single Noacian flood of Genesis. Others thought that fossils grew in place in the rock or had been placed there by Satan to betray humans. Fossils were found that clearly had been parts of plants or animals that were no longer living on earth. This raised a debate concerning the perfection of organic creation if some species had become extinct. Gradually fossils came to be generally accepted as records of ancient life.

Fossils demonstrate two truths about the planet on which we live. First, many species have existed and later became extinct. Second, there has been a succession of plants and animals through time; the communities of life that have existed on earth have gradually changed through time both on land and in the oceans.

THE BEGINNINGS OF PALEONTOLOGY

One of the most intriguing questions that we can ask about fossils is how early scholars came to realize that fossils are the remains of once-living plants and animals that became entombed in rocks. Certainly, this idea about the true nature of fossils did not come easily or quickly into Western thought. This section will deal mainly with developments in western Europe, mainly because we have the best-documented record of the progress of thought in this area. We will also briefly mention early ideas about fossils in China (Fig. 1.1).

The apparent earliest mention of fossils is in writings by early Greeks, perhaps as early as the sixth century B.C. A very early discussion of fossil fishes and marine shells found in the mountains of Greece is attributed to Xenophanes, but his writings on the subject have not survived and credit to him comes from a manuscript prepared 900 years later by Hippolytus.

Another early writer about fossils was Theophrastus, who lived about 300 B.C., who set the study of fossils on an erroneous path that took many centuries

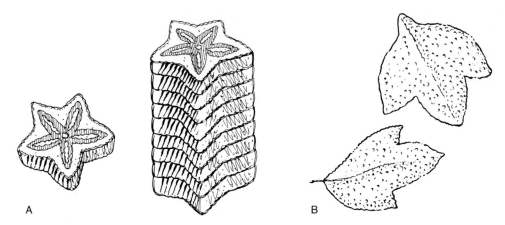

FIGURE 1.1
Two of the earliest known figures of fossils. A. Woodcut of Mesozoic pentagonal crinoid (sea lily) stems from a book by Conrad Gesner written in 1565. Although many earlier works mention fossils, this is the first known book in western Europe to provide illustrations. B. A sketch that is interpreted as two fossil brachiopods called stone swallows by the Chinese. The specimens may represent a Paleozoic spiriferid brachiopod. The illustration is taken from a Chinese book *Pên Tshao Kang Mu*, by Li Shih-Chen, printed in 1596, 31 years after Gesner's work.

to correct. Theophrastus was a student of Aristotle and believed in basic elemental virtues such as fire, water, and air. He attributed bones found in the rocks on the island of Lesbos to a plastic virtue, a characteristic of the rocks themselves that caused the bones to grow within the rocks, thus negating the idea that the bones may have belonged to creatures that were once alive quite independently of the rocks within which their remains are now found.

The Aristotelean point of view held unarguable force among learned men of Europe for many centuries. The naturalistic explanations of early Greeks gave way to mythical or magical explanations that had little relationship to reality. Scholars repeated the words of earlier writers without original thought or criticism. No one took the trouble to go out into the field and examine the rocks and fossils firsthand.

This situation persisted from about 200 A.D. to about 1400 A.D. and has been called the Great Interruption in Western thought. Medieval ideas were so fanciful and basically otherworldly that the very word *medieval* has come to stand for any ideas that are old-fashioned and out of date.

As Western thought emerged from these dark ages, we find that writing about fossils was very uneven in content and understanding. We can see one writer who seems to be on the right track followed a hundred years later by another writer who is still parroting Theophrastus, although in somewhat modified form. Here is an example.

One of the best-written descriptions of fossils during this early time was by Georg Bauer, a German who wrote under the pseudonym of Agricola in the early 1500s. He is known primarily for his works on mining, minerals and ores, and is sometimes called the father of mineralogy, but he also wrote quite detailed descriptions of a number of fossils, especially in one book, *De Natura Fossilium* (1546). You should not jump to the conclusion that this book really concerns fossils because Agricola is using *fossilium* in the original Latin sense of anything dug up from the earth. Fossils to him included stones, ores, crystals, gems, and any other objects obtained from within the ground. Some objects, such as fossil sharks' teeth, he thought had grown within the rocks. Other fossils he thought had been once alive and had been turned to rock, or petrified, by a stone-producing juice within the rocks. However, a hundred years later the Welshman Edward Lhuyd believed that marine fossils resulted from the eggs of sea creatures being washed up onto beaches, dried, then blown by the wind until they fell into crevices in the rock where they hatched and grew into stonified copies of their living counterparts. Thus, it took several hundred years for universal agreement to be reached that fossils were indeed the remains of organisms that were once alive.

Once everyone who was concerned with fossils did accept that these were the remains of organisms, the fossils were traditionally explained as being the results of plants and animals killed off and buried by the great biblical flood. This concept was widely accepted for many years and it resulted in two conclusions: first, that all fossils were of the same age, and second, that all fossils were of relatively recent age. Biblical scholars, typified by Bishop Ussher of Ireland (1664), counted up the number of generations represented in the genealogical chapters of the Old Testament and assigned a reasonable number of years to each generation. They carried this back all the way to Adam and Eve and assumed that the earth and all of its life had been formed just a few days before humans first appeared. This resulted in a calculation that the earth was formed in the year 4004 B.C., and for many years this number was printed in the margin of the King James edition of the Old Testament. Bishop Ussher later refined this calculation and decided that the earth was formed on October 29 in 4004 B.C. at 9:30 in the morning.

Geologists and paleontologists who began to seriously study rock strata and their contained fossils rather quickly realized that such a recent age for the earth was impossible and that the earth must be very old indeed to explain many of the phenomena that they observed. For instance, fossiliferous rocks may be several thousands of feet thick in any one locality; it is unlikely that a single flood event could have produced such a thick sequence of sediments. Rocks have also been deformed into high mountain ranges that were later destroyed by erosion; enormous amounts of time are necessary in order to wear down such lofty peaks. The entire assemblage of fossils that is found in lower (and presumably older) rock layers is completely different from the fossils found in higher (younger) rock layers; this evidence refutes the idea that all of these animals could have lived at the same time. The proponents of the flood as a major feature of earth history countered the rising skepticism of scientists with many ingenious arguments, and some aspects of their point of view still linger today among those who do not fully appreciate the fundamental principles of geology.

Once fossils became appreciated in a rational, modern way, a greatly accelerated rate of discoveries and advances became possible. The French anatomist George Cuvier was able by 1800 to compare the bones of extinct animals with those of modern ones and to reconstruct in detail what the extinct animals had looked like. About this same time the English canal engineer William Smith found that the fossils were distinctive and different in each rock layer, so that the fossils could be used to identify rocks over wide areas of England. He also prepared one of the very first maps showing the distribution of rock layers over the surface of southern England.

Smith's discoveries were quickly followed up by various British and continental geologists who demonstrated that distinctive assemblages of fossils could be recognized over most of western Europe in certain rock formations and signified that these formations were all of the same age. This research rather quickly spread to the United States where it was shown by the 1830s that specific rock sequences in New York State were the same age (that is, contained the same fossils) as rock sequences in Europe. Thus, the way was paved for unraveling the sequence of fossils on all of the continents and establishing a worldwide time scale for reconstructing earth history.

Much of the initial work on fossils was undertaken by persons who had little formal training in geology. Such courses were not part of university curricula. Merchants, doctors, and noblemen all took part in this work. However, as more and more fossils were gathered, the first natural history museums in Europe were created in the eighteenth century; here collections of fossils from many different areas could be assembled and studied. Gradually universities began to include geology and paleontology in their offerings.

In the United States much pioneer work on fossils was undertaken by a series of state-organized agencies, normally called state geological surveys, which started in the 1830s. The impetus for creation of these organizations was mainly economic—to find valuable coal or ore deposits—but much work on fossils was undertaken because of their value in matching up rocks of the same age in different areas. Almost the entire North American continent was unknown territory as far as fossils were concerned, and many important discoveries of new life forms were made. Even our third president, Thomas Jefferson, was intrigued with fossils. Jefferson was interested in many aspects of natural history and was sent a collection of large bones of extinct animals, including mastodons, from Big Bone Lick, Kentucky, on the Ohio River between Cincinnati and Louisville. Jefferson thought that these elephantlike animals might still be alive within the interior of the continent, and when he sent out the Lewis and Clark expedition in 1804, he cautioned them to keep an eye out for such animals.

Future federally sponsored expeditions to the West, beginning in 1850, resulted in many important discoveries of fossils. Most spectacular among these discoveries were the world-famous Mesozoic dinosaur beds of several western states and Canadian provinces and vast areas where fossil mammals, especially of Cenozoic age, were found.

This aspect of paleontology—exploration and discovery—is still an important and valuable part of the science. Many parts of the world's continents have not been adequately explored for fossils even to this day. We have only superficial knowledge of fossil assemblages in many parts of Central and South America. Alaska was poorly known until the oil exploration of recent years. And many parts of eastern Asia, especially China, still have not been adequately surveyed. We will surely continue to make important new discoveries of fossils in the years ahead.

GEOLOGIC TIME

Time is perhaps most comprehensible in terms of how we measure this dimension. Years, months, days, minutes, and seconds are all familiar concepts to us. We deal with them so often that we feel comfortable talking and thinking about them. But millions of years? Billions of years? We have a vague uneasiness in trying to grasp the enormous span of time expressed by such numbers. This is because such lengths of time are far beyond our own experience. We can grasp the idea of centuries in terms of lifespans, generations, and years; and even a few centuries or a few thousand years do not really make us shake our heads. But billions of years? We know that the earth is very old and that life has existed on this planet for a very long period of time. How are we to grasp the significance of such long intervals? One way is to study the events that have taken place during these intervals of time. By relating the succession of events to a scale of time measured in a familiar unit, such as years, we can gradually become accustomed to the idea of what we generally call the **geologic time scale**.

Geologic time differs from any other kind of time only with regard to its span, or duration. There is no essential difference between this kind of time and any other, except duration, and perhaps the ways in which we measure or estimate the duration of geologic time.

Geologic time is unique because it fills a special **interval** of time and has a unique **duration**. The historian is only concerned with time as far back as written historical records are preserved—a few thousand years at most. The archeologist or anthropologist is concerned with time only as far back as the records (bones and artifacts) of humans are preserved. For many years these scientists only dealt with the latest 1 million years of earth history, but recently human or humanlike fossils have been found in rocks as old as 3 million years, tripling the duration of the framework within which the study of humans is conducted. Astronomers, on the other hand, work with durations of time that are much longer than those usually thought of in geologic time. The origin and evolution of the universe involved lengths of time that are probably 10 to 20 billions of years. We do not know exactly how old the universe is, but it is clearly much older than our planet.

The geologist and paleontologist are concerned with an interval of time that is intermediate between the very long times with which the astronomer has to deal and the few millions of years, at most, with which the anthropologist works.

As we shall see, the earth is judged to have come into being about 4.5 billion years ago, although no rocks that old have ever been found. It is unlikely that we will ever find a rock on the earth's surface that was present during the initial formation of the planet. There have been too many changes in the earth's crust for such a rock to have a chance of survival. We have found rocks, however, that can be accurately and confidently dated as being 3.96 billion years old. We know that the earth must be at least that old. The fossil remains of very primitive organisms have been found that are 3.5 billion years old. Again, considering the changes that have taken place on the earth, it seems unlikely that evidence of life will be found in rocks that are very much older than 3 billion years, and we are probably quite lucky to have discovered the few scraps of information that we do have about this very early life. Thus, as far as a time scale for fossils is concerned, we begin about 3 billion years ago.

How close to the present day do we come with our time scale? We said at the start of this chapter that fossils are the preserved remains of past life, or ancient life. How old must a shell or a bone be in order to be judged a fossil? Unfortunately for the beginning student in search of neat, concise answers we cannot provide one to a question like this. A humorous, but wrong, answer is that if the organism no longer has any odor of decay (i.e., doesn't smell), then it is a fossil. Fossils may be only a few thousand years old, but generally not a few hundreds of years old. The remains of an animal that died and was buried and preserved during historical time is generally not considered to be a fossil, but rather the remains of a modern organism. Historical time differs from place to place, and again, no precise age can be given to set the upper limit of prehistory within which we study fossils.

How Do We Tell Geologic Time?

There are really two different answers to this question. We can measure the **absolute ages** of units of rocks in years by analyzing naturally occurring radioactive elements that are found in certain rocks and minerals in minute quantities. This method of radioactive age dating is comparatively new, having been started in the early 1900s and expanded and refined since then. Long before this method was developed, geologists and paleontologists had worked out ways to determine the **relative ages** of different rocks (Rock A is older or younger than Rock B) and had also observed several phenomena that were used to make estimates about the age of the earth and some of the rocks that occur in the earth's crust. Here we will trace these developments historically, beginning with early attempts to determine the relative ages of rocks and fossils.

Sedimentary rocks that contain fossils are usually deposited as horizontal layers of different kinds of rocks: sandstones, shales, limestones, and others (Figure

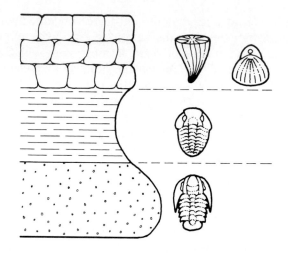

FIGURE 1.2
A vertical sequence of sedimentary rocks with rock types indicated by conventional symbols. A sandstone (stipple pattern) is the oldest (lowest) rock layer; a shale (horizontal lines) is intermediate in age; and a limestone (vertical lines) is the youngest. Fossils found in each layer are shown at the right. Superimposition indicates that the sandstone and shale with trilobites are of a different, older age than the limestone with horn corals and brachiopods.

1.2). These layers form a succession of rock types in any given area. The rock layers are stacked up like a deck of cards, one on top of the other. Scientists have realized for several centuries that the rock layer at the bottom of the "deck" is the oldest, first-formed layer and that the layer at the top is the youngest. This principle is called **superposition** and simply says that in an undisturbed sequence of layered rocks an older layer is always below a younger layer. This principle was understood and applied in western Europe in the 1600s and 1700s before any studies of rocks were undertaken in North America. Geologists gradually came to realize two things: that certain distinctive kinds of rocks were confined to specific parts of the rock sequence, and that different layers of rocks commonly contained unique suites of fossils. These early studies had an economic motive, such as coal production. European geologists studying coal-bearing rocks found that beds of coal could not be located above or below a certain sequence of rocks. This sequence of rocks, which also contained shales, sandstones, and a few thin limestone beds, came to be called the **Carboniferous**, after the carbon of which coal is mainly composed. Other rocks in England and France that included thick beds of chalk were eventually called the **Cretaceous**, coined from the Latin word for chalk. These early geologists realized that beds of chalk would not be found above or below a certain rock interval and that coal beds were found below (were older than) those that contained chalk. As the sequence of rock layers was gradually worked out over larger and larger areas, including parts of France, Germany, and Great Britain, it was learned

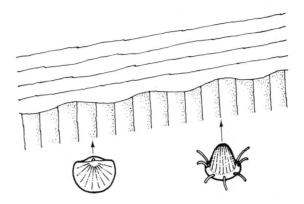

FIGURE 1.3
The rock layers below are standing vertically. They were originally horizontal sandstones that later were tilted and eroded during mountain building. Because they are vertical, it is difficult to tell which layers are younger and which are older. The two kinds of fossil brachiopods shown were found in the layers indicated. We know from study of rocks of the same age that are still horizontal that the spiny brachiopod on the right is younger than the nonspinose one on the left. Therefore, the top of the section of vertical rock layers is to the right. The younger, slightly tilted rocks above the vertical layers are clearly younger than vertical beds, having been deposited on top of them. The surface between the two sets of rock layers records an interval of time when mountain building and erosion took place and no rock layers were laid down. There is a gap in the fossil record at this site, and the surface separating the two sets of rocks is called an unconformity.

that certain kinds of fossils were confined to certain rock horizons, and this observation was used to piece together observations from scattered rock outcrops over increasingly larger areas.

Not all of the sedimentary rocks so studied were still in their original horizontal positions. In some areas the rocks had been folded, broken, and displaced during mountain building (Figure 1.3). Some layers were distorted until they were upside down. Fossils came to be increasingly useful in unraveling these complex disturbances of the earth's crust. In other areas it was found that thick sequences of rocks were missing; older rocks with distinctive fossils were found next below much younger rocks. These layers might have been found elsewhere to be separated by several hundreds or thousands of feet of fossil-bearing sedimentary beds. As these complex relationships were unraveled, a **relative time scale** gradually came into being. It is relative because it says only that any object or event is older or younger than something else. Fossil-bearing rocks in western Europe, by the 1860s, were divided into three great **eras** of geologic time: the **Paleozoic, Mesozoic,** and **Cenozoic** (Table 1.1). These names mean respectively, "ancient life," "middle life," and "young or recent life." The Paleozoic was further divided into six **periods** of time. The oldest three, **Cambrian, Ordovician,** and **Silurian,** were all named from rocks in Wales. Cambria is the ancient Roman name for Wales, and the Ordovices

TABLE 1.1
The geologic time scale. (ages in millions of years)

Eras of Time	Periods of Time	Epochs of Time	Age (millions of years)	Duration (millions of years)	Major Biological Events
C E N O Z O I C	Quaternary	Recent	0.01	1.6	Extinction of large land mammals in northern hemisphere and South America. Evolution of human beings. Rapid shifts in marine and terrestrial communities in response to four major glaciations.
		Pleistocene	1.6		
	Tertiary	Pliocene	5	64	Extensive radiation of flowering plants. Extensive radiation and evolution of mammals. Co-evolution of insects and flowering plants. Dominance of gastropods and pelecypods in the oceans.
		Miocene	24		
		Oligocene	37		
		Eocene	58		
		Paleocene	66		
M E S O Z O I C	Cretaceous		144	78	First flowering plants. Extinction of terrestrial, marine, and aerial reptiles. Extinction of ammonoid cephalopods. Radiation of primitive mammals.
	Jurassic		208	64	Gymnosperms (cycads, conifers, ginkgos), ammonoid cephalopods, and dinosaurs dominant. Radiation of marine reptiles; first birds; flying reptiles.
	Triassic		245	37	Depauperate marine faunas; dominance of ammonoid cephalopods and mammal-like reptiles. Origin of mammals and dinosaurs.

Era	Period	Subdivision	Age (millions of years)	Duration (millions of years)	Life
PALEOZOIC	Permian		245	41	Extinction of trilobites, blastoids, many other marine invertebrates. Dominance of mammal-like reptiles. Decline of amphibians. Evolution of fusulinids.
	Pennsylvanian		286	34	Origin of reptiles. Evolution of fusulinids. Algal-sponge reefs and banks. Extensive coal-swamp forests. Many primitive insects.
	Mississippian		320	40	Echinoderms and bryozoans dominant in the oceans. Amphibians on land. First appearance of coal-swamp forests.
	Devonian		360	48	Extinction of many marine groups. Oldest land vertebrates. Many corals, brachiopods, and echinoderms. Extensive radiation of land plants and fishes.
	Silurian		408	30	Oldest land life: land plants, scorpions, and wingless insects.
	Ordovician		438	67	First diverse marine communities. Dominance of brachiopods, bryozoans, corals, graptolites, and nautiloid cephalopods.
	Cambrian	Tommotian	505 / 570	65	First vertebrates (jawless fishes). First metazoans with skeletons. Dominance of trilobites. Marine faunas of low diversity. No known land life.
PRE-CAMBRIAN	Proterozoic	Ediacaran	570–2500	3920	Origin of life, prokaryotes, eukaryotes, and metazoa.
	Archean		2500–4500		

11

and Silures were two of the early Celtic tribes of this area. **Devonian** is next youngest and named for Devonshire, England. The **Carboniferous** is divided into the Mississippian and Pennsylvanian periods in North America, after the rocks of the upper Mississippi River and those of western Pennsylvania, but this terminology has never been accepted worldwide. The final and youngest period in the Paleozoic, the **Permian**, is named for the town of Perm, just west of the Ural Mountains in Russia.

The Mesozoic era is divided into three periods beginning with the **Triassic**, so named in Germany where rocks of this age occur in a three-fold, or tri-fold, sequence based on color: the Red, White, and Brown Trias. The **Jurassic** is named from the Jura Mountains of northeastern France and adjacent Germany and Switzerland, and the chalk-bearing **Cretaceous** forms the top of the Mesozoic sequence. The Cenozoic Era is divided into two periods, the **Tertiary** and **Quaternary**. The Tertiary is a holdover from a very old eighteenth-century scheme in which all rocks were primary, secondary, and so on, although the latter two terms were soon abandoned. The Tertiary Period is further divided into **epochs** of time: **Paleocene, Eocene, Oligocene, Miocene**, and **Pliocene**. These epochs were initially designated on the basis of the percentage of extinct species of mollusks that rocks of these ages contained. Thus, the Eocene contained clams and snails, 96 percent of which were thought to be extinct, and the youngest Pliocene epoch included rocks in which 90 percent of the species found are still alive. This method was abandoned, however, when it was discovered that many of the mollusks were not extinct but had only been locally exterminated in western Europe. With changes in climate (generally a shift toward cooler conditions during the Tertiary), many of the animals had migrated southward and were found to be alive and well off the west coast of Africa. The Cenozoic Era is divided into the previously mentioned Tertiary and Quaternary, two periods of quite unequal duration. The Tertiary Period ranges from 66 to about 2 million years ago. The last 2 million years of earth history is called the Quaternary, which, like the Tertiary, is a holdover from an archaic scheme of terminology, in this case meaning the fourth series of rocks. The Quaternary is divided into the **Pleistocene**, commonly called the Ice Age, and the **Recent**, encompassing the time since the last major retreat of larger glaciers. In North America there were four major advances and retreats of ice during the Pleistocene, and five in Europe. By the 1880s the relative geologic time scale had been developed to its present state. Rocks older than the Cambrian period that were thought to lack fossils were grouped together into the **Precambrian** (an older name is **Azoic**, meaning "without life." The Precambrian is divided into the Archean and the Proterozoic.

Now that we have briefly reviewed the historical development of the time scale, the names and sequence should not seem as meaningless and arbitrary as they do at first glance. It took scores of geologists working over several lifetimes to fit together the jigsaw puzzle of rocks exposed on the earth's surface into a meaningful and coherent time sequence. And they did it with no clear conception of the ages of the rocks with which they worked.

FIGURE 1.4
Different kinds of clocks.
Those used to measure the
age of the earth, above; those
used for ordinary time,
below.

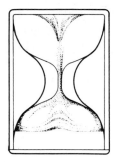

Hourglass clocks:

    rate of cooling of earth's surface;

    rate of salt accumulation in the
      oceans;

    rate of sediment accumulation in
      the oceans.

Early attempts to estimate the age of the earth were based on irreversible
processes, here called hourglass clocks.

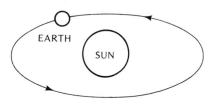

Rotational clocks:

    revolution of the earth
      about the sun;

    rotation of the earth
      on its axis.

Clocks for measuring time in ordinary terms are based on
regularly repeating phenomena that recur at definite
intervals of time.

How Old Is the Earth?

While geologists were untangling the rock record and reaching a decision that
Cambrian rocks were the oldest ones that contained fossils, other scientists were
attempting to solve the riddle of just how old the earth really was. They were
trying to find some kind of natural, reliable clock that they could use to measure
time. Our common household clocks and calendars are based on periodic, regularly
recurring events: the rotation of the earth on its axis and the rotation of the earth
around the sun. But no such regularly recurring events were known that would
reach far enough back in time to be useful for measuring the age of the earth.
Instead, scientists attempted to find some one-way, irreversible series of events
that could be calibrated in terms of years (Figure 1.4). One such attempt was made
in the 1880s by the English scientist Lord Kelvin, who was primarily a physicist
rather than a geologist. He assumed that the earth had originally been a molten
mass and had cooled and solidified to its present state. He was able to measure
the heat flow of rocks at the earth's surface as about 40 calories per year per square
centimeter. He also knew that the temperature of the earth increases with depth,
in deep mines and wells, in the amount of about 2 centigrade degrees per 100
meters. By measuring the heat conductivity of rocks with knowledge of the
temperatures at which commonly occurring rocks will melt, he came up with the

estimate that from 20 to 40 million years ago the earth's surface would have been too hot to have supported life. Kelvin's dating method was ingenious, but we now know that his estimate was too low by far. Why was he so in error? The principal mistake in his calculations is based on the fact (of which he could not have been aware) that much of the heat generated within the earth is not residual heat left over from a time when the earth was molten, but rather is heat generated by decay of radioactive elements within the deep layers of the earth. This heat flow is and has been relatively constant over very long periods of time.

A second attempt along this line was by an Irish geologist, John Joly, in 1899. He reasoned that earth's ocean waters were fresh when first formed and that the salt content, specifically the amount of sodium ions dissolved in ocean waters, was the result of these dissolved substances being carried into ocean basins by rivers. By measuring the amounts of these ions in river waters, estimating the total flow of river waters and the total salt content of the ocean, he theorized that it took 90 million years for the oceans to acquire their present saltiness. Like Kelvin's, Joly's estimate was far too low. Again, the error arose simply from lack of sufficient information. We now know that cubic miles of salt have been precipitated from the oceans during earlier geologic times. Many of these thick salt beds are buried deep in the earth's crust and have only been discovered in recent decades through drilling of oil wells. Thus, much of the sodium that was brought down by rivers was later precipitated and removed from the oceans, causing Joly to seriously underestimate the age of the earth.

Several geologists made yet a third try to establish the earth's age, utilizing known rates of sediment accumulation. They reasoned that if we take the thickest known section of rock of each geological period of time from the Cambrian to the Quaternary and add them all up, we arrive at a thickness of 30 to 100 kilometers. If we then assume that the sediment that formed all of these rocks accumulated at a rate of about 1 centimeter of sediment per year, we are in a position to establish an estimate of the age of the earth, at least back to the beginning of the Cambrian. The simple computation of this estimate will be left to the student, who will readily find it, again, to be far too low. What is wrong this time? The answer is that there has been no place on earth where sediment has accumulated continuously at a steady rate over such long periods of geologic time. Even where rocks of a geological period are thickest, there have been episodes of erosion of these rocks or times when sediment simply did not accumulate. Thus, estimates based on this method are certain to be too low.

Such was the state of the art of setting the age of the earth until this century. However, things changed when **radioactivity** was first discovered by the French physicist Henri Bequerel in 1895. It took only a few years for scientists to realize that the natural decay of certain unstable, or radioactive, elements offered a method for determining the age of the rocks that contained the elements. The first rock date was published by Sir Ernest Rutherford, a Canadian, in 1905. As analytical methods were refined, additional absolute rock dates accumulated until finally in

1930 the first true time scale for geology, in millions of years, was published by Sir Arthur Holmes of Scotland. Geochronologists continue to refine these dates as new techniques and instruments become available.

All of the early dates were determined using the radioactive element uranium. Uranium occurs naturally in rocks only in radioactive form; that is, the nuclei of the atoms of this element are unstable. They contain excess energy that is released in an immutable fashion completely independent of any surrounding atoms, or of temperature or pressure. As this energy is released, the character of the uranium nucleus is altered and the radioactive **parent element** is said to decay to a **daughter element**. Ultimately a stable nuclear configuration is reached, and, in the case of uranium, the nuclei go through a complex series of changes and are finally naturally converted to a stable form of the element, lead.

The lead so produced has a distinctive number of particles in the nucleus that have no electric charge, known as **neutrons**. This number of neutrons is different from the number of neutrons in atoms of lead that are not produced by radioactive decay. Thus, there are several forms of lead atoms called **isotopes,** all of which have the same number of protons (positively charged particles) and electrons (negatively charged particles) but which differ in the number of neutral particles in the nucleus. These forms of lead vary in weight because of the differing numbers of neutrons, and they are spoken of as having different **mass numbers**.

Radioactive uranium also occurs as different isotopes with different mass numbers. Two isotopes of uranium that occur in nature are U^{238} and U^{235}. Each of these isotopes decays to a different isotope of lead, Pb^{206} and Pb^{207}, respectively. Ordinary nonradioactive lead is Pb^{204}. This decay of uranium is known to occur at a constant rate, so that if one started with a given quantity of radioactive uranium 238, for instance, and that uranium produced, through decay, a known amount of lead 206, one could readily calculate the length of time it took for that amount of uranium 238 to produce that quantity of lead 206. The rate of decay of a radioactive element is expressed in what is called a half-life. A **half-life** is the amount of time needed for one-half of the atoms of the radioactive element to decay. The other atoms are still radioactive, but within another half-life, one-half of those atoms will also decay, so that in two half-lives, only one-quarter of the original atoms will remain. Some radioactive elements have extremely short half-lives—a millionth of a second, for instance. Others have very long half-lives, like uranium 238, with a half-life of 4.5 billion years (Table 1.2). Small amounts of naturally occurring uranium were found in rocks and dated using this technique. It became immediately obvious that the dates obtained were very much older than any of the earlier guesstimates.

All of the older dates obtained from radioactive materials used **lead-uranium**. Thus, the number of available dates grew slowly as uranium minerals are rare. Later, other minerals, such as zircon, which may contain only a few parts per million of uranium, were used as new and more delicate instruments were designed to measure the radioactivity. The number of dated rock samples is now very large,

TABLE 1.2
Naturally occurring radioactive elements used in dating rocks.

Parent	Daughter	Half-life in years
Uranium 235	Lead 207	700×10^6 (1 million)
Uranium 238	Lead 206	4500×10^6
Thorium 232	Lead 208	$14{,}000 \times 10^6$
Potassium 40	Argon 40	1300×10^6
Rubidium 87	Strontium 87	$50{,}000 \times 10^6$
Carbon 14	Nitrogen	5730

especially since other naturally occurring parent-daughter combinations were discovered, such as **potassium-argon** and **rubidium-strontium.**

The result of all this age-dating is that we are now able to tie the old relative geologic time scale of eras and periods to an absolute scale of years based on radioactive dates. In order to do this accurately, an igneous rock layer that is being dated radioactively must be related to fossil-bearing beds just above or below. We now know that the era boundary between the Cenozoic and Mesozoic corresponds to about 66 million years ago, sometimes abbreviated as 66 MYP (million years before present) or Ma (mega annum). The Paleozoic-Mesozoic boundary is about 245 million years old and the base of the Cambrian is about 570 million years old. The periods of the Paleozoic range from 65 million years in duration for the Cambrian to about 30 million years for the Silurian. The Mesozoic periods are of about the same order of magnitude in years. The oldest rock so far dated is about 3.96 billion years old; thus the Precambrian ranges from this age to about 570 million years, covering over three-fourths of known earth history.

KEY TERMS

absolute age	fossil	Oligocene
Archean	geologic time scale	Ordovician
Cambrian	half-life	Paleocene
Carboniferous	isotope	paleontology
Cenozoic	Jurassic	Paleozoic
Cretaceous	lead-uranium	Pennsylvanian
Devonian	mass number	period
element	Mesozoic	Permian
Eocene	Miocene	Pleistocene
epoch	Mississippian	Pliocene
era	neutron	potassium-argon

Precambrian	Recent	superposition
Proterozoic	relative time	Tertiary
Quaternary	rubidium-strontium	Triassic
radioactivity	Silurian	

READINGS

All of the modern introductory texts in physical geology and in historical geology include chapters on the geologic time scale and radioactive dating. Most cover the details of radioactive decay series in more detail than is possible in a general paleontology book like this. You are urged to consult one or more of these texts for additional information. Five books, all paperback, that specifically deal with time are the following:

Berry, W.B.N. 1968. *Growth of a Prehistoric Time Scale Based on Organic Evolution*. Freeman. 158 pages. Primarily historical in treatment, this book traces the growth and development of the relative time scale in considerable detail. Radioactive dating is not covered.

Eicher, D.L. 1968. *Geologic Time*. Foundations of Earth Science Series. Prentice-Hall. 150 pages. This book includes discussion of the historical development of the time scale, geologic time in relation to stratigraphy, the study of rock strata, and radioactive dating.

Harland, W.B., and others. 1990. *A Geologic Time Scale*. Cambridge Univ. Press. 131 pages. A comprehensive reference book that details all of the modern geologic time scales now in use.

Needham, J. 1954. *Science and Civilization in China*, Vol. 3. Cambridge Univ. Press. 620 pages. The authoritative study on science in ancient China. Volume 3 contains information on Chinese studies of fossils.

Plants and Animals of Long Ago (filmstrip). 1986. National Geographic Society. 2 film strips (101 frames). Developed for use in primary schools through grade 4.

Toulmin, S., and Goodfield, J. 1965. *The Discovery of Time*. The Science Library. Harper Torchbooks. 279 pages. Written by two historians of science, this book examines gradual recognition of time as an entity and covers early geological and biological discoveries that expanded our concept of time. Very well written and an important book in the history of science.

2

The Organization of Life

Life on this planet is today, and has been in the past, incredibly diverse. Somewhere between 1.5 and 5 million living species have been described, and more species are being named every day. It is impossible for any person to keep up with this enormous body of literature and to understand all species in any detail. If we add to the large number of living organisms the many kinds of extinct fossil species that have been found, estimated at 225,000, the number is just that much more overwhelming. In order to make sense of this complexity, we need some ordering principles to arrange the seeming chaos of named species into a few major categories that will reduce the complexity to a level that can be more readily grasped. This is one of the purposes of the classification of life.

Some of the categories of life are well known and easily understood by everyone—plants and animals, insects, reptiles, ferns, flowering plants, and bacteria—although many may be hard-pressed to precisely define each of them. What is the basis for such classifications? The answer is **evolution**. Each unit of classification, large or small, is judged to contain organisms that are more closely related to each other in an evolutionary, ancestor-descendant sense than they are to organisms contained in an equal but different unit of classification. For example, two kingdoms of organisms that are commonly recognized are the plant and animal kingdoms. All the organisms included in the plant kingdom are more closely allied to each other than to any organisms that we may place in the animal kingdom.

In order to make the evolutionary relationships among organisms explicit, an ordered ranking of categories is used; small categories denote a small series of quite closely related organisms, and increasingly more comprehensive, and thus larger, categories include more and more organisms that are more and more remotely related. Such a scheme is called a **hierarchy. Kingdoms** of life are the largest, most comprehensive, and broadest units in the hierarchy of life. The standard scheme for the organization of categories is as follows:

Kingdom
    Phylum
        Class
            Order
                Family
                    Genus
                        Species

The **species** is the fundamental, underlying unit in this hierarchy. A series of interbreeding populations that share anatomical traits are included in a single species. All of the species that are more closely related to each other than they are to other species are grouped together into a **genus**. The genera (plural of *genus*) in turn are grouped into a series of families, and so on, in increasingly comprehensive categories up to **phylum** and kingdom. For example, consider the common domestic cat. The scientific name of the species is *Felis domesticus*. The first name is that of the genus, which includes other species, such as the bobcat, lynx, and mountain lion. These are all species of the genus *Felis*. This genus is included in the **family** Felidae, which contains other genera of cats (e.g., *Smilodon*, the extinct saber-tooth cat). This family, along with others, is encompassed in the **order** Carnivora, the flesh-eating or carnivorous mammals. This order and many others are grouped together into the **class** Mammalia, those animals that have hair and suckle their young. Mammals, along with reptiles, fishes, birds, and others, are grouped together into the phylum Chordata, all members of which have spinal cords running along their backs. This phylum and others are then classed in the kingdom Animalia, consisting of life forms that are typically large, multicellular, and not able to manufacture their own food. This kingdom contrasts with the kingdom Plantae, which also contains large, multicellular organisms, but these can generally make their own food through photosynthesis. A third kingdom, the Protista, is made up of small organisms, either one-celled or simple aggregates of cells, that may or may not be photosynthetic. Thus we have three of the four broadest categories of life that are commonly recognized today, all members of which have a cartilaginous notochord running along their backs (Figure 2.1).

KINGDOM MONERA

The fourth kingdom of life is the kingdom Monera. It contains the simplest and least highly organized kinds of life. These consist of the bacteria and the

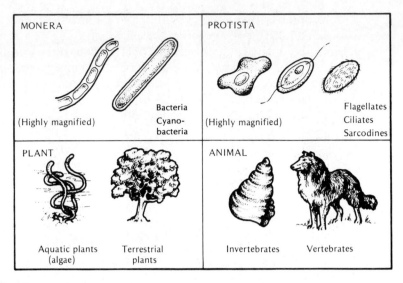

FIGURE 2.1
The four kingdoms into which all life is divided. Only the monerans are prokaryotic; the other three kingdoms are eukaryotic.

cyanobacteria (called blue-green algae in older books), which differ from all other life in that they lack a well-differentiated complex nucleus within the cell. The DNA that is present is not separated from the rest of the cell contents by a nuclear membrane (Figure 2.2). The **monera** are so distinctive that they are commonly set aside from all other life in what might be called a superkingdom. Thus we recognize a fundamental dichotomy of all life; the two main groups are

1. **Prokaryotes:** meaning "before a nucleus" and including the bacteria and cyanobacteria.
2. **Eukaryotes:** meaning "true nucleus" and including all other life.

FIGURE 2.2
The basic dichotomy of life. Prokaryotes are the smallest and simplest forms of life. They lack an organized nucleus, DNA is not organized into chromosomes, and the cells are quite small. All other life is eukaryotic, with a nucleus surrounded by a membrane.

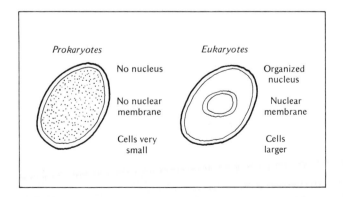

FIGURE 2.3
A strew slide of fossil radio-larians, a common form of protist preserved as fossils. The specimens are Miocene in age from the southwest Pacific and highly magnified. (Courtesy of Joyce R. Blueford, U.S. Geological Survey.)

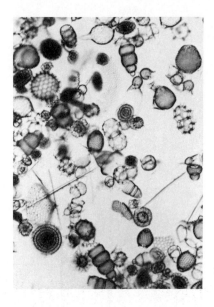

The two groups will be treated here as informal divisions. Thus, at the highest levels, the basic classification of life looks like this:

Prokaryotes
    Kingdom Monera

Eukaryotes
    Kingdom Protista
    Kingdom Animalia
    Kingdom Plantae

At this point we need not concern ourselves further with the monerans. They consist of two phyla (plural of *phylum*) only: one containing the bacteria and the other, the cyanobacteria.

PROTISTAN KINGDOM

The kingdom Protista, or protists, consists of several phyla that are composed of mainly one-celled organisms (Figure 2.3). Some of these are photosynthetic and some are not. There are three main groups (phyla) based primarily on the methods by which the organisms move about. One group is characterized by a whiplike flagellum that beats rapidly for locomotion. In another group the organisms are covered with small, hairlike cilia that beat for movement. These are called, respectively, the Flagellata and the Ciliata. A third group includes the common amoeba which moves by slow extensions of the cell. This phylum is called the Sarcodina.

FIGURE 2.4
Levels of organization in the
animal kingdom.

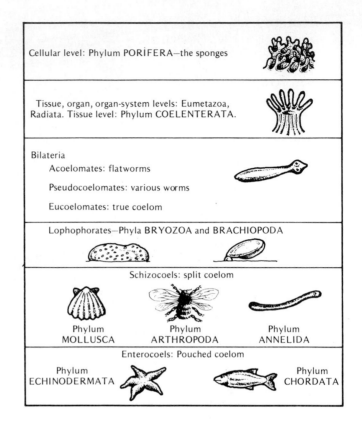

Cellular level: Phylum PORIFERA—the sponges

Tissue, organ, organ-system levels: Eumetazoa,
Radiata. Tissue level: Phylum COELENTERATA.

Bilateria
    Acoelomates: flatworms

    Pseudocoelomates: various worms

    Eucoelomates: true coelom

Lophophorates—Phyla BRYOZOA and BRACHIOPODA

Schizocoels: split coelom

Phylum
MOLLUSCA

Phylum
ARTHROPODA

Phylum
ANNELIDA

Enterocoels: Pouched coelom

Phylum
ECHINODERMATA

Phylum
CHORDATA

ANIMAL KINGDOM

Kingdom Animalia is very diverse; it includes, for example, both sponges and man.
Thus there is a wide range in the complexity of organization of the various
organisms contained within this kingdom. In order to explain how the major
groups of **animals** are related to one another, it is necessary to introduce
subdivisions between the phylum and kingdom levels. Several features of animals
are used to divide these levels into major groups. First is the level of organization
of the cells that make up the body (Figure 2.4). The cells may be largely independent
of each other and not much differentiated or specialized. This is called a **cellular
grade of organization**. All of the protistans have this low level of organization
and, among animals, the sponges are at or near a cellular level. Next more complex
is a **tissue grade of organization** in which groups of cells are organized into tissues
for specific functions. The coelenterates (corals, jellyfish) are at this level. Groups
of different kinds of tissues may be arranged into organs and these organs into
large, complex organ systems, such as the digestive, nervous, and excretory systems
of higher animals. All animals above the coelenterates are at the **organ** or **organ
system level of organization**.

A second characteristic that is used to arrange the animal phyla is symmetry. If an animal body can be divided into two equal halves in only one way it is said to have **bilateral symmetry**. Your body and that of all other vertebrate animals is bilateral. If several planes can bypass through the body, each of which divides the organism into two equal parts, the animal is said to have **radial symmetry**. Many corals and echinoderms have radial symmetry. If an indefinite number of planes provide two equal parts, this is called **spherical symmetry**, and, finally, if no plane of symmetry exists, the animal is said to be **asymmetric**. Generally speaking, those animals that are sessile (attached or unable to move about) are asymmetric or have spherical or radial symmetry. Those animals that are motile (able to move about) generally have bilateral symmetry. There are numerous exceptions to this rule.

A third characteristic has to do with the number of primordial layers of cells that are present in early larval stages. These **dermal layers** during growth develop into the various tissues and organs of animals. The simplest animals lack differentiated dermal layers. More advanced ones may have an outer **ectoderm** (outer skin) and an **endoderm** (inner skin). Still more advanced forms have a third, middle layer, the **mesoderm**.

A fourth important characteristic has to do with the presence or absence of an internal fluid-filled cavity in the body, called a **coelom**, and how that coelom is formed. The coelom forms from the mesoderm in advanced animals and is lacking in less complex forms. A **pseudocoelom**, not completely formed by mesoderm, is found in some animals. Finally, segmentation is important. This is the dividing up of an animal, from head to tail, into a series of sections that repeat important body parts.

These fundamental features of animals are used to formulate a classification of animals between the phylum and kingdom levels (Table 2.1). For the sake of simplicity we will avoid giving names to the rank names of the various categories, although **subkingdom, grade**, and **superphylum** have been used for these.

The animal kingdom is first divided into two basic categories, the Parazoa and the Eumetazoa. The former includes the sponges, which are at a cellular or incipient tissue grade of organization (Figure 2.5). All of the other animals are true (eu-) **metazoan**, or multicelled, animals. The eumetazoans are next divided into two major groups based primarily on symmetry: the Radiata and the Bilateria. The former consists primarily of the coelenterates (corals, jellyfish, and sea anemones), which have a radial symmetry. Virtually all other animals have bilateral symmetry and are, by and large, active animals with a clearly distinguishable head and tail. The radially symmetrical echinoderms are an exception to this symmetry rule, and many of them, like the coelenterates, are sessile animals fixed to the sea floor.

The bilateral animals are next divided into three groups based on the presence, absence, or degree of development of a fluid-filled cavity, or coelom, within the body. Bilateral animals without a coelom (acoelomate) include the flatworms and tapeworms and are not important as fossils. Several phyla are pseudocoelomate (having a false coelom). The cavity between the gut and the body wall is only

TABLE 2.1.
The animal kingdom, to the phylum level.

Parazoa: Cellular grade of organization. Symmetry radial, spherical, or absent.

   Phylum Porifera: The sponges. Skeleton of spicules.

Eumetazoa: Tissue, or organ, or organ system grade of organization.

   Radiata: Primarily radial symmetry. Tissue grade of organization. Mesoderm incipient, basically two germ layers.

      Phylum Coelenterata or **Cnidaria:** The corals, jellyfish, sea anemones.

   Bilateria: Symmetry bilateral and some with radial symmetry (the echinoderms). Mostly with mesoderm, and a coelom is developed in most.

Acoelomata: No body cavity. Includes the flatworms and ribbon worms that are not important in the fossil record.

Pseudocoelomata: Body cavities partly lined with mesoderm. Includes several phyla of worms and rotifers not important in the fossil record.

Eucoelomata: Body cavity a true coelom.

   Lophophorata: Two phlya characterized by a special circular structure around the mouth with ciliated tentacles for food gathering, called a lophophore.

      Phylum Brachiopoda: Skeleton of two valves. Very important fossils.

      Phylum Bryozoa: The moss animals. All colonial, most with a skeleton. Important fossils.

FIGURE 2.5
A large slab on which are preserved the siliceous skeletons of many Devonian glass sponges, *Dichtyophyton*, from New York. The actual spicules have been dissolved or replaced by pyrite (iron sulfide). The slab is about 1 m high. (Courtesy of Field Museum of Natural History, Chicago.)

TABLE 2.1 *continued*
The animal kingdom, to the phylum level.

Schizocoela: Coelom originates from a split in the mesoderm.

Phylum Mollusca: Unsegmented except for one small primitive group (Monoplacophora). Commonly bilaterally symmetrical. Mostly with calcareous exoskeleton secreted by a mantle that surrounds vital organs. Includes the clams (Class Pelecypoda or Bivalvia); the snails (Class Gastropoda); and the cephalopods, as well as other, less important, classes.

Phylum Annelida: The annelid worms. Segmented, with a round, elongate body; lack jointed appendages. Tiny chitinous jaws (scolecodonts) found as fossils as well as rare, carbon-film impressions of body.

Phylum Arthropoda: The jointed-leg animals; conspicuously segmented. Includes crabs, shrimp, barnacles, spiders, and insects. The extinct trilobites are the most conspicuous fossil group.

Enterocoela: Mesoderm-lined coelom formed from outpocketings of the embryonic gut.

Phylum Echinodermata: The spiny-skinned animals. Conspicuous five-sided radial symmetry; unique water vascular system. Includes sand dollars and sea urchins (echinoids), starfishes, sea cucumbers (holothurians), and crinoids. In addition to the five living classes, there are as many as sixteen extinct fossil classes.

Phylum Chordata: Elongate, rodlike cartilaginous notochord runs along the back for support. Main nerve cord dorsal (top); heart ventral (bottom). Most adults with gill slits, a bony vertebral column (backbone), or both. The most important groups are the Hemichordata, which include the extinct graptolites, and the Vertebrata, which include all fishes, amphibians, reptiles, birds, and mammals.

partly lined by mesoderm; the remainder is formed of ectoderm and endoderm. The pseudocoelomates include several groups of worms that are not important as fossils. The true coelomate animals (eucoelomate) have a fluid-filled cavity completely lined with mesodermal tissue. This includes the remainder of the animal phyla. Eucoelomates are divided into three major groups: the lophophorates, the schizocoels, and the enterocoels.

The **lophophorates** consist of two phyla that are judged to be rather closely related because they have in common a conspicuous, distinctive structure called the lophophore. This is a ring of tentacles close to the mouth. The tentacles are covered with cilia and mucus and serve primarily for food gathering. The two phyla of lophophorates are the Brachiopoda and the Bryozoa, both very important in the fossil record. Representatives of the two groups are quite different in appearance. The brachiopods are all relatively large, have two external valves or shells that enclose the animal, and consist of solitary individuals. The brachiopods and clams are the only two larger animals with two valves or shells that are very common as fossils. The bryozoans, commonly called moss animals, are all colonial,

FIGURE 2.6
The two different modes of
forming the coelom in the
eucoelomates. In schizocoels,
proliferating mesodermal cells
divide to form the fluid-filled
cavity, the coelom. In enterocoels, pouches from the side
of the gut expand to form the
coelom.

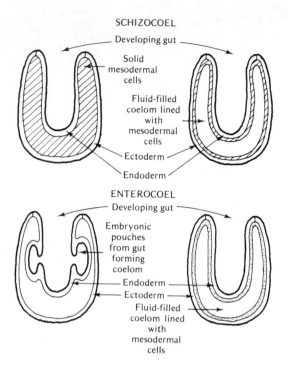

and each individual of a colony is microscopic. The common skeleton of the colonial is typically calcified and readily preservable as a fossil.

The **schizocoels** consist of three major phyla: the Mollusca, the Annelida, and the Arthropoda. In early larval stages, the coelom is formed by splitting (schizo-) of a solid mass of mesodermal cells to form the internal coelomic cavity (Figure 2.6). Superficially, the three phyla do not look very much alike. The mollusks, except for one primitive group, have an unsegmented body, whereas the annelid and arthropod body is conspicuously segmented. Mollusks are very important fossils, especially three groups—the clams, the snails, and the cephalo-pods. All have external calcareous shells that are readily preserved. The annelids (worms) do not have a hard exoskeleton and are not common fossils. They have legs that are not jointed, distinguishing them from the arthropods. Phylum Arthropoda is the most diverse group of animals alive today, mainly because it includes hordes of insects. Although crabs, lobsters, shrimp, and so on are very common today, they are relatively rare fossils (Figure 2.7). Insects are rare as fossils because they do not have a mineralized exoskeleton and are not marine. The most common fossil arthropods are the trilobites, which are exclusively Paleozoic forms that have been long extinct. Another group of microscopic arthropods, the ostracodes, are also exceedingly common fossils although easily overlooked. These have a calcareous external shell composed of two halves, or valves. In this respect they are like the larger brachiopods and clams.

FIGURE 2.7
A fossil arthropod, *Eryon*, related to modern lobsters and crayfish from Jurassic rocks of Germany. Note the segmentation of the body and division of the legs into segments. The specimen is 9 cm long. (Courtesy of Field Museum of Natural History, Chicago.)

The final major group consists of the **enterocoels**. They are all marine, and the calcareous plates have spines or nodes on them. The coelom in these animals is not formed in the same way as in the schizocoels. Instead, in early larval stages, small pockets form in the sides of the developing gut. These pockets pinch off from the gut, enlarge, and become lined with mesodermal tissue, thus forming a true coelom. Two phyla are included here: the Echinodermata and the Chordata. The first group is sometimes called spiny-skinned animals. They are all marine and many have calcareous spines with nodes or bumps on them. The echinoids (e.g., sand dollars, sea urchins) and the starfishes are the most familiar living forms. This phylum is unusual in that it is divided today into twenty-one different classes. Of these only five are still alive, and of the extinct sixteen groups, all but one are confined to rocks of the Lower and Middle Paleozoic.

The final animal phylum is the Chordata. This includes two groups of interest to paleontologists. The hemichordates are simple, wormlike forms called acorn worms. These are thought to be related to an extinct group of colonial animals (graptolites) that flourished in the Lower and Middle Paleozoic. The other group, the vertebrates, are of course very important animals today and have been in the past. Apart from insects they are the most generally known group of animals. The vertebrates are especially characterized by having a bony spinal column. The fishes are the most primitive and earliest known of the vertebrates. They are followed in evolution by the most primitive land tetrapods (four-legged), the amphibians, which include toads, frogs, and salamanders. These in turn gave rise to the reptiles: snakes, lizards, turtles, and alligators. From the reptiles arose the birds and the mammals, both warm-blooded vertebrates. All of these groups have long and important fossil records, and the tracing of their development through time is one of the most important stories to be told in paleontology.

TABLE 2.2.
The plant kingdom, to the phylum level.

Phylum Algae: Photosynthetic aquatic plants in fresh or salt water, single-celled or multicellular.

> *Subphylum Chlorophyta:* The green algae, with same photosynthetic pigments as the higher land plants.
>
> *Subphylum Chrysophyta:* The yellow-green algae. Includes several groups, one of which, diatoms, is important as fossils. Mostly unicellular; many with a silica skeleton.
>
> *Subphylum Phaeophyta:* The brown algae. Mostly all marine; rarely with preservable hard parts. Not important as fossils.
>
> *Subphylum Rhodophyta:* The red algae. Exclusively marine, usually multicellular, and with a complex calcareous skeleton that is commonly preserved as a fossil.

Phylum Mycophyta: The fungi. Nonphotosynthetic land plants that derive food from the soil and decaying organic matter. Includes mushrooms, molds, rusts, and mildew. Not important as fossils.

Phylum Bryophyta: The mosses and liverworts. Photosynthetic land plants that lack vascular tissues and readily preservable hard parts. Not important as fossils.

Phylum Tracheophyta: The vascular land plants. Fossil record extends from the Silurian to the Recent. Includes a host of different varieties that are today dominated by the flowering plants.

PLANT KINGDOM

We now have reviewed the Monera (bacteria, blue-green algae), the Protista (flagellates, ciliates, and sarcodines), and the Animalia, consisting of about twenty phyla, only about eight of which are very important as fossils. Now we come to the final major group of life, Kingdom Plantae (Table 2.2). The single most important generalization that one can make about **plants** is that they do, by and large, manufacture their own food. They engage in photosynthesis, the production of reasonably complex organic molecules used for food—sugars and carbohydrates—from simple inorganic starting materials—water, carbon dioxide, and traces of minerals in solution. In order to combine these materials into food, energy is required, and plants use sunlight as a source of energy. Such organisms are called **autotrophs** (self-feeding), in contrast to **heterotrophs** (different-feeding), which cannot manufacture their own food but which instead capture organic material in one way or another—by predation, scavenging, and so on. More specifically, plants are **photoautotrophs**, because they use light (photo-) as a source of energy. Bacteria include another kind of autotroph called **chemoautotrophs** that can produce chemical reactions, such as $SO_4^= \rightarrow H_2S$, to release energy that is used to synthesize food.

Not all plants are photosynthetic. The principal group that relies on external food sources is the fungi (mushrooms, molds, and mildews). These are so distinctive that some botanists feel they should be assigned a separate kingdom of their own. Besides the plants, many of the protists are also photosynthetic, although none of the animals are. Thus, the method by which food is obtained is one of the basic criteria for the splitting of life into kingdoms.

There are several ways in which to split plants into two primary groups. First we could divide them into plants that live in water and those that live on land. Water-dwelling plants generally are called **algae**, the Latin word for seaweed. As it turns out, such a division is useful for the algae but not very useful for the land plants, which are sufficiently diverse in structure as to resist being easily lumped together. Another criterion that could be used is similar to that which divides the protists and animals, the former mainly one-celled, the latter multicellular. If we divide plants up in this way, we find that all of the land plants, with few exceptions, are multicellular, but the algae include many multicellular forms—the most spectacular example is the giant kelp seaweeds of the Pacific Ocean—as well as unicellular forms. A third way to divide up plants is in terms of what is called their vascular structure. *Vascular* means circulatory. Your blood system of arteries and veins is your vascular system. No water-dwelling plants or algae have a vascular system; it is unnecessary because they are immersed in water. They either do not have to transport fluids through a large body, or they accomplish such internal transport in other ways. Most land plants do have a vascular system, but one important group does not. This comprises the Bryophytes—mosses and liverworts. These are quite primitive land plants that are generally small and confined to moist habitats, where a water-conducting system is not a necessity.

For the purpose of our discussion, we will divide up the plant kingdom as follows. We will consider the algae, all water-immersed plants, as one group; then we will divide the land plants into three groups: the nonvascular bryophytes, the fungi, and all of the higher land plants. This latter group are called the **tracheophytes** (Figure 2.8) a distinctive kind of cell found in their vascular systems.

Algae

Algae live in both marine and fresh water. They are most common close to the water surface, where they receive ample sunlight. They cannot live in the dark parts of the deep ocean. Many algae are floaters and contain oil, fat, or gas inclusions that make them about the same density as the water. The four major groups listed in Table 2.2 have names based on the color of the algae. Thus we have green, yellow-green, brown, and red algae. These color differences are due to photosynthetic pigments that generally absorb different wavelengths of light. We tend to think of most plants as being green because in land plants one pigment, chlorophyll, is very abundant and tends to mask other photosynthetic pigments of different color contained in land-plant leaves. In the autumn, when the chlorophyll degrades, the other photosynthetic pigments provide fall colors. The green algae

FIGURE 2.8
Common kinds of plants preserved as fossils. The green and yellow-green algae are commonly single cells and small. Red algae are commonly multicellular and complex. Yellow-green algae (diatoms) have a siliceous skeleton; red algae, a calcareous skeleton. Major groups of land plants include spore-bearing ferns as well as the seed-bearing gymnosperms (conifers) and flowering plants (angiosperms).

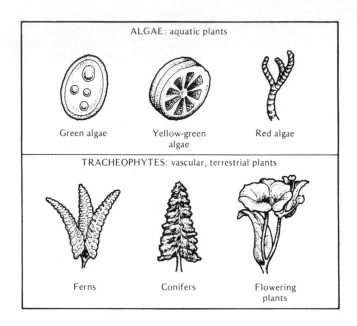

(Chlorophyta) are mostly unicelled or simple filaments of cells. They generally do not have hard skeletal parts and thus are not especially common as fossils. They are important in two ways. First, they are thought to have been the ancestors of the land plants. One reason is that chlorophyll is the dominant pigment in both groups. Secondly, green algae have been identified in the late Precambrian rocks of Australia, thus making them the oldest known eukaryotic fossils (see Chapter 5).

The yellow-green algae (Chrysophyta) include one group that is very conspicuous as fossils. These are the diatoms. They have a skeleton composed of opaline silica ($SiO_2 \cdot nH_2O$) that is in two parts fitting together like a pillbox. These have been found as fossils in the Mesozoic and younger rocks, but not with complete certainty in the Paleozoic. It is unclear whether they came into existence in the Mesozoic, whether they were present but lacked a skeleton in the Paleozoic, or whether they were present in older rocks and the microscopic skeleton simply was not preserved. They are so abundant in some Miocene rocks in California that they compose much of the rock, which has economic value in cleansing powders, toothpaste, and other products.

The brown algae (Phaeophyta) are very common today but have a scanty fossil record because they rarely, if ever, become skeletonized. These include the giants among the algae, marine kelps, which may be 200 feet long.

The final group, the red algae, or Rhodophyta, are important as fossils. Many of these have a calcareous complex skeleton (a reflection of their complex structure) that is readily buried and fossilized. Red algae are common on modern coral reefs, and their remains have been identified in many fossil reefs. They generally are not unicells, but rather are multicellular, and are marble- to fist-sized.

Bryophytes and Fungi

The bryophytes (mosses) and the fungi have little importance to the paleontologist. They do not have hard parts that are easily preserved. Thus they have a scanty fossil record.

Tracheophytes

The final group of plants is the tracheophytes, the vascular plants. These plants had aquatic ancestors and made the successful transition from living in the water to living on dry land—one of the most severe and difficult changes in habitat that is possible. They are all characterized by specialized tissues in the plant body. These vascular tissues are primitively in the center of the main plant body but later, secondarily, may be situated near the surface of the body. The central core of vascular tissue provides for uptake of water and dissolved materials from the soil and their distribution throughout the plant. The tissues comprising the central core are called xylem. Other vascular tissues, called phloem, transport manufactured food from photosynthetic areas to other parts of the plant.

The tracheophytes are very diverse. They include small, very primitive plants that have not yet developed true leaves or true roots as well as the most advanced, youngest, and most conspicuous of the vascular plants—those with flowers. A very great majority of the plants with which you are familiar and which you see all around you are flowering plants. Other vascular plants include the ferns, conifers, and several other smaller groups that are much reduced in variety today but were important parts of the landscape earlier in earth history.

KEY TERMS

algae	eukaryotes	monera
animal	evolution	order
autotrophs	family	phylum
bilateral symmetry	genus	plant
class	grade	prokaryotes
coelom	heterotroph	protists
cyanobacteria	hierarchy	radial symmetry
dermal layer	kingdom	schizocoels
enterocoels	lophophorates	species
	metazoa	tracheophytes

READINGS

You may consult any modern college-level textbook in biology, botany, or zoology for a discussion of the organization of life. These texts will also include one or more classifications of life. Do not expect these texts to present identical classifications; each one will differ from the others in some fashion.

Blackwelder, R.E. 1963. *Classification of the Animal Kingdom*. Southern Illinois Univ. Press. 80 pages. A detailed listing without illustration or explanation of the animal kingdom. One of many different classifications of animals.

Bold, A.C. 1960. *The Plant Kingdom*. Prentice-Hall. 106 pages. An excellent short discussion of the major features of monera, algae, and higher plants.

Lane, N.G. 1990. "Census of Past and Present Life." Journal of Geological Education 38:119–122. A summary of the numbers of known living and fossil genera or species of life.

Margulis, L., and Schwartz, K.V. 1988. *Five Kingdoms: An Illustrated Guide to the Phyla of Life on Earth*, 2nd ed. Freeman. 378 pages. Descriptions and illustrations of eighty-nine phyla of life.

Parker, S.H., ed. 1982. *Synopsis and Classification of Living Organisms*. McGraw-Hill. 2 vols., 1232 pages. A detailed compilation of all known living organisms. Tabulations of numbers of species can be extracted from the text.

3
Rocks and Fossils

We have seen how fossils fit into a time framework and how they may be of widely different ages. Fossils also are considered within the context of a physical framework provided by the rocks within which they are found. This chapter is devoted to a brief discussion of the relationship between rocks and fossils.

FOSSILIZATION

How and why does any living thing, after it dies, become a fossil? The odds against any specific individual becoming a fossil are very great indeed. In the first place an organism must ordinarily have some part of its body that does not easily decay. Such parts are called **hard parts**, or a **skeleton**. In vertebrate animals the parts most readily preserved as fossils are the bones and teeth. In invertebrate animals the skeleton is commonly an external shell of some kind that is composed of minerals, or definite chemical compounds. The most common such skeletons are composed of the mineral **calcite**, or calcium carbonate ($CaCO_3$). Other kinds of invertebrate skeletons may be composed of **silica** (SiO_2), which is the same as quartz; however, skeletons commonly have water added to the silica, forming opaline silica that is not nearly as stable as is quartz. A third type of skeleton is a **chitinophosphatic** shell. This consists of microscopic layers of chitin, a complex organic molecule, and calcium phosphate, or the mineral apatite ($CaPO_4$). In plants the hard parts most commonly preserved contain the organic material cellulose,

FIGURE 3.1
The fates of hard parts.

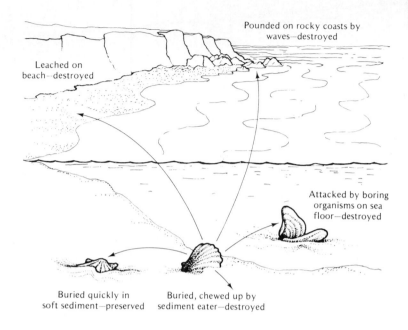

which is quite resistant to decay. Woody parts of plants and the leaves are the most readily preserved. Although the great majority of fossils consist only of the preserved hard parts, there are instances in which traces of soft tissues may also be preserved. Such examples are quite rare and depend upon exceptionally favorable circumstances of preservation.

Just having hard parts is not enough to ensure **fossilization**. All hard parts are subject to decay by physical or chemical processes. A shell may be washed onto a beach and broken up by waves. Sunlight and bacteria will eventually crumble a bone exposed on dry land. We know that most wood and leaves in a forest rot and decay. In order for any hard parts to be preserved as fossils, they must become buried. Such burial is accomplished by **sediments** within which the skeleton becomes entombed. Sand, or silt, or mud and clay may be washed over the skeleton, sealing it off from water or air. Burial cuts down on bacteria, the principal agents of decay, and helps prevent destruction of the skeleton. The more prompt the burial after the death of an animal or plant, the greater the likelihood that the specimen will eventually become a fossil. Thus, rapid burial is a prerequisite for fossilization.

Sediments accumulate in areas that are called **sites of deposition**; that is, places where sand or mud is deposited (Figure 3.1). Such sites are the most likely places for skeletons to be buried. Any plant or animal that lives near, in, or on a site of deposition has a greater chance of being preserved as a fossil than has an organism that lives far from such a site. Thus the leaves of a tree growing along a lakeshore are more likely to be preserved than are the leaves of a tree living on a hillside where erosion takes place.

If burial by sediments is so important in fossilization, then the environment in which any plant or animal lives has much to do with whether or not hard parts are preserved. Plants and animals that live on dry land, rather than in water, are less likely to be preserved. Sediments accumulate over much larger areas in the oceans, where there is less erosion taking place, than they do on land; thus, marine plants and animals have a better chance of being fossilized than do terrestrial organisms. A clam that lives in a burrow in the mud has already buried itself and is a prime candidate for fossilization. Our fossil record of birds and flying insects is quite poor, partly because these animals commonly do not live close to sites of deposition.

PRESERVATION

Once a skeleton has been buried and removed from a zone of decay, it still has a long way to go before becoming a fossil. Sediment must accumulate more and more deeply over the skeleton, protecting it from later erosion. As sediments pile up, they place an increasing weight on the layers in which a potential fossil is buried. The pile of sediment squeezes out water between grains of sand or mud. With deeper and deeper burial, the temperature of the sediment increases. These changes, called **diagenesis**, alter the soft, unconsolidated sediment into a solid rock. The mineral grains of the sediment are bound together by cement. The sediment is compacted as water is driven out and the end result is a hard, firmly bound together **sedimentary rock**, so called because it is formed of what was initially loose particles of sediment—sand, silt, or clay.

There are many different kinds of sedimentary rocks, based on the original materials of which they are composed and the conditions of burial. In general we can recognize two major types of sedimentary rocks: those of **detrital** or **clastic origin** and those of **chemical origin**. The former consists of those rocks formed by the transport of discrete particles to a depositional site where they are buried. The particles may be moved by water, ice, or wind. Different kinds of detrital rocks are recognized based on the sizes of the sedimentary particles. Thus, the rock **sandstone** is composed predominantly of sand-sized particles (Figure 3.2), **siltstone** is made of silt-sized particles, and **claystone** of still smaller grains. **Mudstone** is a mixture of clay and silt. One of the most common sedimentary rocks is **shale**, composed of clay and silt that have been compacted so that they split into thin layers. The coarsest detrital rocks—consolidated gravels—are called **conglomerates**.

Chemical sedimentary rocks are those formed of particles that have been chemically precipitated. Rock salt is an example. The most abundant such rocks, and the ones of most interest to paleontologists, are **limestones**. These are formed of calcium carbonate or lime that is either a fragment of a fossil, and thus was biologically secreted, or was chemically precipitated out of water. Only a few limestones are completely unfossiliferous. Most are marine but some are freshwater in origin. There are many different kinds of limestone, too numerous to expand on here, based on differences in texture and the components that make up the rock.

FIGURE 3.2
Relative sizes of common sedimentary particles, highly magnified. Sand grains range from 2 mm to $\frac{1}{16}$ mm; silt particles are $\frac{1}{16}$ to $\frac{1}{256}$ mm in diameter; clay particles are smaller still.

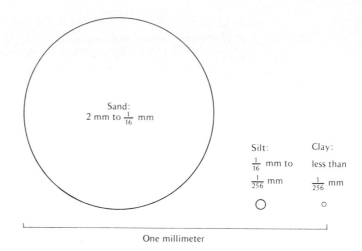

Sand:
2 mm to $\frac{1}{16}$ mm

Silt:
$\frac{1}{16}$ mm to
$\frac{1}{256}$ mm

Clay:
less than
$\frac{1}{256}$ mm

One millimeter

There are two other main types of rocks: **igneous** and **metamorphic**. The former consists of rocks that were once molten, such as lava from a volcano. Obviously such rocks are rarely going to have fossils preserved in them, although some terrestrial plant and animal fossils are known that were preserved by being buried under a lava flow. Metamorphic rocks are those that form either from igneous or sedimentary rocks by alteration of the rock due to high temperatures and pressures. If a metamorphic rock is formed from a sedimentary rock, such as marble from a limestone, or slate from a shale, and if the rock is not too altered, fossils may be preserved. Fossils in metamorphic rocks are relatively rare and are commonly quite poorly preserved.

As sedimentary rocks are formed, they undergo conspicuous chemical and physical changes. It is not surprising then that the skeletons of organisms preserved in such rocks also undergo many changes from their original state. If a rock is quite young, say 1 to 3 million years old, contained fossils may not have been altered too much, if at all. For example, such fossils may closely resemble shells you would pick up on a beach, except that any coloration would probably be leached out or faded. On the other hand, many (but not all) fossils that are older than a few million years show evidence of having been altered in one or several ways. Therefore, preservation has two basic categories: **original preservation** and **altered preservation**. Some of the most unusual examples of fossil preservation are those found in an unaltered condition.

A **mummy** is a specimen of a once-living organism in which some of the soft tissues are preserved, usually muscles, skin, and tendons. Both artificial and natural mummies are known, the former produced especially by the early Egyptians, who used sodium salts and tars to preserve humans as well as a wide variety of other animals such as snakes, bulls, cats, and beetles. All of the internal organs and brain were removed during the initial stages of mummification and preserved separately.

FIGURE 3.3
The carcass of an adult male mammoth found in 1900 on the bank of the Berezovka River in eastern Siberia. The individual was about 35 years old, with a height of about 9 feet. The geologic age is about 44,000 years. The mount is a dermoplastic reproduction of the original. (Courtesy of I. M. Kerzhuer, Zoological Institute, Leningrad, USSR.)

Natural mummies are ones that occur through some unusual happenstance of nature. The most common such fossils are frozen mummies preserved in the permafrost regions of Alaska and Sibera (Figures 3.3 and 3.4). The most common frozen mummies are mammoths, although mastodons, horses, bison, wooly rhinoceros, musk oxen, moose, and other animals have been found on rare occasions. At least some animals seem to have been suddenly entombed when they became stuck in soft thawed sediment during a summer or fell into a melted out ice crevasse. Preserved stomach contents have been examined, and it has been possible to do albumin and collagen typing on some of the preserved tissues. These tests confirm that the mammoth is much more closely related to the living Asian and African elephants, which are two different genera, than any of them is to the extinct mastodon.

The tundra of Siberia has yielded tons of fossil ivory in the form of mammoth tusks for several centuries. There has been an ages-long trade of this ivory from Siberia to China where the tusks, generally known as dragon bones, are ground up and used in Chinese medicine. In Alaska there are reports that fossil mammoth hair was so common in Pleistocene gravels that it clogged the dredges of placer gold miners.

In addition to frozen mummies, dessicated mummies have been found. In these mummies the flesh has been preserved by drying out all water, thus rendering

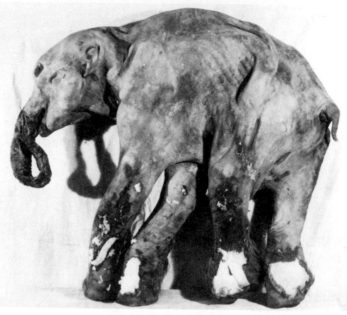

FIGURE 3.4
The mummified carcass of a male baby mammoth found in 1977 in Mgadan Province in eastern Siberia. The specimen was found in a sandpit at a depth of 6 feet in permafrost ground; it is emaciated, about 8 months in age, and slightly over 3 feet tall. The geologic age is about 40,000 years, based on radiocarbon dating. (Courtesy of I.M. Kerzhuer, Zoological Institute, Leningrad, USSR.)

the material unsuitable for bacterial action. Some of the most spectacular dessicated mummies are of extinct ground sloths preserved in caves or other cavities in the southwestern United States. The hair, skin, and dried tissues of one of these animals was found in a fumarole cave in a volcanic crater in New Mexico (Figure 3.5). In some instances, the mummies occur in caves that also contain piles of sloth dung up to a meter high that yield much information about the plant diet of these animals (Figure 3.6).

Even natural human mummies are known. A very old human burial in Egypt has been discovered that contained a mummified person that had not been treated with the later mummification procedures of the ancient Egyptians (Figure 3.7).

The oldest known mummy, of late Cretaceous age from Wyoming, is of a duck-billed dinosaur that was suddenly suffocated and preserved underneath a dense cloud of hot volanic ash. This specimen gives us an unparalleled view of what dinosaur skin was like and also includes valuable information on muscles and tendons in a bipedal dinosaur (Figure 3.8).

What are some of the things that can happen to a fossil as it is buried deeper in the earth and sediments gradually lithify into rocks (Figure 3.9)? The most

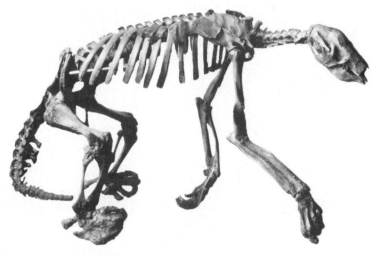

FIGURE 3.5
Skeleton of an extinct ground sloth, *Nothotheriops*, from a fumarole at Aden Crater, New Mexico. Small patches of skin and hair are attached to the extremities of the legs. (Courtesy of Paul S. Martin, University of Arizona, and Peabody Museum, Yale University.)

FIGURE 3.6
View of the interior of Rampart Cave in northwestern Arizona showing piles of fossil ground sloth dung on the floor of the cave. The dung has been radiocarbon dated at 11,000 years old. (Courtesy of Paul S. Martin, University of Arizona.)

FIGURE 3.7

A natural human mummy that is several thousand years old, from a predynastic burial in the sands of Egypt. (Reproduced by courtesy of the Trustees of the British Museum.)

FIGURE 3.8

One of the rare mummies of dinosaurs and one of the oldest mummies known. The mummy includes the skeleton with preserved skin, muscles, and tendons of the ornithopod dinosaur *Edmontosaurus annectens*, of Cretaceous age from the Kirtland Formation in San Juan County, New Mexico. (Neg. no. 35607 courtesy Department of Library Services, American Museum of Natural History.)

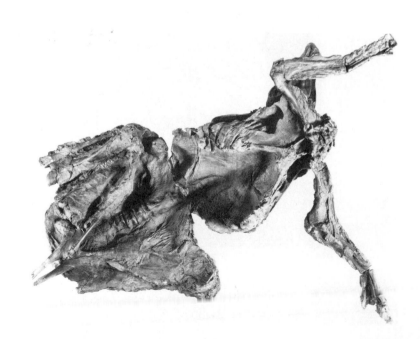

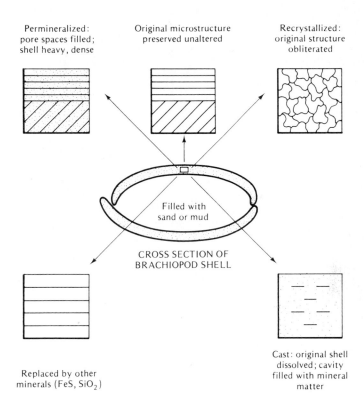

FIGURE 3.9
Different modes of preservation of fossil shell material. Small section of brachiopod shell is enlarged to diagrammatically show the various ways in which hard parts may be preserved.

Permineralized: pore spaces filled; shell heavy, dense

Original microstructure preserved unaltered

Recrystallized: original structure obliterated

Filled with sand or mud

CROSS SECTION OF BRACHIOPOD SHELL

Replaced by other minerals (FeS, SiO_2)

Cast: original shell dissolved; cavity filled with mineral matter

common type of alteration is that in which the hard parts have all microscopical pore spaces filled in by minerals precipitated from surrounding waters. This process is called **permineralization** and makes the hard part heavier and denser than it was originally. The open cell spaces in bones and in wood that become filled in to form petrified wood or bone provide an example (Figure 3.10). Even the external shells of invertebrate animals (e.g., clam shells) which seem to the naked eye to already be solid material, have microscopic pore spaces that fill up with mineral matter. Virtually any fossil that is more than a few million years old and that looks like it may be unaltered will turn out to have been permineralized. You can easily test this by hefting a fossil and a modern shell of similar size and shape in each hand. The fossil will feel heavier than the shell of Recent age.

Much more drastic things can happen to a skeleton than simply having its pore spaces filled up. The fossil may be buried under a sufficient thickness of rocks such that the pressure, or chemical changes in surrounding water in the rock, causes the original layers of the shell to become **recrystallized**. Virtually all shells and bones have a detailed microscopic structure dependent on the way the mineral crystals of the skeleton were deposited by the animal that grew the skeleton. These microstructures are commonly clearly visible in permineralized shells, but they become obliterated when the fossil material has recrystallized.

FIGURE 3.10
A piece of petrified wood of Tertiary age about 20 cm across. The open woody cells of the plant tissue have been filled in with silica precipitated from percolating water in the sediments, forming a permineralized fossil. (Courtesy of Rebecca Lindsay, Children's Museum, Indianapolis.)

The fossil may also be **replaced** (Figure 3.11). In this process the original material is dissolved by chemical action and other minerals substituted, so that a replica of the original shell is formed of foreign material. By far the most common replacing material is silica, which usually replaces skeletons composed of lime. The process takes place on a one-to-one volumetric basis; that is, a given volume of original material is replaced by the same volume of silica. Thus the size and shape of the fossil is not disturbed. In those instances involving limestones, the silicified fossils may be selectively etched out of the lime with dilute acids. Exquisitely detailed fossils that are virtually impossible to chip out of the solid rock may be obtained in this way. The next most common replacing material is the mineral **pyrite**, or iron sulfide (FeS_2). Fossils replaced by pyrite (fool's gold) are especially attractive as they have a shiny golden color. The pyrite is generally formed in organic-rich shales, with the iron and sulfur being formed in the process of decay of organic matter.

Another condition of preservation is the formation of molds and casts. Let us suppose that you surround a shell with potter's clay and then fire it in an oven, simulating formation of a rock through diagenesis. Now you break the clay open and there is an impression of the shell on the inside of the clay. Such an impression is called a **mold** (Figure 3.12). Now let us imagine that you somehow dissolved away the shell that you placed in the clay. The molds of the exterior of the shell are still in place, and you now fill the vacant space up with plaster of paris. You have formed a **cast** of the shell. Both casts and molds occur in nature (Figure 3.13). Shells or bone may be dissolved after they have been buried, and the vacant spaces may or may not later become filled with foreign material. Crystals may form in the spaces or mud may be squeezed into the openings. Molds and casts

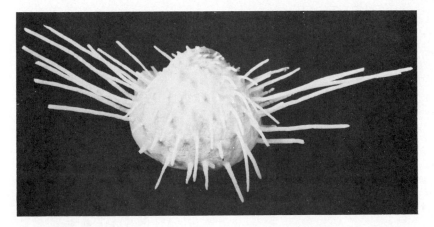

FIGURE 3.11
A brachiopod shell of Permian age from the Glass Mountains of West Texas. The original calcareous shell was selectively replaced by silica in the rock. Dissolving the limestone matrix away with acid released the shell with its long, delicate spines. The specimen is enlarged. (Courtesy of National Museum of Natural History.)

FIGURE 3.12
Molds of a large starfish, *Devonaster*, a small ophiuroid starfish, several brachiopods, and other fossils. The original shell material has been removed by solution, leaving behind an imprint of the fossil in the rock. The specimens are from Devonian rocks of New York. The slab is about 10 in. across. (Courtesy of National Museum of Natural History.)

FIGURE 3.13
Natural cast of the stem of an extinct, primitive tree (lycopod, *Sigillaria*) from the Pennsylvanian of Illinois. The distinctive scars on the cast are places where leaves attached when the plant was alive. (Courtesy of Field Museum of Natural History, Chicago.)

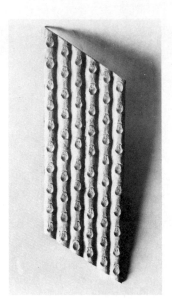

are especially common in coarse-grained rocks such as sandstones. Water may be easily percolated through these rocks, dissolving shells away.

There are many other ways in which fossils may be preserved (Figure 3.14). Insects are known from fossil amber, the hardened resin of certain conifers (Figure 3.15). The insects were trapped on the sticky surface and then engulfed in more resin. Black shales, rich in organic matter and deposited on a sea floor where

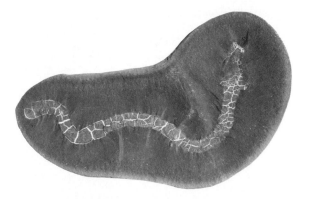

FIGURE 3.14
Preserved impression of the soft parts of a holothurian (an echinoderm that lacks hard parts except for microscopic spicules) of Pennsylvanian age from Illinois. The specimen, about 10 cm long, is preserved in an ironstone nodule that presumably formed soon after the animal died, preventing decay. (Courtesy of Field Museum of Natural History, Chicago.)

FIGURE 3.15
A caddis fly preserved in amber, the hardened resin of certain kinds of coniferous gymnosperms. The specimen is much enlarged and is of Oligocene age from East Prussia. The wing venation, hairs, bristles, and eye facets are all preserved. (Courtesy of National Museum of Natural History.)

oxygen was depleted and decay could not take place, may have preserved on them the outlines of the entire soft body of an animal. All of the light, more mobile organic compounds were driven off during compaction and diagenesis, but carbon molecules are very stable, remaining behind to record the fossil as a thin carbon film on the rock (Figure 3.16). Such a process is called **carbonization.**

The hard or soft parts need not be preserved in order for an object to qualify as a fossil (Figure 3.17). Recall that in the first chapter we defined a fossil as *any* evidence for the former existence of life. Footprints of ancient animals are sometimes preserved in what were once soft muds or sands (Figure 3.18). These provide indirect evidence of life. Many marine worms and crustaceans live in burrows in

FIGURE 3.16
Carbon film of an extinct arthropod belonging to a group called eurypterids. These were especially common in Silurian time. The specimen is about 25 cm long. (Courtesy of Field Museum of Natural History, Chicago.)

FIGURE 3.17
Two halves of a fish coprolite that has been preserved in a nodule of Pennsylvanian age from Illinois. A coprolite is the preserved feces of an animal, and coprolites are common fossils at some localities. (Courtesy of Field Museum of Natural History, Chicago.)

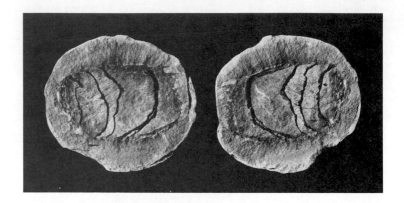

FIGURE 3.18
Footprints of two fossil vertebrates, much reduced, of Permian age from Arizona. Can you tell the direction in which each of the animals was moving and which one walked over the area first? (Courtesy of National Museum of Natural History.)

the sea floor. Their living burrows may become filled with sediments after they die and the outlines of the burrows preserved; thus, the burrows are fossils. These are called **trace fossils** because they provide only indirect evidence of the former presence or activity of the animal.

Even certain chemicals may truly be called fossils. When the chlorophyll of a green plant decomposes, it breaks down into organic compounds. Two of these that are very stable are pristane and phytane. These chemicals have been recovered from rocks over 2 billion years old, and they provide indirect evidence of the early existence of photosynthetic organisms. Thus, they qualify as **chemical fossils**.

In some respects it is amazing that so many different kinds of plants and animals have been preserved in the fossil record as we do find. On the other hand it is quite true that our record of life is not perfect. Given the vicissitudes of burial and alteration of fossils, our record is really pretty good; and occasionally we do find truly remarkable fossils, such as fossil mummies, insects in amber, or body outlines preserved in black shales. Our knowledge of fossils is still very uneven in a geographical sense. We are only now beginning to get much information about fossils contained in rocks of the ocean basins. A deep-sea drilling project called JOIDES (Joint Oceanographic Institutions for Deep Earth Sampling) that has been going on for several years has provided us with samples of deep ocean rocks from many areas. We know very much more about fossils preserved in rocks that are exposed on land because they are so much more easily collected. Even on land, we know a lot more about the fossils of western Europe and North America than we do about those of many other parts of the world. This is mainly because most paleontologists are trained in these two areas, later living and doing research within them. The paleontology of parts of South America, Africa, Asia, and Antarctica is still in an exploratory state. Finally, our knowledge of fossils collected on land areas is largely limited to those rocks that are exposed at the surface today. Many rock layers have been removed by erosion and their contained fossils destroyed. Other rocks are completely buried under younger ones and may be accessible only in mines or by taking cores of the rocks with oil well drilling rigs. Despite all these difficulties, paleontologists have put together a remarkably complete record of life on earth. We are missing many small details, and we continue to argue about the causes of some of the phenomena exhibited by fossils, but the history of life is certainly well documented in its broad outlines.

ENVIRONMENTS OF DEPOSITION

While life today is very diverse, all species of plants and animals are restricted to certain habitats and environments. On the broadest scale, there are very few animals that can live both in water and on dry land. We expect that the same situation applied to ancient life—that fossil plants and animals were restricted to specific environments. The rocks in which we find specific fossils may or may not provide a record of the environment in which the ancient organisms once lived. Only if an individual died and was buried at or near its living site would it provide a faithful indicator of environment. We know that the hard parts of organisms can be transported far from their place of origin. We find, for instance, logs of ancient tree trunks that were rafted down rivers out into the sea and eventually sank far from their living sites. Waves and currents may move empty shells from one environment to another in the ocean. The paleontologist must always be aware of the possibility of fossil transport; determinations of whether or not it has taken place are prerequisite to valid interpretations of ancient environments.

Not only are the remains of life different from place to place in rocks, but obviously the physical aspects of the environment differ also. The floor of the ocean today is covered by very different kinds of sediments from place to place. A coastal bay may have sandy beaches and bars, while the center of the bay may be floored by mud, and rocky shorelines may have coarse gravel seaward. All of these sediments may be buried and changed into sedimentary rocks, yet they will all be of the same age and will be in close proximity to each other. If we look at the animals living on or in a sand bar, a muddy bottom, or a gravel bed in a single bay, we find that each environment is occupied by a different suite of animals. Both the physical and biological characters of sedimentary rocks provide clues to the environments recorded by any layer of fossiliferous sedimentary rock. We commonly have very specific knowledge of what the sea floor was like at a given place and time since rocks reflect and record the nature of that sea floor. However, we may have to make inferences or do especially detailed studies if we hope to answer questions about what the temperatures may have been, or the salinity, or the depth of the water, in an ancient bay. Reconstruction of ancient environments is called **paleoecology**. In order to be successful, such research must usually take into account all data that can be obtained by study of both the rocks and the fossils they contain. Studying one without the other may lead to results that are inconclusive or to misinterpretations.

THE TIME-SPACE PROBLEM IN PALEONTOLOGY

It is very important in paleontology to be able to demonstrate whether or not different rocks and fossils in different areas are of the same age or of different ages. If a paleontologist studies only a single outcrop of rock, or a single vertical section of rock, the problem is relatively simple—superposition alone determines the age relationship. However, just as soon as rocks and fossils are studied over an area, whether it is a square mile or several states, the problem arises as to whether or not rocks and fossils in different areas are contemporaneous or perhaps of different ages. Questions about time relations over an area are referred to as the **time-space problem**, and they cannot be answered by the study of fossils alone. Several aspects of sedimentary rocks must be considered.

Stratigraphy

The study of stratified (layered or bedded) rocks is called **stratigraphy**. A stratigrapher attempts to decipher the sources, distribution, limits, and environments of deposition in sedimentary rocks. In any stratigraphic study the first step is to divide up a rock sequence into working units that can be measured and described as to their physical characteristics and fossil content. Imagine a large hill with rocks exposed from top to bottom that consist of beds of shale, sandstone, conglomerate, and limestone. Each type of rock must be delimited and the sequence of kinds of rocks worked out. Depending on its thickness, each rock type occupies a certain area on the hill and has a definite interval of altitude on the hill, from 200 to 250 meters above sea level, for example.

Now let us suppose that there are several other hills in the immediate area, each of which also has a rock sequence exposed. The stratigrapher wants to match

FIGURE 3.19

Use of formations to con-
struct a geologic map. At top
is a section of rocks with
three mappable rock units
(formations). Map below
shows distribution of these
three rock types over an area.
Bold lines are streams; light
lines indicate the positions of
the boundaries between the
formations. Numbers give
heights above sea level. Note
that the oldest formation, the
sandstone, occupies the val-
ley floor and that the youngest formation, the limestone,
occupies the highest ground.

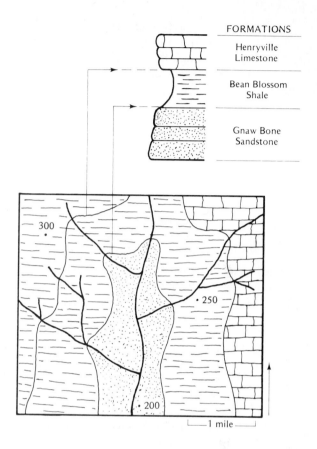

FORMATIONS

Henryville
Limestone

Bean Blossom
Shale

Gnaw Bone
Sandstone

300

• 250

• 200

└—1 mile—┘

up these rocks from one hill to another. How is this accomplished? The technique
that is used is called **geologic mapping**. A geologic map shows the distribution of
different rock types over an area and thus differs from a highway map, for instance,
which shows the distribution of roads over an area (Figure 3.19). The sequence
of rocks on the hill must be divided up into mapping units. Such rock units are
called **formations**. They are based on physical and fossil characteristics of the rocks.
Each formation has definite upper and lower boundaries at which there are changes
in rock characteristics. Every bed of rock must be contained within a formation,
but not in two formations. There can be no gaps or overlaps. Formations are given
geographic names such as a St. Louis Limestone, a Chattanooga Shale, a Topeka
Limestone. These mapping units are named after towns, streams, mountains, or
other geographic features.

   As an example, let us suppose that at the base of our hill we can see 20
meters of exposed sandstone; directly above this is 10 meters of limestone, followed
by 15 meters of shale. The transition from one rock type to another is quite abrupt,
so that the change from sandstone to limestone forms a visible line on the side of
the hill. That line marks the boundary between two formations, and it can be

accurately marked on a map. The line between the limestone and the shale can also be so delimited. Once the positions of these boundaries have been established on one hill, then the different rock types can be traced to an adjacent hill with the lines indicating the boundaries between formations carried along from one hill to another. In this way, a geologic map showing the geographic distribution of each formation can be constructed. If the mapping is done over a large enough area, the map may then show the geographic limits of different formations. Some units, such as the limestone we have mentioned, may thin and disappear so that the shale is directly above the sandstone. Other new formations may appear in the rock section.

The original horizontality of sedimentary rocks is one of the assumptions made in stratigraphy. We assume that the layers of sediments were laid down on top of one another in horizontal position. However, there are exceptions to this rule. Ancient sand dunes that become sandstone commonly have the beds of sand inclined to the horizontal. Beds that were originally deposited at an angle to the horizon are said to be cross-bedded or cross-stratified. If beds that were originally horizontal are now tilted, various internal forces within the earth have been active in the area where the tilted beds occur and mountain building has taken place.

We also assume that individual beds had lateral continuity. Even though a specific bed may only be apparent in certain places, we assume that the bed was originally present over an entire area. The bed may have been removed by erosion in some places, then buried below other rocks, and may not be exposed at the surface.

At this point you may well ask what all of this has to do with fossils. The process of matching up sequences of rocks from one area to another is called **correlation**. In making a geologic map from one hill to another, a stratigrapher may correlate or match up the shale formation and the limestone formation over an area. This is done largely on the basis of the physical characteristics of the rock—sandstone versus limestone versus shale. The stratigrapher may then ponder, "I know that the shale unit is the same rock unit over this area, but is the shale of the same geological age over the entire area? Perhaps the muds that formed the shale may have been deposited earlier or later in some areas than in others. Perhaps while mud was being deposited here, limy sediments were still being deposited over there."

There are several ways in which this problem can be solved. If the shale contains one or more thin beds of a clay rock called **bentonite**, which is altered volcanic ash, the observer can assume that the ash was deposited virtually instantaneously in geologic terms, and contemporaneously over the entire area; thus, bentonite rock would serve as a time marker. Other very distinctive kinds of rocks that occur in thin beds may also be assumed to be time markers. Another solution would be to study the fossils in the rocks. All species of the fossils have definite time spans, from their time of origination to their time of extinction. Some species lived for relatively short periods of geologic time, others for long periods. By carefully collecting fossils through the rock sequence, a paleontologist can

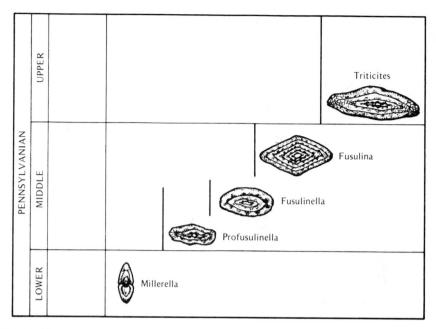

FIGURE 3.20
Use of fossils in biostratigraphy. The time range of each genus of extinct protozoans called fusulinids is shown, with diagrammatic cross sections of each type. These fossils mark a sequence of fossil zones in rocks of Pennsylvanian age that is used in biostratigraphy.

carefully pinpoint the level at which any species first appears and then disappears. By plotting these vertical (time) ranges through the rocks, he or she can establish a pattern of species distribution in time that can be compared with similar patterns in adjacent rock sequences. This process is called **biostratigraphy**, the study of the biological content of rocks and the use of fossils to correlate rocks in time (Figure 3.20). An interval of rock, whether it is sandstone, limestone, or whatever, that contains a distinctive assemblage of fossils is called a **fossil zone**. One or more of the species may be confined to that interval, or the interval may be characterized by a distinct suite of concurrent species.

In order for a fossil to qualify as a good **zonal fossil** it generally should have a reasonably short geologic time span—a few million years at most. The fossil brachiopod *Lingula* is known from the Ordovician period to the Recent, a span of 500 million years. Obviously *Lingula* is not a good zonal fossil. In addition to short duration, a zonal fossil ideally should be reasonably common and easy to collect, it should be distinctive so that it is readily recognized, and, most important, it should be of some plant or animal that was not strongly controlled by environment. A fossil brachiopod may have lived only on soft, muddy bottoms and thus may only be found in shales and mudstones. A fish, on the other hand, may have swum around over all kinds of bottoms—sandy, muddy, or gravelly. Generally speaking, those plants and animals that lived in the water, swimming or floating, make better zonal fossils than do animals that lived on the bottom and were controlled by bottom conditions.

Fossil zones can be arranged so they completely divide up any rock sequence. Thus, Jurassic rocks are divided worldwide into seventy-three zones, based largely on fossil ammonoid cephalopods. Because the duration of this time period is 64 million years, we can discriminate time intervals of less than a million years by using these fossils. The system of using fossils for correlation is founded on the principle of faunal and floral succession. This was recognized in the eighteenth century by William Smith, a British surveyor and canal digger, who first recognized the orderly succession of fossils. Succession is the product of organic evolution and occurs as species replace each other through time and older plants and animals become extinct. Because the communities of life change through time, we can use fossil succession to characterize and date formations of rock. We can do this not only on a local basis but can also correlate and match up rocks that are widely separated on different continents.

Biostratigraphy is an important function of the paleonotologist. In order to make environmental analyses of fossils it is usually necessary to be quite sure that the rock specimens being studied are of the same age. Biostratigraphy may be approached on very different geographic scales. On one hand, a paleontologist may study sequences of fossils on two adjacent hills; on the other, he or she may attempt to correlate rocks from New York State and western England.

Facies

If you think for a moment about the face of the earth today, you will realize that the sediments that are being deposited right now are not all alike. On some ocean bottoms sand is being deposited, on others muds are accumulating, and on still others there are gravels or loose limy particles from broken shells. On land, rivers may deposit sand on their beds and, during times of flood, fine muds over the flood plains. All of these different kinds of sediment are contemporaneous; that is, they are all being deposited at the same time. If these sediments accumulate and form sedimentary rocks, then the rock record for this year will be very diverse at some time in the future. We would expect that the same situation held in the past, that different kinds of sediments were being deposited at the same time in different areas.

The principle we have just described is called **uniformitarianism**, a fancy word that means that the same processes that are operating today also took place in the past. This is sometimes summed up in the phrase, "the present is the key to the past." There have certainly been differences in the rates at which events occurred in the past and today, as well as differences in emphases of certain processes, but the basic mechanisms are thought to have been the same. For example, there have been times in the past when there were immense coastal

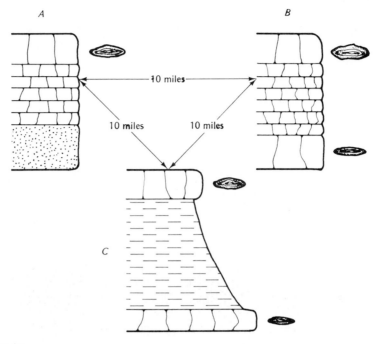

FIGURE 3.21

Facies and correlation. Three stratigraphic sections of rock 10 mi equidistant from each other are shown. The limestone at the top contains a particular species of a fusulinid fossil, *Fusulina*. The lower limestone contains another species of *Fusulina* at sections *B* and *C*. The thin-bedded limestones in the middle of the section at *A* and *B* are a facies of the thicker shale at section *C*. The basal sandstone at *A* may or may not be a facies of the lower limestone at *B* and *C*. (See Figure 1.2 for standard symbols of rock types.)

swamps, the plant debris of which formed extensive coal beds. Such coastal swamps are much restricted today, but the basic process of coal formation is judged to be the same.

In an ancient bay, mud could have been deposited in the center of the bay and this mud could have given way to sandy muds and, finally, sands, proceeding from the center of the bay to its shores. When changed into rocks, these muds and sands would have become shales and sandstones that intergraded laterally with each other. Such changes, when they occur within a formation or mappable rock unit, are called **facies** changes, which means changes in distinctive aspects of the rock (Figure 3.21). Not only would the physical aspects of the sedimentary rocks have changed, but we would expect the fossil content of the rocks also to have changed. We would not expect to find many of the same animals living on a muddy bottom as we find on a sandy bottom. Such lateral changes in fossils are called **biofacies** changes.

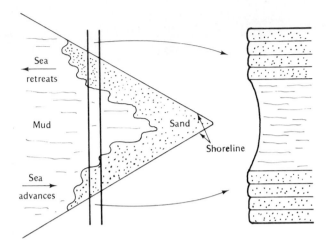

FIGURE 3.22
Diagram of an advance and retreat of the sea on the left, with sand being deposited close
to shoreline and mud offshore. The rocks that result from this transgression and regression of the sea are shown on the right. The section is located in the area shown by the
double column on the left. Shale is in the middle of the section, indicating maximum advance of the sea, with nearshore sandstones above and below.

Facies changes in rocks and contained fossils are one of the main clues that
paleontologists use to solve the time-space problem. We know that the face of the
earth is changing constantly. Rivers shift their course, lakes fill up with sediment
and are destroyed, the sizes of shallow marine bays shrink or expand, sea level
may rise or fall. These kinds of changes are what cause different rock types to be
deposited in the same area. At times in the past, North America was almost 75
percent covered by shallow seas; these seas retreated, only to flood parts of the
continent again. During a time of flooding, offshore muds may gradually be
deposited on top of nearshore sands. During a retreat of the sea, the opposite may
occur; nearshore sands may be laid down on top of offshore muds. This would
produce a rock sequence of sandstone overlain by shale, in turn overlain by
sandstone (Figure 3.22). Some of the shale would be younger than the sandstone
below and older than the sandstone above, but not all of it. Some shale would be
of the same age as the sandstones, and these would be facies of each other. Sorting
out age differences versus facies differences is one of the primary tasks of
paleontologists using biostratigraphy and fossil zones. It should be obvious why
fossils that were not strongly controlled by environment would be most helpful in
determining age relationships. Fossils that were strongly influenced by environment
when they were alive would, on the other hand, be the most helpful tools in
characterizing the ancient environments that were present in the past. Such fossils,
called **facies fossils** because they are strictly controlled by environment, are of

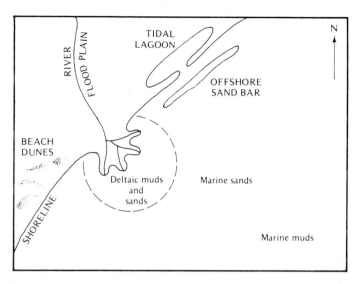

FIGURE 3.23
A simple paleogeographic map showing reconstruction of ancient environments over an area. The map is based on interpretations of the different kinds of rocks that now occupy this area. Fossils contained in these rocks aid in interpretation. Marine pelecypod shells were found in the central, southern, and eastern areas; fossil leaves and mammal bones were found in the northwestern part of the map. Because the differing kinds of rocks laid down in the different parts of the map area are all of the same age, the various rocks are facies of each other.

special interest in paleoecological interpretations. Good zonal fossils may be of little interest in environmental studies.

One of the goals of paleontology is to reconstruct the past as accurately and in as much detail as possible. A paleoecological study may reveal where and for how long particular environments were present over an area. It may be possible to construct **paleogeographic maps** showing the ancient distribution of past environments (Figure 3.23). Such maps might show ancient shorelines where land met the sea, areas of coastal sand dunes or offshore sand bars, river channels, deltas, islands, ancient coral reefs, and so on. The plants and animals that lived on, in, and above these environments can be analyzed in terms of communities, associations of organisms that once lived together. Communities and environments can be traced forward or backward in time, revealing changes that have occurred, shifting of environments, and deletion or addition of organisms to communities as they evolved. Such comprehensive analyses must be based on careful study of both the physical and biological aspects of the rocks. Neither can be neglected. The age relationships of the rocks must be evaluated and time-space problems solved; otherwise, rocks that are of different ages may be judged erroneously to be contemporaneous facies of each other.

KEY TERMS

biofacies	facies fossil	pyrite
biostratigraphy	formation	recrystallization
calcite	fossilization	replacement
carbonization	fossil zone	sandstone
cast	igneous rock	sedimentary rock
chemical fossil	limestone	shale
chitinophosphatic	metamorphic rock	silica
clastic	mold	skeleton
conglomerate	mudstone	stratigraphy
correlation	mummy	trace fossil
detrital	paleoecology	uniformitarianism
diagenesis	paleogeography	
facies	permineralization	

READINGS

Introductory college-level texts in physical geology and for historical geology include discussions of sedimentary rocks, environments of deposition, facies, and correlation. The latter books also discuss the uses of fossils in correlation, biostratigraphy, and preservation. These books are designed for introductory geology courses and provide much more detail than is possible to cover in a paleontology course.

Feldman, R.M., and others. 1989. *Paleotechniques*. Paleontological Society Special Publication No. 4. 358 pages. This is the only up-to-date book in English dealing with all of the various laboratory techniques for working with fossils.

Lockley, M.G., and Rice, A. 1990. *Volcanism and Fossil Biotas*. Geological Society of America Special Paper No. 244. 125 pages. A fascinating account of sudden burial of organic remains by volcanic ash; includes an interesting chapter on plant remains preserved at Pompeii.

Matthews, R.K. 1974. *Dynamic Stratigraphy*. Prentice-Hall. 370 pages. Emphasis in this book is on broad-scale interpretations of sedimentary rocks with emphasis on facies and environments of deposition.

Rixon, A.E. 1976. *Fossil Animal Remains: Their Preparation and Conservation*. Athlone Press. 304 pages. A useful handbook of techniques for working with invertebrate and vertebrate fossils.

4

Origins of the Earth, Oceans, Atmosphere, and Life

W hy consider the origin of the earth in a book about fossils? The answer is that the origin and early evolution of life are thought to have been closely tied to the nature of the early atmosphere and the oceans. These prime features of the earth have not always been in existence, nor were they the same in early earth history as they are today. Obviously, the planet earth formed before the oceans and the atmosphere could have existed; therefore, the process of earth formation is an important part of these interrelated events.

ORIGIN OF THE EARTH

In the first chapter on geologic time, we saw that the earth is indeed very ancient, older than 3.96 billion years. We know that the earth is a **planet** with one moon in a solar system that includes one star, the sun, and planets orbiting around the sun at various distances. Our solar system is a tiny part of an immense nebula of stars represented by the familiar Milky Way, and this galaxy is only a small fraction of the large number of nebulae and galaxies that make up the universe. The origin of the earth cannot be considered in isolation because any theory that explains its origin must also explain the remainder of the solar system. It is highly improbable that the earth formed alone and uniquely, completely independent of its nearest neighbors. Any explanation of earth or solar system origins must take into account and satisfactorily explain a series of important observations that have been made by astronomers:

1. The solar system is part of a nebula.
2. The sun represents most of the mass of the solar system (2×10^{30} kilograms).
3. The planets have most (98 percent) of the angular momentum of the solar system. The angular momentum is mass times velocity times distance. Since most of the mass is concentrated in the sun, the velocities of movement of the planets are great as are their distances from the sun.
4. The earth almost lacks such noble (heavy, inert) gases as xenon, neon, and krypton, at least in the abundance that they are present in the sun.
5. The earth is layered with a thin outer **crust** only a few kilometers thick, and thick **mantle** composed of very heavy dark rock under tremendous heat and pressure, and a liquid or molten outer **core** and a solid inner core, both composed of a material like iron-nickel silicate.
6. Each planet has a different density, and most of them are denser than the sun. The earth has a density of 5.5 grams per cubic centimeter; the sun and Jupiter have densities of 1.4. The differences in density mean that each planet has a chemical composition that is somewhat different from the others, and that each may have formed at a somewhat different temperature. They may also imply that the planets formed at somewhat different times and that there may have been a temperature gradient from the sun to the outer limits of the solar system.

Two contrasting ideas concerning the origin of the solar system have vied for scientists' acceptance over the past 400 years. The two ideas can be expressed in the question, Are the planets the daughters or the sisters of the sun? The older of the two ideas, originally put forward by the French philosopher and mathematician René Descartes, has the planets spinning off into space from a large, hot proto-sun (Figure 4.1). In order to accomplish this feat, a near collision of the sun with another star would have been necessary in order to pull out the material for the planets. A serious objection to this hypothesis is based on the distribution of mass and angular momentum in the solar system. If the planets are indeed daughters of the sun, the sun should still be spinning rapidly enough for it to have a much greater share of the angular momentum and somewhat less of the total mass of the solar system than it does.

The other hypothesis, that the sun and planets are sisters, envisions the solar system beginning as an enormous **nebular cloud** of gases and dust spinning as a spiral arm of the Milky Way nebula. The cloud would have been very large, thirty to forty light years across, with a mass two to ten times less than that of the present solar system, and there would have been only a few atoms of matter per cubic centimeter. As this cloud slowly spun, the force of gravity and magnetic attraction caused particles to coalesce and the cloud to shrink. As dust accumulated into

FIGURE 4.1
Alternate hypotheses for the
formation of the solar system.
A. Planets formed from the
sun. **B.** A swirling dust cloud
with local aggregations of
matter formed a proto-sun
and proto-planets.

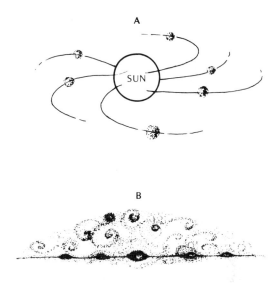

larger particles, they collided and adhered to each other. This coagulation process
continued until a large part of the mass of the nebular cloud was concentrated
near the center of the cloud, mainly by gravitational attraction, with much smaller
aggregations of matter at various distances from the center. The concentration of
mass in the center, with collisions of materials, produced heat, and the proto-sun
gradually warmed up until it reached a temperature at which spontaneous nuclear
reactions could take place, mainly involving hydrogen. At this point the sun
became very much hotter, and the heat drove off vast streams of lighter atoms into
space, creating what is called the **solar wind**. This wind stripped the frozen inert
noble gases and light gases such as hydrogen from the developing proto-planets.
As the earth formed, much of its original mass was dissipated into space, especially
huge quantities of light elements such as hydrogen and helium. It is estimated that
the present mass of the earth is only about $\frac{1}{1200}$ of its original mass. As coalescence
continued, the heaviest materials worked their way to the center of the planet
forming the core, and the lightest rocks floated at the surface forming the crust.
The accretion of particles to form the earth may have occurred very rapidly, some
estimates being as short as 10,000 years.

The dust cloud hypothesis shown in Figure 4.1B is generally accepted by
astronomers. The earth may have initially accreted in a cool condition. As the
earth became larger, the heat released by the radioactive decay of elements within
the rocks became trapped and could not escape at the surface. Thus, the earth
gradually warmed up and became hot to the point where melting occurred. In a
molten state, heavier elements and compounds slowly sank to the center of the
earth, forming a dense, heavy core, thought to be composed of iron-nickel silicate
minerals still partly in a molten state. Lighter materials floated to the top, forming
a thin, light crust. Materials of intermediate density, those forming the bulk of the

earth, made up the mantle. As the earth gradually cooled, the crust solidified. There was no atmosphere, all of the lighter elements having been blown away by the solar wind. There was no water on the surface, the heat of melting having vaporized any free water that may have been present. The earth gradually cooled off, but as it did, it continued to receive an important source of heat. Radioactive elements trapped within the earth give off heat as they decay. This provides a major source of the internal heat of the earth today.

ORIGIN OF THE OCEANS

When the earth was born it was naked, hot, and dry. There was no water and no atmosphere. The earth contained the elements that form the oceans and atmosphere within it, but these took time to be released to form water and air. The inert gases, which do not react readily with other elements, are seriously lacking in the earth's chemical composition today. The earth was stripped of these inert, light elements by the solar wind during the very early periods of its formation. At the same time those gases that are reactive and combine easily with other elements, such as hydrogen, oxygen, nitrogen, and carbon dioxide, were bound into the crystal structures of minerals and so escaped being blown away by solar wind.

By studying stony **meteorites** that have fallen onto the earth's surface, we know that they contain minerals that have water bound into their crystalline structure. The water amounts to about one-half of one percent of the material in such meteorites. This seems like a very small percentage of water, but if all of the rocks of the earth's crust and mantle contained an equivalent amount of water and it were all released on the surface, the amount would fill the present ocean basins twenty times. Scientists believe that water composing the oceans, lakes, and rivers was gradually released as water vapor from the interior of the earth, primarily through the venting of gases by volcanoes and fumaroles and through hot springs. Thus the oceans gradually accumulated and grew in size early in earth history, and much of the water on the earth's surface is very old, having been recycled millions of times.

Measurements of the water vapor content of the gases that come from volcanoes and springs that are active today indicate that the amounts of water emitted, if produced in the same amounts over geologic time, would be sufficient to fill the ocean basins 20 times. One may well wonder then why the world ocean is as small as it is and why all the continents are not covered by hundreds of feet of ocean water. The answer is that a tremendous amount of the water produced from the earth's interior is taken out of circulation in a variety of ways. A great deal of water gets trapped in sediments between grains of sand or silt. It is eventually buried and lost to the water cycle. Other water gets bound into the crystal structure of clay minerals or of other minerals such as gypsum (calcium sulfate) which contains ions of water. So while there is constant input of new water onto the earth's surface, there is also constant loss of water, effecting a balance that seems to have operated at much its present rate for long periods of earth history.

We have now had a brief look at how the earth formed and how the oceans came to be spread across the earth's surface as water was released from the interior of the earth. The stage is set to consider the origin and early evolution of life and the origin and evolution of the atmosphere.

ORIGIN OF THE ATMOSPHERE AND LIFE

During the earliest period of its formation, the earth probably lacked any kind of surrounding gaseous envelope, or **atmosphere**. The solar wind would have removed such light atoms from the earth's surface. After the initial formation of the sun, the solar wind abated, and thus light atoms could accumulate above the earth. We have already seen how the internal heat of the earth produced volcanoes and hot springs, bringing to the surface water vapor that cooled, condensed, and eventually formed the world ocean. Along with the water vapor, other gases were belched out by volcanoes and geysers. These are thought to have been mainly nitrogen (N) and carbon dioxide (CO_2); no free oxygen would have been present. This primitive atmosphere may have contained simple gases such as methane (CH_4) and ammonia (NH_3), although scientists argue about their presence and whether or not they were present in significant amounts.

From this primitive beginning, two highly significant events occurred on the earth's surface: life originated and the atmosphere evolved from a **reducing atmosphere** that lacked free oxygen to one that contained significant amounts of free oxygen. These two events are linked to one another.

The primitive atmosphere contained nitrogen (N), hydrogen (H), mainly tied up in water vapor, oxygen (O), combined in water vapor and in carbon dioxide, and carbon (C) in the latter gas. These four elements are the basic building blocks of all organic compounds and of living things. How were they combined into the very large and complex molecules that characterize life?

Part of the answer lies in an experiment conducted by Stanley Miller, a chemist, in 1953 (Figure 4.2). He put methane, ammonia, hydrogen, and water vapor in a closed flask at ordinary room temperature and pressure. Note that these four materials—or at least the four elements that compose them—are thought to have been present in the primitive atmosphere. No free oxygen was used. He then subjected this mixture to continuous electrical discharges for one week. The liquid phase of the mixture gradually changed color, turning a light yellow. When Miller analyzed the resultant liquid, he found that he had synthesized several **amino acids**. These are the building blocks of life, especially of large protein molecules. Miller had simulated possible conditions in the early atmosphere and oceans on the assumption that storms, and lightning, may have been of frequent occurrence, and that amino acids may have been formed close to the air-water interface. Other experimenters tried somewhat different combinations of starting materials and different sources of energy to synthesize amino acids. It is now known that amino

FIGURE 4.2
Stanley Miller's experiment that produced amino acids, the building blocks of life. Three gases, water vapor, methane, and ammonia were exposed to repeated electrical discharge. This resulted in gradual accumulation of amino acids and other simple organic molecules in a connected boiling water bath.

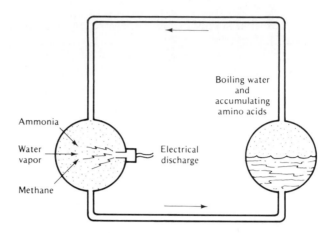

acids can be formed experimentally in this way by using not only electricity but heat, ultraviolet light, sunlight, and radioactivity. Such experiments give us an insight into the very first events on earth that ultimately would lead to life.

Amino acids are constructed out of small molecules that have from ten to twenty atoms each. The initial steps in forming the amino acids seem to have involved several very simple but highly reactive organic molecules such as formaldehyde (CH_2O), acetylene (H_2C_2), and hydrogen cyanide (HCN) (Figure 4.3). These have also been detected in the experiments, and they would have rapidly combined to produce amino acids.

This process of producing organic molecules, such as amino acids, from nonbiological materials is called **abiotic synthesis**. The term means that these materials were synthesized without any living organisms being present. Today amino acids are formed naturally only by living things, but initially, before life existed, these building blocks were synthesized from materials by nonliving energy sources. Abiotic synthesis does not occur today because organic molecules are used up by living organisms, especially bacteria, before complex organization can occur. As amino acids and other simple organic molecules gradually built up in the oceans or in fresh water (we are not sure which), some of them combined into larger and larger molecules, so that proteins or proteinlike giant molecules would eventually have formed. These then would have combined into still larger structures that may have been cell-like. In order for such cells to qualify as being living entities, they would have to be self-replicating, that is, have the ability to reproduce, and they would have to utilize other, external molecules as a source of energy, or food. Exactly how these latter steps of abiotic synthesis took place is still largely a matter of speculation. One intriguing experiment was performed by Sidney Fox, a biochemist, who was able to link amino acids together by boiling them in seawater, thus simulating conditions around an undersea volcano. He found in the boiled mixture tiny microspheres surrounded by a double-walled membrane, some of which had divided. For living systems to develop, amino acids must be assembled

FIGURE 4.3

A potential pathway for the origin of complex organic molecules. We start with a few simple inorganic compounds such as water, carbon dioxide, methane, and ammonia. These may be synthesized into highly unstable, short-lived intermediate compounds that, in turn, may combine into amino acids, the building blocks of proteins and life.

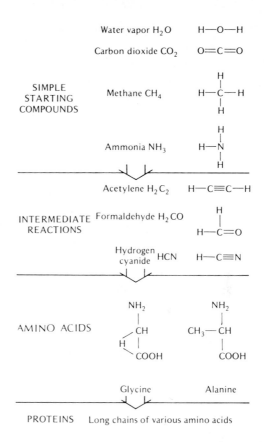

into long chains to produce protein and nucleic acid molecules. How did these very large molecules become assembled? One suggestion is that tiny crystals of clays served as templates to which the amino acids adhered in definite sequences.

All of this synthesis took place in a reducing environment. There was no free oxygen that could be burned by a primitive organism as a source of energy. There was no ozone layer in the atmosphere to shield the earth's surface from the ultraviolet light that would have been lethal to first-formed life. Probably the surface waters of the oceans provided the first ultraviolet shield. However, about 10 meters of water is needed to provide such a shield, so life surely did not originate right at the surface of the water but at some depth.

What would these first kinds of life have been like? They would have been very small (1 to 2 micrometers) and simple (composed of a single cell). They would have been able to survive without the presence of oxygen; in fact, free oxygen would have been a deadly poison for them. How would they have obtained nourishment? They would have been what are known as anaerobic heterotrophs. By **anaerobic** we mean that they survived in an oxygen-free environment, utilizing outside sources for food; hence, they were **heterotrophs** (different food). Organisms that can manufacture their own food (complex organic molecules from simple compounds), such as green plants, are called **autotrophs** (same food).

The waters of the earth at this stage probably contained large numbers of organic molecules of various sorts, formed in the ways we have just described. These first organisms would have utilized the molecules in what has been called a primordial or prebiotic or organic soup. The very earliest kinds of life would have fed from the organic soup that had been formed in water over millions of years. This source of food was definitely limited in amount and was renewed at a very slow rate. Thus, early heterotrophs may have depleted the source rapidly and created the earth's first food shortage. By the time this happened, however, other forms of life had solved the problem of a diminishing supply of organic molecules. They had learned how to take simple compounds that were present in large quantities and combine these materials into more complex organic molecules that could serve as a source of energy (i.e., food) for them.

In order to do this there had to be an external source of energy, since the chemical reactions in which simple materials such as carbon dioxide and water are combined into a sugar require energy to take place. One potential source of energy is that produced when certain molecules are split. For instance, energy is released when hydrogen sulfide (H_2S) is split, and the sulfur is changed to elemental sulfur (S). Another obvious source of energy is sunlight, but in order to use it an organism must have some way of converting solar energy to chemical energy. This is done in green plants by the pigment chlorophyll among others, and in bacteria and some algae by other pigments. Prior to photosynthesis, the original chemical reaction, which did not require sunlight, may have been a fermentation process producing alcohol and carbon dioxide as from sugar:

$$(CH_2O)_6 \rightarrow C_2H_5OH + CO_2$$

Early life forms may have been photoassimilators, using sunlight as a source of energy and converting it into chemical energy in reactions similar to fermentation. This step would have led to organisms possibly resembling certain living bacteria, which are anaerobic photoautotrophs; that is, they get along without oxygen— shun it, in fact—and use sunlight as a source of energy to convert simple compounds into food, in this case, converting carbon dioxide and hydrogen sulfide into formaldehyde and free sulfur:

$$CO_2 + H_2S \rightarrow (CH_2O)_n + S$$

Once energy-converting pigments evolved, the next step was to use two of the most abundant compounds available as a source of food:

$$CO_2 + H_2O \rightarrow (CH_2O)_n + H_2O + O_2$$

This is probably the single most important chemical reaction that has ever taken place on earth. Note that both carbon dioxide and water are very abundant and easy to acquire for the organism. The formaldehyde produced, with conversion by fermentation to sugar, served as a source of food. Water and free oxygen are released as by-products of the reaction. This is the first time that free oxygen was being produced. The oxygen was released into a reducing environment and initially, and for some time to come, it would have combined with other elements to form inert compounds, especially with reduced iron to form iron oxides (FeO, Fe_2O_3). As autotrophs flourished and produced more and more oxygen, the various reduced elements would have all become oxidized and bound up with oxygen. At this point, free oxygen would have begun to accumulate in the atmosphere and to dissolve in water. As we have mentioned, free oxygen would have been a poison to organisms acclimated to a reducing environment. Perhaps many of these organisms became extinct, but others were able to adjust to these different chemical conditions and to survive in an oxidizing environment. When this happened, it marked the beginning of an atmosphere much like our present one.

We have now seen how life may have originated and how our oxygen-rich atmosphere evolved by the development of photoautotrophs. Do we have any evidence in the fossil record to support the series of hypotheses we have outlined? Indeed, there is considerable evidence to support the scenario given here, and this will be the subject of the next chapter.

KEY TERMS

abiotic synthesis	core	meteorite
amino acids	crust	planet
anaerobic	heterotroph	reducing atmosphere
atmosphere	mantle	solar wind
autotroph		

READINGS

Most introductory textbooks in physical geology include chapters on the origin and early history of the earth. There is considerable variation in terms of detail and comprehensiveness. Some of these texts are listed below.

Bernal, J.D. 1967. *The Origin of Life*. World Press. This book explains the possible origin of large molecules on clay templates.

Fox, S.W., ed. 1988. *The Emergence of Life: Darwinian Evolution from the Inside*. Basic Books. 208 pages. Papers from recent symposium on the abiotic origin and synthesis of life.

Origins: Can Man Survive? (videotape). 1987. BBC, Group W Television. 30 minutes. Covers the origin of the universe, cosmology, human evolution, and artificial intelligence.

Schopf, J.W., ed. 1983. *Earth's Earliest Biosphere: Its Origin and Evolution*. Princeton Univ. Press. 543 pages. An important book with chapters by several experts in the field. Includes information on the earth's oldest known fossils from Australia.

5

The Precambrian Fossil Record

Now that we are aware of how life may have originated, we are prepared to ask the question, Is there fossil evidence to support the hypothesis of abiotic synthesis? The answer to this question is yes, there is considerable evidence available, and more evidence is becoming available each year. Documentation of this early fossil record is a relatively recent event in paleontology that has generated much interest on the part of all paleontologists. Prior to 1954 there were almost no fossils known at all from Precambrian rocks, and the few that were known were almost invariably suspect in one way or another. But the mystery of early life began to unravel as geologists and paleontologists learned just what kinds of Precambrian rocks to look in for fossils and how to recognize these fossils.

To a great extent, the work involved the study of **black chert**. Chert is a microcrystalline form of silica, which in its common crystalline form, quartz, is one of the most common minerals on the earth's surface. The black color is due to finely disseminated carbon. These ancient black cherts apparently formed on the sea floor from gels of silica that were precipitated directly on the surface of Precambrian sediments. The colloids of silica were soft and sticky, allowing the bodies of microscopic primitive organisms to become entrapped and preserved in the gels, which later were buried and hardened into cherts (Figure 5.1). Except in the case of the very youngest Precambrian rocks, in which fossils are found in other kinds of sedimentary rocks, most known Precambrian fossils come from black cherts.

FIGURE 5.1

Photomicrographs of Precambrian fossils. Specimen *A* is a pseudo- or false fossil. *E–P* are preserved in black chert. Specimens *B–D* are among the oldest known fossils, consisting of an algal filament (*B*) and spherical cells (*C, D*) 2250 million years old. Specimens *E–H* are 1200 million years old; *I–K*, 1400 million; and *L–P*, 850 million. Magnifications in micrometers (microns) are indicated by bars. (Reproduced, with permission, from *Annual Review of Earth and Planetary Sciences*, vol. 3. Copyright © 1975 by Annual Reviews, Inc. All rights reserved.)

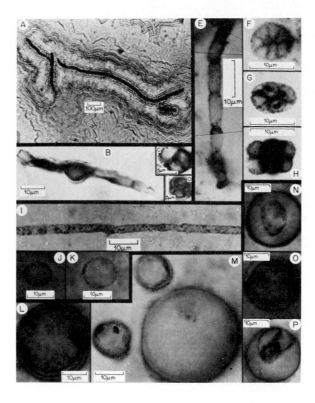

THE OLDEST KNOWN FOSSILS

For almost twenty years the oldest known fossils were thought to be from southern Africa in what is known as the **Fig Tree Group**, which have been dated as being at least 3.1 billion years old (Figure 5.2). In 1982 still-older fossils were discovered in northwestern Australia which are 3.5 billion years old. These fossils are those of bacteria and single celled cyanobacteria. They were found associated with organic structures called **stromatolites** which provide evidence for algal activity. Stromatolites are concentrically layered rocks that look like a cabbage head. The layers are caused by successive growth of thin algal mats, one on top of the other. The mats trap sediment and the algae cause precipitation of the mineral calcite, producing sediment layers that harden into stromatolites. These structures, present sporadically throughout the Precambrian, provide evidence for life and are thus considered fossils, even though they are not the skeletons of organisms but simply sedimentary rock layers produced by the activity of photosynthetic organisms. The organic-walled fossils that have been reported from the Fig Tree Formation are preserved in chert that was associated with stromatolites. It is remarkable that such small, fragile organisms could be preserved for such a long period of time without being destroyed.

FIGURE 5.2
Important occurrences of fossils in Precambrian rocks, ranging from 3.1 to 0.7 billion years ago.

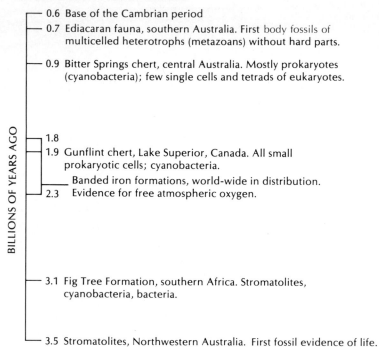

THE PRECAMBRIAN FOSSIL RECORD

0.6 Base of the Cambrian period

0.7 Ediacaran fauna, southern Australia. First body fossils of multicelled heterotrophs (metazoans) without hard parts.

0.9 Bitter Springs chert, central Australia. Mostly prokaryotes (cyanobacteria); few single cells and tetrads of eukaryotes.

1.8
1.9 Gunflint chert, Lake Superior, Canada. All small prokaryotic cells; cyanobacteria.

Banded iron formations, world-wide in distribution.
2.3 Evidence for free atmospheric oxygen.

3.1 Fig Tree Formation, southern Africa. Stromatolites, cyanobacteria, bacteria.

3.5 Stromatolites, Northwestern Australia. First fossil evidence of life.

BILLIONS OF YEARS AGO

The small spheroidal fossils that constitute the oldest known life forms are only about 10 to 20 μm in diameter. These earliest known fossils are entirely consistent with the hypothesis of abiotic synthesis outlined in the previous chapter, because bacteria and cyanobacteria are the simplest and smallest forms of life that are living today. Cyanobacteria are bacteria that can photosynthesize (they were formerly called blue-green algae). Together, they constitute a division of life called the **prokaryotes** (or **monera**), which stands in contrast to all other life, called **eukaryotes**. Prokaryotic cells are considerably simpler than are those of a eukaryote. Prokaryotes lack a membrane surrounding the nucleus; the nucleus is, instead, continuous with the cytoplasm of the cell. Eukaryotic cells also contain other membrane-bound structures called **organelles** which are lacking in prokaryotes. The nuclear material of prokaryotes does contain DNA, the master blueprint molecule for reproduction, but it is not organized into chromosomes as it is in eukaryotic organisms. Thus, prokaryotes are less highly organized and simpler than eukaryotes; their cells are also conspicuously smaller than most eukaryotic cells. They may represent a very early divergent group in the evolution of life and, again, such small, simple cells are what we would expect to find as earliest fossils.

One theory explaining the development of eukaryotic cells from prokaryotic cells is that organelles in the larger cells were initially independent prokaryotic cells that became incorporated into other cells. Instead of being digested or eaten,

FIGURE 5.3
Microfossils from the Gunflint
Chert. These are all of long,
filamentous strands that are
thought to be the remnants
of cyanobacteria. Scale is indicated by 10-micrometer
bars. (Courtesy of J.W.
Schopf, U.C.L.A.)

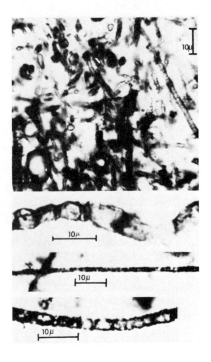

the prokaryotic cells were retained and the two types of cells interacted to form a
mutually beneficial relationship. For instance, the ingested cell might use its
photosynthetic capability to manufacture food for both cells while the other might
take care of metabolic wastes. A relationship of this kind between two independent
organisms is called symbiosis, and organelles may have evolved from an initial
relationship of this kind.

THE GUNFLINT AND BITTER SPRINGS FOSSILS

Australian and African fossils are from Precambrian rocks that are termed **Archean**.
These are the oldest known rocks on the earth's surface and range in age from 4
billion to 2.5 billion years old. Younger Precambrian rocks, which are from 2.5 to
0.6 billion years old, are termed **Proterozoic**. These younger Precambrian rocks
contain many more fossil localities than do Archean rocks, which are usually
highly metamorphosed. All known Archean fossils were formed from prokaryotic
cells and are associated with black cherts and stromatolites. Precambrian rocks of
the Proterozoic have yielded a variety of small organic-walled microfossils. A
variety of forms of cyanobacteria and rare bacteria have been reported from the
Gunflint Chert on the north shore of Lake Superior (Figure 5.3). These are dated
at 1.9 billion years old. The **Bitter Springs Formation** of central Australia, which
is 0.9 billion years old, has yielded microfossils interpreted as remains of green

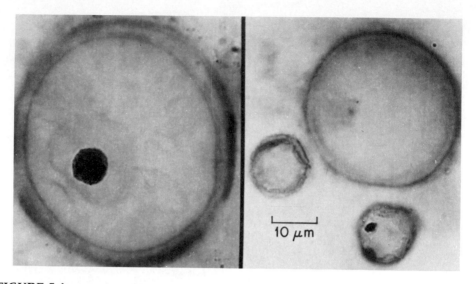

FIGURE 5.4
Precambrian microfossils about 850 million years old that probably record unicellular eukaryotes. These fossils are from the Grand Canyon, Arizona. The dark areas inside the cells are thought to represent degraded organelles typical of eukaryotic cells. (Courtesy of J.W. Schopf, U.C.L.A.)

algae, which are eukaryotic organisms. Thus, by this time in earth history, life with an organized nucleus was present (Figure 5.4). In addition, whereas prokaryotes reproduce asexually, by simple splitting of a parent, eukaryotes reproduce sexually, with exchange of genetic materials. The latter reproductive mode allows for greatly increased genetic variation, and once sexual reproduction was achieved, the rate of evolutionary change may have increased dramatically.

CHEMICAL INDICATORS

In addition to the preserved remains of organisms as evidence for life in the Precambrian, a variety of less direct evidence is available to support the idea that life evolved very early in earth history and had a long evolutionary history in the Precambrian that is still very poorly known. Some of the black cherts we have mentioned have been analyzed chemically for the organic compounds that they may contain. Two of the products that have been isolated are **pristane** and **phytane**. These are compounds that form part of the photosynthetic pigment chlorophyll and are stable degradation products of chlorophyll. The presence of these molecules in early Precambrian rocks is evidence of photoautotrophs and reinforces the interpretation of the Fig Tree and other preserved microfossils as being truly the remains of organisms. Another chemical indicator of the presence of life in these ancient rocks is provided by studying isotopes of carbon. Two stable carbon isotopes are **carbon 12** and **carbon 13,** both present today, although C_{12} is the prevalent form. During photosynthesis, which uses CO_2, the organism takes up proportionately more carbon dioxide in which the carbon is C_{12} than that in which C_{13} is present. Thus, biologically formed organic materials tend to be enriched in the lighter carbon isotope, C_{12}. By studying the ratios of these two isotopes in ancient black cherts, it has been possible to determine that they are consistent with the ratios produced by modern living organisms, thus again substantiating the record of photosynthetic organisms very early in the Precambrian.

BANDED IRON FORMATIONS

Cyanobacteria are photosynthetic, and the carbon and pristane-phytane data accumulated from Precambrian rocks also document the presence of photosynthetic organisms quite early in the Precambrian. We have seen that photosynthesis releases free oxygen to the atmosphere. Does this mean that an oxygen-rich atmosphere was present very early in the Precambrian? It seems that this was not the case, but rather, that it took millions of years for enough oxygen to be produced to oxidize all of the reduced elements that were present in water and in the air. Precambrian iron formations are the principal source of evidence for the timing of these events. These extensive **banded iron formations** of rock are rich in iron oxides. They are valuable sources of iron ore found in the Lake Superior region, Montana, Wyoming, New Mexico, Labrador, Brazil, Venezuela, Australia, southern Africa, and the Soviet Union. The deposits range in age from 1.8 to 2.3 billion years. Enormous amounts of oxygen were required to oxidize all of the iron contained in these deposits, and it wasn't until this iron was oxidized that oxygen was able to begin accumulating in the atmosphere. Thus, our oxygen-rich atmosphere did not even begin accumulating until about 1.8 billion years ago, and it took a billion years to reach its present level of concentration.

EVOLUTION OF METAPHYTES AND METAZOANS

The fossil record indicates that from about 3.5 billion to 0.9 billion years life on earth consisted of simple, small organisms that were single-celled or loose aggregations of a few cells (Figure 5.5). Truly multicellular plants and animals, with differentiation of cells to perform specific functions, had not yet evolved. Both eukaryotic and prokaryotic life forms were present, widespread, and reasonably diverse. The next important event in evolution was for these simple organisms to begin combining into macroscopic entities that contained millions of cells. These larger forms of life we call **metaphytes** (or plants) and **metazoans** (or animals). The first metaphytes were certainly plants that lived in water and, in the vernacular, we would call them algae or seaweeds. The fossil record for such plants is quite poor and none are known from Precambrian rocks, the first records of seaweeds being impressions of the body, or **thallus,** in rocks of Middle Cambrian age at the beginning of the Paleozoic Era.

FIGURE 5.5
A loose colony of cells, probably of cyanobacteria, about 1000 million years old, from South Australia. Highly magnified. (Courtesy of J.W. Schopf, U.C.L.A.)

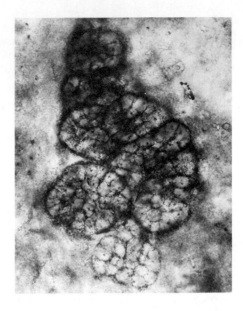

The very first hints of multicelled animals are trace fossils. A variety of burrows and feeding traces have been found in late Precambrian (Proterozoic) rocks in several thick sections of rock especially in Montana and Canada. These traces are sufficiently large that they must have been made by large metazoan animals that could move through the sediment. Such creatures must have been several centimeters in size and probably entirely soft-bodied since no traces of skeletons have been found. The burrow and feeding track traces are up to 1 billion years old, and the oldest ones are quite small and simple. They become increasingly large and complex in younger Proterozoic rocks. In addition to these animal traces carbon impressions that have been ascribed to multicellular algae have been found in rocks up to 1.3 billion years old in Montana. Thus, metaphytes may have arisen up to 300 million years earlier than metazoans.

THE EDIACARAN AND TOMMOTIAN FAUNAS

The oldest known body fossils of animals are known mainly from South Australia where a very famous fossil fauna has been found, named after a small town in the area, Ediacara (Figure 5.6). The fossils of **Ediacaran fauna** occur in a sandstone and are preserved strictly as casts and molds. No skeletal remains have been found, and it is presumed that the animals were soft-bodied. The sandstone in which they occur is several hundred feet below the lowest bed in that area that contains skeletonized fossils of trilobites. Elements of Ediacaran fauna have been found on other continents, most notably in Nova Scotia, Canada, but with nowhere near the abundance and variety of the original Australian site.

FIGURE 5.6
A slab with the impressions of several fossil coelenterates (jellyfish or medusae) from the Ediacaran beds of South Australia. The specimens average about 3 cm in diameter. (Courtesy of Field Museum of Natural History, Chicago.)

Most of the Ediacaran fossils are small, generally only an inch or two in maximum size, but a few are over a foot long. Many have radial symmetry, but a few show a bilateral pattern. One kind has a three-sided symmetry. Most of the specimens have been interpreted as being of simple jellyfish and worms, although none of the fossils is very similar to any living jellyfish or worm. They may very well represent a fauna that has no modern descendants at all.

Thus, at the very close of the Proterozoic Era, just before the onset of Paleozoic time, we have our first good look at what larger animal life was like. In this strange world there were probably not very many kinds of metazoans, and they probably lived in shallow marine waters on sandy bottoms.

Ediacaran fossils are important in several respects. First and foremost they provide oldest evidence for a heterotrophic lifestyle in the usual sense. We have seen that a few bacteria are known from the Precambrian. These organisms were probably heterotrophic, by which we mean that they were not capable of manufacturing their own food, depending instead upon outside food sources. We also mentioned that the very earliest life was probably heterotrophic, feeding on the primordial soup. All of the other fossils known from 3.1 to 0.9 billion years seem to have been photoautotrophs that were capable of synthesizing complex organic compounds which they then used as a source of energy. The Ediacaran fauna were surely advanced heterotrophs in the sense that they had to capture other organisms for food. Presumably, heterotrophs evolved from autotrophs by losing the capacity to photosynthesize and learning how to ingest other life. This may have happened at the unicellular stage of development, but we have no fossil record of Precambrian unicellular heterotrophs.

Ediacaran fossils also shed light on the transition from single cells to multicellularity. Loose associations or colonies of cells are known in the Precambrian. With the advent of heterotrophy, an increase in size may have had a distinct survival value. As size increased and the number of cells in a colony multiplied,

FIGURE 5.7
Four typical skeletal fossils of
Tommotian age. All are phosphatic, quite small (1 or 2
mm in size), and without
clear relationship to living
animals, although some may
be mollusks. (Adapted from
Matthews and Missarzhevski,
1975.)

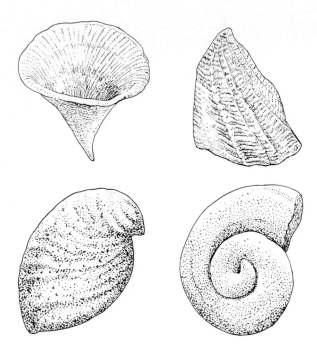

some of the cells would have been completely surrounded by other cells of the
colony and would have been isolated from direct contact with the surrounding
water medium. These interior cells would not have been able to obtain nutrients
or oxygen directly from water or to release waste or reproductive products directly
into the water. Specialization of cell functions may have arisen under these
conditions, with cooperation and communication among cells. This process would
have led ultimately to aggregations of cells specialized to perform specific functions,
thus forming the first tissues. Increasing specialization with increasing size would
have resulted in tissues combining into complex organs and, finally, groups of
organs forming even more complex systems of organs, such as the digestive,
nervous, excretory, and reproductive systems. The Ediacaran fossils clearly indicate
the presence of such systems of organs. Once multicellularity had been achieved,
the evolutionary steps from clusters of cells to tissues and organs to organ systems
may have taken place quite rapidly. At any rate, large complex heterotrophs with
several organ systems were assuredly present by 700 million years ago when the
Ediacaran fauna were alive.

The next most important fossils belong to the very basal rocks of the Cambrian
Period that have been named the Tommotian Stage, from an area in the USSR,
where some of the fossils were first discovered by Russian paleontologists (Figure
5.7). These fossils are the oldest known animals with a hard skeleton, but they
are unique in that they occur below the oldest known specimens of trilobites, the
animals that for years were considered to be the oldest skeletonized creatures and

that were used to define the base of the Cambrian system. We now know that trilobites always occur above the oldest fossil skeletons; thus the lowest trilobites mark the top of the Tommotian Stage.

These earliest skeletons are strange. They are all very small and consist mainly of phosphatic skeletons. The specimens are obtained by dissolving limestones or dolomites in dilute acetic acid; this dissolves the lime but does not affect calcium phosphate and thus releases the fossils from the surrounding matrix. Many of these tiny fossils cannot be assigned to existing phyla of animals with a great deal of certainty. Some look like sponge spicules; others are simple cones that may have been built by mollusks or worms. Still others are coiled like small snails. Once again, these fossils show us that animal life near the start of the Paleozoic Era was really quite unlike anything alive today.

Tommotian fossils have been found not only in the Soviet Union but in the western United States, Great Britain, and elsewhere. Shortly after their appearance in the rock the first trilobites are found and the more typical and much longer known Cambrian fauna begin to appear. This fauna is dominated by trilobites and secondarily by phosphatic brachiopods. A wide variety of other fossils also appears, including verifiable echinoderms and mollusks, but they remain rare.

KEY TERMS

Archean	eukaryotes	pristane
banded iron formations	Fig Tree	prokaryotes
Bitter Springs	Gunflint	Proterozoic
black chert	metazoa	stromatolite
carbon 13	metaphyte	thallus
cyanobacteria	phytane	Tommotian fauna
Ediacaran fauna		

READINGS

Precambrian paleontology is a rapidly expanding field of research that is relatively new. Modern studies date from a classical paper by S.A. Tyler and E.S. Barghoorn in 1954. The literature consists mainly of systematic papers describing new occurrences of Precambrian fossils and of review papers that summarize knowledge to date that has been published in the field. Several review papers are listed below. These have extensive references that provide entry into the literature on Precambrian fossils.

Barghoorn, E.S., and Schopf, J.W. 1966. "Microorganisms Three Billion Years Old from the Precambrian of South Africa." *Science* 152:758–63. A description of the oldest known fossils.

Barghoorn, E.S. 1971. "The Oldest Fossils." *Scientific American* 224:30–42. A popular account of Fig Tree fossils.

Glaessner, M.F., and Wade, M. 1966. "The Late Precambrian Fossils from Ediacara, South Australia." *Palaeontology* 9:599–628. Description and illustration of the oldest known metazoan body fossils.

Glaessner, M.F. 1985. *The Dawn of Animal Life.* Cambridge Univ. Press. 244 pages. A biohistorical study with special emphasis on fossils at or close to the Precambrian-Cambrian boundary and especially the Ediacaran fauna.

Hoffman, A., and Nitecki, M.H. ed. 1986. *Problematic Fossil Taxa.* Clarendon Press. 267 pages. A series of papers on all sorts of skeletal fossils that have uncertain affinities with known major groups of animals, living or fossil. Very well illustrated.

Matthews, S.C., and Missarzhevski, V.V. 1975. "Small Shelly Fossils of Late Precambrian and Early Cambrian Age: A Review of Recent Work." *Journal of the Geological Society of London* 131:289–304. A landmark paper on Vendian and Tommotian fossils with superb photographs of these small, enigmatic fossils.

Schopf, J.W. 1975. "Precambrian Paleobiology: Problems and Perspectives." *Annual Review of Earth and Planetary Sciences* 3:213–45. A recent review of the status of knowledge concerning Precambrian fossils.

6
Organic Evolution and Extinction

Evolution is one of the most provocative theories ever conceived. It has entered into debates not only in science but also in religion, politics, and education. The idea that species are not immutable, fixed packages of life, but can be transformed one into another, has changed our perspective of the world around us. For centuries one of the prime dogmas of natural history was that all life was created on earth solely for the benefit and use of humankind. As the fossil record slowly accumulated, it came to be realized that there were species of plants and animals that had lived in the past but that no longer existed. Even that point took a long time to be established; some persistently argued that these supposedly extinct species would eventually turn out to be living in still unexplored parts of the earth.

During medieval time, all life was viewed as forming a "Great Chain of Being." Each species was a link in this chain, the lowest forms of life being algae, then moving up through higher plants, then low forms of animals, culminating in the final link, human beings. Links closest to each other were most closely related, but each link was immutable and unchanging. This idea was popular for many years and still is evidenced in such terms as "missing links," used for animals that are supposedly or actually intermediate in relationship between two major groups of life.

This view of a static world was thoroughly upset by the theory of Charles Darwin. In his monumental work *Origin of Species*, published in 1859, he presented many detailed examples demonstrating that species had arisen from each other and that evolution had occurred.

Evolution provides the underlying foundation for all of paleontology. The basic processes of evolution are essential to understanding the history of life. There have been many books written on the subject, yet here we must condense our discussion to a single chapter. There are three main areas of evolutionary theory and practice that concern us in the study of fossils. First is the fact that the sequence of occurrence of fossils in rocks of different ages and their transition from one form to another through time provide the material evidence for the evolution of life that has indeed taken place on earth. This aspect of evolution we will not discuss in this chapter because most of the book is devoted to discussion of the documentary record of evolution provided by fossils. The other two important facets of organic evolution are (1) the genetical basis of evolution—the ways in which individual plants and animals are constructed that has resulted from the evolutionary process and (2) the mechanism of evolution—the ways in which a species of organisms, composed of many individuals and populations, may change through time into another species.

THE BASIS OF EVOLUTION

We know that the offspring of any plant or animal inherits many of its characteristics from the parents. If the offspring is the result of sexual reproduction, it exhibits a blend of the characters of each parent, being exactly like neither the male nor female parent but combining traits of each, and thus having some features that are uniquely its own. The science of how inheritance takes place is called **genetics**.

The transfer of heritable characters from parent to offspring is effected by the nuclei of the sex cells, the sperm and the egg. Each nucleus contains protein strands of nucleic material called **chromosomes** that have very large molecules of deoxyribonucleic acid, commonly abbreviated to **DNA**, in them. These molecules are the transmitters of genetic information from the parent to the offspring; they ensure that the offspring will develop most of the same characters that are possessed by the two parents. Each parent contributes the same number of chromosomes to the first-formed cell of the offspring from which all other cells of the young develop, thus ensuring that each parent will have equal representation in the genetic materials of the new generation.

A DNA molecule is composed of two helically spiral strands, each of which is composed of a linear chain of sugar and phosphate molecules (Figure 6.1). The two spiral strands are held together by crossbars composed of four kinds of nitrogenous bases called **adenine, thymine, cytosine**, and **guanine**. It is the arrangement of these four bases that provides the genetic code of the chromosome that is passed on from one generation to the next. The bases are arranged in very definite sequences along the helix, two bases composing each cross-link. Because of the chemical structure of the bases, adenine can only form a crossbar with thymine and cytosine with guanine. When the helix unravels to form a single strand, that strand contains all the information needed to build another compatible new strand to make up a new double helix in the offspring. If the parent strand

FIGURE 6.1
The molecular structure of
DNA. Unraveling of the spiral
helix and splitting of the pairs
of cross-linking bases provides a pattern for duplica-
tion of the structure. A unit
length of the DNA molecule
that includes about 1500
pairs of bases constitutes a
gene, a single hereditary unit.

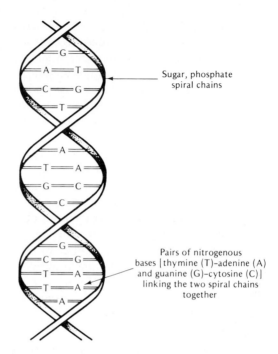

Sugar, phosphate
spiral chains

Pairs of nitrogenous
bases [thymine (T)–adenine (A)
and guanine (G)–cytosine (C)]
linking the two spiral chains
together

has adenine at a specific site, the new crossbar can only be completed by thymine.
The sequence provides a code for the construction of proteins and enzymes of the
new individual, ensuring that the genetic features of the parents will be passed on
to the offspring.

The synthesis of new molecules, proteins and enzymes, is carried out by
another kind of large, complex molecule called ribonucleic acid, or **RNA**. This
molecule is shorter than DNA, is single stranded, and is linear rather than helical.
It chemically copies a single DNA strand, using the compatible opposite member
of each of the two pairs of nitrogenous bases. That is, where the DNA strand has
cytosine, the RNA molecule will insert guanine.

If the matching of these bases is so exact, one may very well ask, How do
any possible changes ever occur in the DNA molecules? Why are not all the genetic
materials of all organisms exactly the same? Changes in the sequences of
nitrogenous bases, called **mutations**, can and do occur in several ways. One or
more pairs of bases may be deleted from the structure, two or more pairs may be
inverted in their order, or one or more pairs of bases may flip-flop from one strand
to the other. If only one or a few of the bases are involved, the mutation is a minor
one and commonly cannot be detected. The subunit of the chromosome that results
in transfer of a detectable heritable characteristic is called a **gene**. We now think
that each gene consists of as many as 1500 pairs of bases; thus a change in only
one or two bases may have very little effect on the gene. We do know that large
changes, mutations that affect a conspicuous part of the chromosome, are almost

invariably lethal or severely deletirious to the organism. The gradual buildup of many very small mutations over a long period of time is not necessarily lethal or disadvantageous but may, in fact, result in genetic changes that are advantageous to the organism. It is the gradual accretion of these many quite small mutations that are considered to be the underlying foundation of evolution.

The mutations themselves do not cause evolution. The physical expression of genetic features is called the **phenotype** of the individual. The genetic factors that result in the phenotype are called the **genotype** of the individual. If there is sufficient change in the genotype, through a series of small mutations that are neither lethal nor disadvantageous, a change in the outward appearance, or phenotype, will result. This process results in phenotypic variation. All populations are variable in one or many of the aspects of the individuals that make up the population. Evolution operates at the population level. It does not take place on or in a single individual, much less on the genes of that individual. Thus, mutations provide an important source of heritable variation within populations. Mutations are reasonably common in individuals of a single, interbreeding population. Yet those individuals remain part of a single species. The mutations do not result in evolution; they provide some of the phenotypic variation exhibited by the population.

THE MECHANISM OF EVOLUTION

Evolution takes place within populations of plants and animals. Although each natural population is made up of a collection of interbreeding organisms of a single species, it is not the genetic makeup of each individual that is important in evolution but rather the genetic composition of the entire population.

Variation

The first aspect to be considered is **variation**. Every species is made up of one to many local interbreeding populations. Individuals within a population show a certain amount of variability—great in some instances, small in others. Some of this variation is the result of interaction between individuals and their surrounding environment and is due to factors such as food supply, living habitat, temperature, and rainfall or sunlight. This type of variation is not heritable and is not passed on from parent to offspring, so it is of little interest here. Some of the observed variation is genetic in origin and is due to the somewhat different genetic makeups of the individuals in the population. As far as is known, there is no natural population of organisms in which the genotypes and phenotypes of all individuals are identical. All populations include some genetic variability among the members, even though phenotypic variation may be slight or absent in rare cases. Even differences in sex are examples of genetic variation in populations.

The variations that are of most interest in the study of evolution are mutations. There are relatively sudden changes in one or more genes that may or may not produce noticeable phenotypic changes. Many of the mutations that are known in organisms are either lethal, killing the offspring, or they render the offspring sterile, so that the mutation is not passed on to succeeding generations. There are some mutations, however, that represent small changes in phenotype; they are heritable, and are passed on to succeeding generations. Nonlethal mutations in which offspring are not rendered sterile apparently occur in all natural populations, but at a slow rate, producing changes in phenotype that become noticeable only after generations, as the mutation is passed through increasingly greater numbers of the population in which it occurs.

Even if genetic variability produced by mutations can occur in viable offspring, this does not necessarily mean that the mutation confers any degree of advantage on those individuals that carry the mutation. Any mutation may be advantageous, disadvantageous, or neutral as far as the fitness of the individual carrying it is concerned. This brings us to the next aspect of the evolutionary mechanism—natural selection.

Natural Selection

Given that all or most populations display genetic variability, not all individuals are equally likely to cope successfully with their environment to exactly the same degree. Especially important in this respect is success at reproduction. Those plants and animals that are successful breeders and produce viable offspring pass on their genetic characters to the next generation. Some scientists have argued that the physical body of an organism is simply a vehicle designed by genes to transmit the genes to the next generation. Many organisms die upon successful completion of reproduction. Annual flowering plants die after setting seed. Male black widow spiders are eaten by the females after mating. Any interaction between the individual and the environment that tends to prevent or diminish reproductive success is lethal or disadvantageous to the genes of that organism. The plant or animal may die before it reaches breeding age, it may be unsuccessful at attracting a mate, or it may be unable to establish and defend a breeding territory. Other disadvantageous aspects may relate to feeding, migration, predation, and so on. Conversely, one or several mutations may result in individuals that are more successful at breeding than are other members of their population. They may be better able to establish a breeding territory, utilize their food resources, or resist predation. These individuals will pass on their genetic features to an increasingly larger proportion of the population with each succeeding generation. The particular mutations that confer these advantages will spread and become fixed into more and more individuals. This process, in which those individuals that are less fit to cope with certain aspects of the environment are diminished in breeding capabilities and those that carry advantageous genotypic and phenotypic characters are enhanced in their ability to reproduce, is called **natural selection**. Some are advantageous, and are selected for; others are disadvantageous, and are selected against. Over a sufficiently long period of time, typically many generations, this process may result in a gradual shift of the genetic makeup of the population. This does not mean that disadvantageous mutations are necessarily eliminated from the population. They may survive but generally not in very large numbers. Neutral mutations may persist and spread but generally at slower rates than advantageous mutations, and without significant selective action on them.

Natural selection results in a slow, continuing shift in the genetic makeup of a population. Given enough time, this change could result in a younger population having a sufficiently different pool of genes to make it significantly different genetically from an ancestral population that lived several hundred generations earlier. Since the individuals of the past generation are long since dead, there is no direct way to test for the genetic compatibility of the two populations to see if viable offspring could be produced from their interbreeding. If viable offspring could be produced, we would consider the two populations to be the same species. If they could not, then the populations would be labeled two distinct species, and we would say that evolution had occurred. Change in genetic makeup through time is commonly thought to have occurred in fossils, but we cannot test this hypothesis directly. Instead, we infer that such changes have taken place by studying observable phenotypic changes over many generations. If the observed changes in the physical aspects of the fossils are sufficiently great, then the paleontologist may infer that a new species developed directly from an ancestral species. The younger and older populations should be uniformly distinguishable. This problem, the need to make inferences concerning genetic composition in paleontology, is commonly known as the species problem.

The gradual change through time of one species population to another, phenotypically distinct species population is called **phyletic evolution**. Phylogeny (hence *phyletic*) is the study of ancestral-descendant relationships among organisms. Notice that phyletic evolution as we have described it does not result in any multiplication of species. Older species A gradually evolves into younger species B as there is a gradual shift through time in certain morphologic characters used to define the species. This process is sufficiently slow that it cannot be detected in living populations.

For many years phyletic evolution has been considered an important process in the evolutionary history of past life. There is a current controversy concerning the importance of phyletic evolution in the fossil record. Some paleontologists argue that phyletic evolution has recurred repeatedly and is an important mode of evolution. Others believe that the common instance is for a fossil lineage to persist for long periods of time without any significant changes in morphology. New species then arise rather abruptly as a result of what is called **allopatric speciation**. In order for these speciation events to occur, a segment of the population must be geographically isolated, as we will presently discuss. Perhaps both evolutionary modes are important. We are not yet in a position to evaluate the relative importance of phyletic evolution and abrupt species events. Phyletic evolution is also called phyletic gradualism because this kind of evolution leads to true speciation, or splitting of lineages through time. Sudden allopatric speciation is called **punctuated equilibrium**. In this process a species will show little or no appreciable change over a long period of time and then abruptly split into two species. The long periods of equilibrium that many species exhibit are a time of **stasis** and are interrupted or punctuated by species events.

Isolation

We have discussed two of three important aspects of evolutionary mechanisms—variation and natural selection. The third, and most important, facet of evolution is **isolation**, especially geographic isolation. Almost every species consists of a series of local populations of interbreeding individuals. These populations may be in contact with each other at the edges of each local geographic range. As long as there is contact and migration back and forth between local populations, the entire set will have a genetic makeup that is more or less uniform. Certainly there will be variability among local populations, and somewhat different kinds or strengths of natural selection. But these differences will not be sufficiently strong to prevent an individual of one local population from interbreeding successfully with an individual of another local population. There will be a certain amount of coherence to the gene pool of the entire species.

Let us consider a hypothetical small population of a species that somehow becomes established in an area in which it is effectively isolated from all other local populations of the species (Figure 6.2). The migrants, the founders of the new isolated unit, will have only a small sample of the total genetic variability of the species. Furthermore, the environmental conditions of the new area will not be exactly like those of the area in which the rest of the populations live. Phenotypic or genotypic variables that are rare or absent in the parent populations may prove to be advantageous under new and different circumstances, and the processes of natural selection may also be different. Over a number of generations, the isolated population will show differences both in genotype and in phenotype from the parent population.

Now with time and changes in climate or habitats, it is possible for migration to occur between the members of the old species and members of the isolated population. If the changes that have taken place in the isolated population are sufficiently great, then its members will not reproduce successfully with members of the old species. A new species will have been created. Failure to produce viable offspring is a critical test of speciation. In order for sufficient genetic and physical changes to take place in an isolated population, it must be genetically isolated from the parent populations for a sufficiently long period of time for these changes to build up. If the founder population is rather atypical genetically, and if the new area that it invades is sufficiently different from the old one, these changes may take place relatively rapidly.

As we have already mentioned, allopatric speciation is the name for this process of formation or splitting of species by geographic isolation. *Allopatry* (strange country) means living in different areas. *Sympatry* (same country) means living in

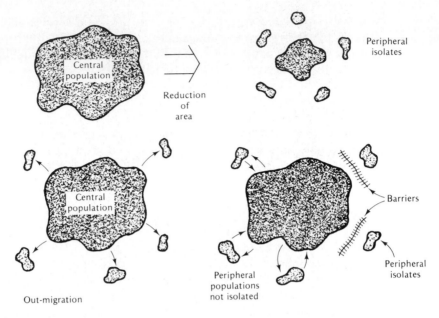

FIGURE 6.2
Hypothetical formation of peripheral isolates. Above: Area occupied by a species shrinks in size, leaving behind peripheral isolates in perimeter of original large area. Below: Peripheral populations are established, then later isolated by formation of barriers.

the same area. Some biologists today believe that new species cannot arise sympatrically, in the same area as the parent species, because of the mixing and blending of genotypes. The process of allopatric speciation can be seen taking place in living species. It is apparently a considerably faster process than phyletic evolution, which cannot be observed in a few generations in living organisms. Biologists would say that all speciation, as far as they can tell, arises through allopatry. Recall, though, that some paleontologists hold that speciation can also occur through phyletic evolution, that is, slow, gradual changes in genotype and phenotype through time. Phyletic evolution may not, however, result in multiplication of species; allopatric speciation does result in two or more species living at the same time when originally there was a single species.

Now we need to look into the ways in which isolation of populations may occur. Every species has a certain ability for dispersal, either as adults or in early larval stages. Yet very few species have a worldwide distribution. Dispersal is limited by ecological factors—differences in moisture or rainfall, in temperature, or in food sources. Most species ultimately run up against some kind of geographic barrier, such as a river or a mountain range or, for terrestrial species, an ocean, that limits dispersal. Obviously there are very many kinds of ecological and geographical barriers that prevent ubiquitous species. The founding of new, isolated populations outside the normal range of a species results when a few migrant individuals are able to surmount one or more barriers and establish themselves outside the normal range of the species. They thus become **peripheral isolates** of the species and may diverge enough genetically to become new species. Where isolation is complete and long-enduring, many new species may arise, as for instance in the fauna and flora of oceanic islands far removed from mainland areas that provided the founder population. Not only may peripheral isolates arise from migration out from a central population area, but they also may be the result of residual populations left behind and isolated when the normal range of a species contracts and shrinks due to changes in environment in intervening areas (see Figure 6.2).

The evolutionary mechanisms that we have sketched consist of three primary features: variation, natural selection, and geographic isolation. Taken together these processes are the proximal causes of evolution at the species level. We have described how a new species may arise from a different ancestral species. The evolution of new species is the foundation of evolution, but it does not encompass all of evolution. We will turn now to those processes that operate on groups of species, resulting in major changes in the diversity of larger groups of plants and animals, remembering all the time that evolution at the species level constitutes the warp and woof from which these larger designs are made.

PATTERNS OF EVOLUTION

We have now seen how and why new species may arise through time. The development of new species is at the heart of evolution. When we examine the fossil record, however, we see much more than just a random assortment of new species appearing here and there at different times in different ages of rocks. Clusters of species may be closely related and grouped into genera. Different genera in the same or related families may, over the course of time, show remarkably similar or different trends in morphologic change that can be interpreted in terms of different or similar ways of coping with the environment. These changes can represent adaptation to the environment in various lineages and, when traced through time, yield a variety of patterns that the course of evolution has followed. In this section we will discuss some of these patterns and what they mean.

Adaptive Radiation

The process of speciation has as one of its prime features the splitting off of one group from another in such a way that new species may arise. This process produces multiplication of species as well as divergence of species. Species diverge not only in their morphological characteristics but also in their environmental requirements. The degree of environmental divergence may be very slight, involving a single food source, for instance, or slight differences in a preferred living site. There may also be a drastic shift in adaptation to the environment. If a major new zone of adaptation becomes available to a species, innumerable ways of life may

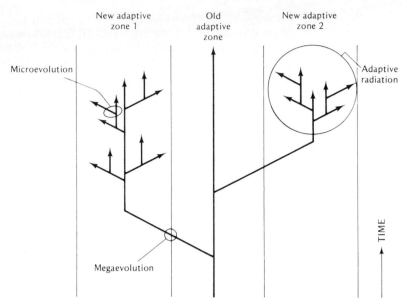

FIGURE 6.3
Diagram showing theoretical relationship of the major processes of evolution. Shift from one adaptive zone to another results in megaevolution; radiation within a new adaptive zone results in adaptive radiation, or macroevolution; origin of new species within a radiating group is microevolution.

open up to be exploited by a rapidly developing series of descendant species. Such a process is called **adaptive radiation**, or **macroevolution**, and is basically one in which a new species and genera gradually adjust to new ways of life (Figure 6.3).

There are many good examples of adaptive radiation in the fossil record. The gradual development, spread, and diversification of shelled marine invertebrates in the early Paleozoic is one example. The gradual dominance of reptiles on land and in the sea and air in the Mesozoic is another. The tremendous diversification of flowering plants after their origin in the Cretaceous is still another. Adaptive radiation takes place relatively slowly, over millions of years. The pattern of divergence may be likened to a branching tree. Each major branch represents a broad environmental setting, and small side branches record specific adaptation to microhabitats within that larger setting. The process of adaptive radiation produces new genera, new families, and even new orders of plants and animals. This process does not happen abruptly.

The fossil record shows that many episodes of adaptive radiation occurred both in the oceans and on land since the Cambrian Period. Examples include the Cambrian radiation of trilobites and the extensive expansion of mammal species after the dinosaurs and other large reptiles became extinct during the early Cenozoic Era. Mollusks from shallow marine waters also underwent adaptive radiation after

FIGURE 6.4
A specimen of a Recent *Lingula*. The chitinophosphatic shell consists of two valves above, with long fleshy stalk called the pedicle extending below from between the two valves; the pedicle anchors the shell in a soft bottom. (Courtesy of National Museum of Natural History.)

many bryozoans, brachiopods, and stalked echinoderms became extinct at the close of the Paleozoic Era. In many instances, adaptive radiation occurs after a species has become extinct because many habitats are vacant and are available for new exploitation.

Adaptive radiation can be contrasted with other evolutionary patterns, some of which are more abrupt and others much slower. On the one hand, we have examples in the fossil record of fossils that persist for millions of years with barely any perceivable change. A classic example is *Lingula*, a shallow-water brachiopod that has a shell in the Ordovician that is not detectably different from that of living *Lingula* (Figure 6.4). *Lingula* is adapted for living in sandy, intertidal environments in a burrow. Apparently it is very well adjusted to this microhabitat; it has managed to flourish continuously in this condition for a very long time. The possibility always exists that *Lingula* has evolved in its soft parts but not in its shell. Such evolution would be virtually undetectable in the fossil record.

Megaevolution

The opposite condition to that seen in *Lingula* is found where new major groups of fossils appear rather suddenly in the rock record. The sudden appearance of a new group has commonly been called **megaevolution** or, sometimes, explosive evolution (see Figure 6.3, page 86). These occurrences seem to take place rapidly in geologic terms, over a few hundred or thousand or a million or two years. Examples include the sudden appearance of land-based amphibians at the close of the Devonian, the appearance of land plants near the end of the Silurian, and the abrupt origin (but not the spread) of flowering plants at the beginning of the

Cretaceous. In most cases new major groups occupy a new major environment and have made an important shift in habitat. One of the single most difficult transitions that can be made on earth is from living in water, salt or fresh, to living on dry land. Only a few groups of plants and animals have been able to make this change successfully.

How do these relatively sudden evolutionary changes arise? Megaevolution probably takes place in relatively small populations that are geographically restricted so as to be peripheral isolates, as described for allopatric speciation. The odds of having members of such populations preserved as fossils are very small indeed and, even if preserved, the chances of a paleontologist finding fossils of the organisms making the transition are equally small. Thus, there is very likely to be a gap in the fossil record where and when the transition occurs, producing so-called missing links. Once the major step has been taken and the transition made, from water to dry land for instance, the plants or animals may then be quite successful, spread rapidly, proliferate, and be much more likely to be preserved as fossils. Thus, we tend to catch such changes not in the act but just afterward, creating the impression of a very sudden, abrupt appearance of the new group.

An important condition of megaevolutionary changes is thought to be a process called **preadaptation**. The plant or animal that makes the habitat transition has evolved a series of features in response to better adaptation to the environment in which it lives. These changes are such that they also happen to fit the organism to a successful life in a new habitat. Thus, the plant or animal is preadapted to make the transition. In the Devonian, lobe-finned fishes evolved sturdy, stumpy paired ventral fins that may have allowed them to push themselves from one freshwater pond to another. This fin condition had tremendous survival value for these fishes to remain as fishes. But the paired fins also were relatively easily converted to limbs for walking around on dry land. Thus, the fishes were preadapted for life on land, and, given the right set of circumstances, one or more small populations of lobe-fins may have rapidly made the transition from fins to limbs. This, along with other modifications, converted them into amphibians. The important thing to realize about preadaptation is that the evolutionary changes that take place do so because they have clear adaptive value for the *old* way of life. Evolution does not occur from anticipation of major changes in lifestyle. At the right time and place, however, certain unique features of an organism may enable it to make such a change. These major transitions seem to take place relatively rapidly. Given the time factor plus small populations and geographic restriction, our knowledge of just precisely how and when such shifts took place is often blurred.

Megaevolution commonly leads to adaptive radiation. The process of forming an important new way of life may allow the organisms to exploit a variety of new habitats, thus producing divergence and adaptive radiation. The relationship may not be a direct one and may take place quite slowly. Although mammals first appeared in the Triassic Period, they did not undergo detectable radiation until the Cretaceous, and conspicuous divergence did not take place until the beginning of

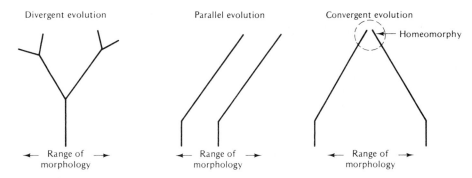

FIGURE 6.5
Patterns of evolution. In divergent evolution on the left, new species constantly diverge or differ in morphology; in parallel evolution (center), two or more series of unrelated species display a series of similar morphologic trends; in convergent evolution (right), two unrelated species come to be more and more morphologically similar, resulting in homeomorphy. The two species in convergent evolution may be synchronous (live at the same time) or heterochronous (develop the same morphology at different times).

the Cenozoic Era after many reptiles had become extinct. For many millions of years after they originated, mammals were not an especially diverse group of animals.

Parallel and Convergent Evolution

In addition to various patterns of divergence, we can also recognize other patterns in the fossil record that are either parallel or convergent. Let us suppose that we have two different lineages or stocks gradually evolving through time. A sequence of morphological changes may appear in each lineage that duplicates similar changes in the other lineage. Such sequences of changes are called **parallel evolution** (Figure 6.5). Prime examples are found in several groups of fossil invertebrates. Fossil **graptolites**, extinct colonial animals thought to be related to chordates, radiated into several distinct lineages in the Ordovician. Each lineage can be traced because of distinctive, conservative characters of the **thecae** (small chitinous structures that housed each individual of the colony). The thecae are arranged in linear branches. In each stock the number of branches gradually decreased during the Ordovician, from thirty-two to sixteen to eight to four, two, and finally to a single branch in the Silurian (see Figure 10.11, page 161). This happened in two or three distinct and separate groups of graptolites. It seems clear that these lineages were genetically similar enough to allow for the possibility of branch reduction in each group. The several stocks must also have been subjected to similar pressures of natural selection that caused them to undergo such changes.

A similar pattern can be seen in the evolution of an extinct group of shelled cephalopods called **ammonoids**. When these animals first appear in the Devonian,

FIGURE 6.6
Increase in complexity of ammonoid septa, from simplest at the bottom to most complex at the top. All but one septum has been omitted for clarity. These changes occurred in several different evolving lineages of ammonoids.

the shell is divided up into chambers by very simple shelly partitions called **septa**. During their evolution, the septa became progressively wrinkled at the edges and more and more complex (Figure 6.6). This increase in complexity occurred in several different and distinct families and superfamilies in the Paleozoic—an outstanding instance of parallel evolution. Ammonoids underwent a conspicuous adaptive radiation during the course of the Mesozoic Era and, again, different groups display increasing complexity of the septa. Toward the close of the Mesozoic, several different groups underwent a reversion of this trend, producing septa that were progressively simpler in structure, so that some of them looked much like some of their remote ancestors from the Paleozoic. These changes are likewise best interpreted as responses to repeatedly similar selective pressures. The change to simpler sutures signals adaptation to different conditions of life that made a complex suture disadvantageous. The ammonoids did not revert back to old genetic programs that were available to their Paleozoic ancestors; they moved forward toward simpler types under the pressure of selection that we find difficult to specify for these extinct animals. Evolution does not reverse itself in the sense of going back to simpler, older life form.

In **convergent evolution** two or more stocks not only develop a sequence of the same structures, as in parallel evolution, but these structures come to look very much alike in response to adaptation to a similar habitat. The convergence may be so close that the hard parts of the animals may eventually bear such a striking resemblance that their distant relationship can be revealed only through careful

FIGURE 6.7

Analogous and homologous structures in animals. The wing of an insect (*A*) and the bones in the wing of a bird (*B*) are analogous in that both serve the same function—flying. The bones in the wing of a bird (*B*) and the bones in the forelimb of a reptile (*C*) are homologous but not analogous; the bones have the same origin but serve different functions.

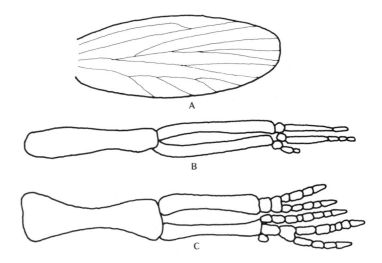

study. Such creatures are said to be **homeomorphic** (of the same form). There are numerous examples of homeomorphs among various groups of fossil invertebrates, especially among some brachiopods, ammonoids, and crinoids. The homeomorphs may occur in the same time interval and be found together in the same bed of rock. If they are contemporaneous, they are called **synchronous homeomorphs** (*synchronous* means at the same time). Alternatively, the homeomorphs may be of quite distinctly different ages. For instance, there are a few Jurassic ammonoids that are virtually indistinguishable from unrelated Cretaceous ammonoids. These are called **heterochronous homeomorphs** (existing at different times).

One of the classical examples of convergent evolution is the completely independent evolution of wings in those groups of animals that successfully invaded the aerial environment. Insects, birds, reptiles (pterosaurs), and mammals (bats) all became successful aerial creatures through evolution of the wing. Yet the details of wing structure in each of these groups are conspicuously different (Figure 6.7). The wings of these groups are said to be **analogous structures**; that is, they all serve the same function of flying. Yet the bones that make up the wing in the vertebrates and the veins of the insect wing do not all have the same origin; that is, they are not **homologous structures**. Thus, we may find wholly unrelated structures modified in similar ways to perform the same function. Such structures are analogous but not homologous. The reverse may also be true; for example, the wing of a bat, the foreleg of a dog, and a person's arm are all homologous structures because they all have the same origin, but obviously these structures are not analogous, one serving for flying, one for walking and running, and the third for grasping objects.

You will note that the preceding discussion does not refer to the evolution of an entire animal but rather to certain morphological parts of that animal. Thus, it is generally not correct to speak of one individual as being analogous or homologous with another entire specimen of a different species. Members of different species, genetically isolated from each other, cannot be completely alike in all of their functions, nor will all of their similar structures necessarily have exactly the same origin.

In speaking about the evolutionary aspects of fossils, careful distinctions must usually be made as to whether the concern is with an entire population, with a complete individual of that population, or with a specific structure borne by an individual. This is especially true when using such descriptive pairs of words as *primitive* and *advanced* or *generalized* and *specialized*. Sometimes these words are applied to whole organisms and sometimes to particular morphological features. *Primitive* and *advanced* have time connotations. A primitive species or structure may be older than a younger, advanced form or trait. Alternatively, one may refer to the condition of a structure as being relatively simple (primitive) or quite complex (advanced). In either case, the use of such qualifiers may be no more than a supposition and subject to interpretation. *Primitive* and *advanced* do not imply fitness or unfitness. Every species that we find as a fossil was obviously a successful animal for its time, or we would not find it preserved in the rocks. It would probably never have become common enough to have a remote chance of being buried and preserved if it were disadvantaged in any way.

Generalized and *specialized* may refer to an entire species or population and the way in which it exploits its environment or to an individual and the overall nature of its organization or to a specific part or system of that individual. A species may occupy a narrow microhabitat that permits little variation in environmental factors, the species thus being specialized for that particular lifestyle. On the other hand, a species may be able to live in a wide variety of environments as long as certain requirements are met within a broad range of variation. A generalized individual may display a set of characters that places it close to the ancestral stock of an evolving lineage, whereas descendant individuals in younger species may display evolved characters that are judged to be specialized for specific functions. Note again that in this case a time factor and ancestor-descendant relationships are implied. Finally, a given structure may be generalized or specialized, such as the five-toed flat foot of a mouse in contrast to the specialized single toe of a horse that is raised up on its tip.

Despite whether we consider a group of fossils or a particular structure to be primitive or generalized, that population or that feature evolved through a process of natural selection that operated on the organism and its parts to make them successful in the role they played. We recognize that just because a species became a successful animal through evolution does not necessarily mean that it was equally successful from that time forward. We know that environmental conditions on the earth have changed time and time again over millions of years. Some organisms have been able to cope with such changes and have evolved into successful descendants. Others have not been able to make such adjustments and have become extinct. Thus in a sense, extinction is the foil of evolution.

EXTINCTION

Evolution of new life forms is, at least, partly related to the extinction of old life forms. The diversity of life on earth has apparently remained relatively stable for long periods of time. Thus, as new species are added by evolution, old species fade away through **extinction**. The fossil record shows that extinction events tend to occur relatively suddenly. There have been many such events, but some are much more important than others. Recent evidence seems to indicate that extinction may occur on a cyclical basis, roughly every 26 million years.

The two most important types of extinction events largely determined the two era boundaries between the Paleozoic and Mesozoic and between the Mesozoic and Cenozoic. The former affected mainly marine bottom-dwelling organisms. Although major groups of brachiopods, bryozoans, corals, stalked echinoderms, and ammonoid cephalopods became extinct during the Paleozoic-Mesozoic event, land plants and animals and marine fishes were much less affected. The Mesozoic-Cenozoic extinction event was quite different. Both land animals, especially the dinosaurian reptiles, and marine animals, the ammonoid cephalopods, were drastically affected. Other large reptiles, such as the marine plesiosaurs and ichthyosaurs, were wiped out, along with the flying reptiles, and pterodactyls. Recent evidence attempts to link this event with the impact of a large asteroid on earth, which created a dust cloud that reduced sunlight and affected photosynthesis and primary production of plants, resulting in disruption of food webs that led to the extinction of these species. Evidence for such a collision is found in unusually high concentrations of the rare element **iridium** in clay layers of many scattered localities at the Cretaceous-Tertiary boundary. The iridium is believed to have come from the asteroid. This theory is not yet universally accepted and further research is being vigorously pursued. The cyclical nature of the pattern of extinction events is thought to be related to periodic collisions of the earth with asteroids and other bodies in the solar system.

CYCLES IN EARTH HISTORY

Many of the features that we can see in the rock record and in the history of life seem to repeat themselves in a cyclic pattern. Paleontologists have discovered many different **cycles**, but the explanations for these observed rhythms are often difficult to prove. In this section we will review some of the most widely discussed cycles, indicate their nature and duration, and how they may affect the fossil record.

The longest rhythms observed on earth seem to be those associated with worldwide glaciations and the rise and fall of sea level. The period at which major ice ages reoccur, or repeat themselves, on earth is about 200 million years. We have either just witnessed or are still in the midst of one such ice age, which we call the Pleistocene Epoch of geologic time. Near the close of the Paleozoic Era, mainly in the Permian Period, there was an ice age that especially affected the southern Gondwana continents which were massed together over the south rotational pole of the earth. This ice age occurred somewhat more than 200 million years ago, centered on 250 Ma. There was another major ice age during the Devonian Period, about 400 million years ago, one at the end of the Precambrian, and two in the Proterozoic. Thus, ice ages seem to have occurred in a cycle, but explanations for the cause of the cycle are nebulous.

The rise and fall of sea level is one of the truisms of geology. The rock record documents that the continents have been flooded by shallow seas time after time during earth history. These flooding episodes are called **transgressions**, and the subsequent withdrawals of the sea are called **regressions**. Obviously, these events affect very strongly the record of life that we find within the rocks in any given area. Careful study of the record of transgressions and regressions, especially by oil company geologists, has led to the hypothesis that these events are cyclical on a grand scale, and exhibit smaller cycles as well. The largest cycles, those of longest duration, are called first-order cycles, and they approximate the glacial cycles, lasting about 200 million years. Occurring within these big cycles are shorter cycles called second-order cycles, within those are still shorter cycles called third-order cycles, and so on. The second-order cycles are of about 10 to 80 Ma in duration. Thus, they approximate the average length of many periods of geologic time in the Paleozoic and Mesozoic eras. Third-order cycles are of 1 to 10 Ma long. Each cycle consists of a relatively slow invasion of land areas by the oceans followed by a relatively rapid withdrawal or regression of the sea, although there is some dispute about whether this pattern of slow transgression and fast regression really exists. A basic cause of such cycles may be changes in the volume of the ocean basins. If plate tectonic forces within the earth result in shrinking of the ocean basins, the surplus water would spill over onto the continents. The rise of major new mid-ocean ridges might result in such volume reductions in the basins. Deepening of the ocean basins would result in regressions. What might cause such cyclic events deep within the earth? We really do not know; there is no very good explanation for these major repetitive phenomena that we can clearly see recorded in rocks.

One source of demonstrable cycles lies in various astronomical observations. Astronomers can see, or at least calculate, a variety of cycles ranging from sunspot cycles that have a period of 11 years to ones 300 Ma long. These cycles are well recognized. What is not clear is exactly which astronomical cycles might affect life on earth. For instance, the longest such cycle is the galactic orbit, the length of time it takes for our solar system to make a single revolution around our galaxy. That cycle has a period of about 300 Ma. As the solar system moves through the galaxy, it oscillates with respect to the midgalactic plane on a period of about 30 Ma. These two cycles approximate the first- and second-order sea-level change cycles, or at least they are the right order of magnitude. But it is far from clear how such long-term astronomical events might have any effect at all on the earth.

There are several shorter-term cycles, however, that have definitely affected the planet earth. These cycles, taken together, are thought to have had direct and profound effects on earth history, especially its climate. The cycles are commonly called the **Milankovitch cycles**, after the Yugoslavian physicist Milutin Milankovitch, who theorized in the 1920s and 1930s that these cycles were the direct cause of the continental glaciers of the Pleistocene.

The first of these significant shorter cyclical events relates to the eccentricity of the earth's orbit. We know that the earth's orbit around the sun is not a circular path but rather follows an ellipse. Thus, twice a year the earth is farthest from the sun and twice a year it is closest to the sun. (These far and close positions have absolutely nothing to do with our seasonal summer and winter climate; the earth is actually closest to the sun during our Northern Hemisphere winter.) This orbit does not stay the same each year. Some years the ellipse is shorter and fatter; sometimes it is stretched out longer and leaner. The period of orbit varies from 100,000 to 400,000 years.

The second important short cycle relates to the obliquity of the earth's plane of orbit. The earth's rotational poles are not perpendicular to the plane of the orbit that the earth makes around the sun. Instead, the north and south rotational poles lie at an angle of about 23 degrees to that plane. However, because the earth wobbles on its axis of rotation, this angle changes very slightly and very slowly. The angle is more or less oblique at different times. This cycle of variation has a period of about 40,000 years. When the earth's rotational axis is at a greater angle, there is a slight tendency for winters to be more severe and summers hotter. More equitable climates result when the angle of obliquity decreases.

The third significant astronomical cycle is called the precession of the equinoxes. This has a period of about 21,000 to 26,000 years. Today we celebrate the winter and summer equinoxes (the shortest and longest days) in the Northern Hemisphere on about December 20 and June 20. However, these times gradually shift through the calendar. In another 10,500 years the shortest day of the year in the Northern Hemisphere will be on June 21, and June, July, and August will be our winter months. Thus, the equinoxes slowly progress (precess) through the fixed calendar that we use.

Now, given these three kinds of cycles with their different periods, there will be times when all three cycles reinforce each other and other times when they completely cancel each other out. You can imagine them as represented by three different sine curves of different amplitude and period.

Milankovitch postulated that the combined effect of these cycles affected world climate in the past. When the cycles reinforced each other, the earth experienced climatic extremes that resulted in a very slight lowering of the worldwide mean annual temperature by only a couple of degrees. This would be sufficient to trigger the onset of large-scale glaciation. Slightly more ice would accumulate each year than would melt. As the glaciers slowly grew in size, they would help lower temperatures as well. When the cycles gradually shifted, so that they did not reinforce each other but instead canceled each other out, climates would become less extreme or more even, and the glaciers would begin to retreat. The overall cycle between more extreme and less extreme climates is about 100,000 years.

Thus, we believe that very subtle, long-term shifts in the basic relationships of the earth to the sun have caused the ice ages. It is still not clear why these major ice ages should only recur every 200 Ma or so.

EVOLUTION VERSUS CREATIONISM

In recent years there has been considerable controversy concerning the Darwinian theory of evolution and what is called scientific **creationism**. The latter idea assumes the the story of creation in the Bible is literally and scientifically true and thus deserves equal treatment in science classes, especially at the primary and secondary levels.

Not all Biblical literalists necessarily accept that creation occurred in six twenty-four-hour days because a day can be interpreted to be much longer in duration. The creationists' claim for the scientific equality of their views has been rejected by federal courts in Arkansas and Louisiana, which have ruled that creationism is essentially religious in nature, not scientific, and therefore should not be taught in science classes. This section examines some of the more important claims and counterclaims of Darwinian scientists and creationists.

Any scientific theory must be testable as to its truth or falsity. New evidence may result in modification or rejection of a theory. In this sense, creationism is not scientific because it cannot be falsified or modified with new data. One must either accept the premise of the literal truth of the Bible or reject that premise.

The enormous span of time for life history on earth causes problems for creationists who interpret the six days of creation in Genesis as sequential twenty-four-hour days. They believe that the earth is only a few thousand years old. Efforts have been made to discredit the reliability of absolute radioactive dating, but these efforts have generally been rejected by scientists as lacking any evidence other than statements in the Judaic testaments. Other dating methods, such as the length of time needed for light to arrive from distant stars, support the idea that the universe is at least 8 to 13 billion years old.

Creationists try to disprove evolution by trying to find evidence against the succession of fossil faunas and floras. One way to do this is to discover co-occurring fossils that could not have lived at the same time, according to paleontologists. This effort has concentrated on finding fossil evidence for the presence of humans on earth in rocks that paleontologists agree are much too old to yield such evidence. The evidence has commonly taken the form of fossil human footprints or sandal prints. Thus, it is claimed that in Utah a sandal print has been found in Cambrian rocks that also contain trilobites. Creationists interpret this evidence to mean that the rocks that contain these fossils were all deposited during the Noah's flood episode of the Old Testament. According to creationists all of the plants and animals that we know are now extinct were living on earth prior to the flood but were wiped out by that flood, and there was no room for the dinosaurs on the ark. A Cretaceous bed in Texas has yielded what are said to be human footprints. If this is true, these occurrences would falsify our current view that man and dinosaurs could not have co-existed and that Cretaceous rocks are 60 million years old. Geologists who have examined these occurrences have found that the so-called footprints and sandal prints are not fossils at all but are inorganic structures, impressions of concretions, or fragments of dinosaur toe impressions in the rocks.

Finally, creationists believe that one species cannot be transformed through organic evolution into another species. Thus, they attempt to discredit intermediate fossil forms that seem to bridge the gaps between major life groups. They deny that *Archaeopteryx* is a transitional form between dinosaurs and birds and argue that *Ichthyostega* does not indicate the origin of amphibians from lobe-finned fishes. These arguments have been effectively countered by detailed anatomical rebuttals from paleontologists.

In summary, creationism is an unscientific approach to understanding ancient life on earth because it cannot be falsified or modified. One either accepts creationist ideas as an article of faith or one does not. Many church leaders oppose attempts to teach scientific creationism in the schools or to equate it with evolution. Gradual, rather than sudden creation, is accepted by some; others see life in terms of theistic or God-related evolution. An overwhelming amount of paleontological evidence and modern biological evidence contradicts creationist arguments.

In an entirely empirical way, faunal succession and age-dating of rocks are used daily by successful multibillion dollar corporations in the search for fossil fuels and minerals. For instance, ancient organic reefs buried deep within the earth are prime sources of petroleum. How they could use creationist ideas of the history of the earth to find oil and gas is difficult to imagine.

KEY TERMS

adaptive radiation

allopatric speciation

analogous structures

convergent evolution

creationism

cycle

DNA

extinction

gene

genotype

homeomorphic

homologous structures

iridium

isolation

macroevolution

megaevolution

Milankovitch cycles

mutation

natural selection

parallel evolution

peripheral isolates

phenotype

phyletic evolution

preadaptation

punctuated equilibrium

stasis

sympatry

variation

READINGS

The processes of evolution and extinction have been written about extensively. Discussions range from the very simple to the quite complex. Each of the books listed here places emphasis on a somewhat different aspect of evolution or extinction.

Berggren, W.A., and Van Couvering, J.A. 1984. *Catastrophes and Earth History*. Princeton Univ. Press. 461 pages. A variety of essays on different types of catastrophic events that may have shaped the history of the earth, including climate and life.

Elliott, D.K. 1986. *Dynamics of Extinction*. Wiley. 294 pages. An evenhanded and thorough biological treatment of extinction.

Evolution. (videotape) 1986. Hawkshill. 39 min. Covers the history of scientific thought on evolution, the concept of natural selection, and evidence for evolution.

Geotimes. 1990, August. The four articles on pages 13–20 of this issue present conflicting views on the Cretaceous-Tertiary (or K/T) boundary, cataclysms, bolide impacts, and extinction.

Hoffman, A. 1989. *Arguments on Evolution: A Paleontologist's Perspective*. Oxford Univ. Press. 274 pages. A series of essays on the fossil record of evolution, with emphasis on macroevolution and megaevolution.

Kauffman, E.G., and Walliser, O.H., eds. 1990. *Extinction Events in Earth History*. Lecture Notes in Earth Sciences No. 30. Springer-Verlag. 431 pages. Comprehensive discussion of all hypothesized extinction events, including cyclic phenomena.

Lewin, R. 1983. "What Killed the Giant Mammals?" *Science* 221:1036–1037.

Nitecki, M.H., ed. 1984. *Extinction*. Univ. of Chicago Press. 335 pages. A collection of thirteen technical papers on various aspects of plant and animal extinction.

Origins: The Mutation Machine (videotape). 1987. Films for the Humanities. BBC, 27 minutes. Evolution in relation to mass extinctions, very early fossils, and molecular RNA evidence.

Out of the Past (videotape). 1985. Films for the Humanities. Grandes Iberia. 3 cassettes, 20 minutes each. The importance of fossils in the theory of evolution, emphasizing plants and tetrapods.

Runnegar, B., and Schopf, J.W. 1988. *Molecular Evolution and the Fossil Record*. The Paleontological Society, Short Courses in Paleontology No. 1. 167 pages. Presents newest information on the study of organic molecules of all kinds and their use in deciphering the evolutionary history of life.

Stebbins, G.L. 1981. *Processes of Organic Evolution*. Concepts of Modern Biology Series. Prentice-Hall. 183 pages. A short paperback with emphasis on genetic control of evolution.

7
Continental Drift and Plate Tectonics

$\mathbf{I}$f you ask the man or woman on the street if the positions of the oceans and continents have always been the same as they are now, the answer will probably be yes. The continents are the largest land masses and because of their size most people think they must surely be fixed in position. This has always been the traditional viewpoint, even among most paleontologists and geologists. In the 1960s, however, the old idea of fixed continents and oceans was seriously challenged. Today most earth scientists think that the continents have moved about over the face of the earth.

Why should we concern ourselves about the crust of the earth in a course on paleontology? One of the aspects of ancient life that interests us is the past distribution of life—topics such as migration, barriers, isolation, and faunal and floral provinces. In order to discuss ancient-life geography, also called **paleobiogeography** (see Chapter 8), we have to assume some model for the distribution of land and sea. In the 1940s we would have chosen a model of a fixed earth crust, with oceans and continents forever in their present positions. But with new techniques that have been developed to study the ocean floor, we have obtained much new evidence that has led to a scientific revolution in our thinking about the stability of the crust. We now can support the view that the crust is in constant movement and that the continents and oceans have not always had their present positions or outlines. Here we will first discuss the century-old idea of continental drift, then present a synopsis of the driving force for drift, or sea-floor spreading. Finally, we will discuss the unifying theory that brings these two ideas together—plate tectonics.

CONTINENTAL DRIFT

Soon after North and South America were discovered and their eastern coasts roughly mapped, several scientists noted in the 1600s that there was a similarity of outline between the east coasts of North and South America and the west coasts of Europe and Africa. These early scientists, including Francis Bacon, simply noted the similarity in outline on both sides of the Atlantic; they did not propose that the continents had separated, forming the Atlantic Ocean. This idea of drifting continents was first expressed in the 1850s by two European scientists, Richard Owen and Antonio Snider. It was not until 1912, however, that the theory of **continental drift** was elaborated and much evidence gathered to support it. For the most part, this work is credited to a German meteorologist, Alfred Wegener. The theory of continental drift depended largely on geological and paleontological observations that had been made in the southern hemisphere. The theory was strongly supported by geologists working in South America, Africa, and Australia, especially by a South African geologist, Alexander DuToit. A number of European geologists also supported the theory, but some North American geologists were especially reluctant to accept drift.

What was the evidence for and against in this controversy concerning continental drift? The old idea about the close fit of North and South America with Europe and Africa was used, of course. The continents were not matched at their shorelines, but at the edges of the continental areas, including submerged continental shelves and slopes. The best fit was obtained by matching at a depth of about 2000 meters, halfway down the continental slopes. Additional support for the tearing away of the continents from each other was gained when the nature of the Mid-Atlantic Ridge was ascertained by early oceanographic expeditions. This high ridge, which only reaches the surface of the ocean at Iceland, the Azores, and two small islands in the South Atlantic, has an outline that matches the edges of the Atlantic on either side (Figure 7.1). It was suggested that the ridge records the scar left behind where the continents separated.

Rocks and fossils of the southern hemisphere were found to have several unique features that suggested drift. Rocks of Permian age were found to contain evidence for ancient continental glaciers. Ice had scoured and scratched pre-Permian rocks, then had melted, leaving behind tillite deposits. In South America the ice had moved from east to west, as determined by the direction of the grooves made by the ice, and would have had a source in the present Atlantic (Figure 7.2). In South Africa ice movement was from west to east, again from the Atlantic. Evidence for these large, ancient glaciers was also found in Australia, India, and Antarctica. It was argued that large ice sheets could not have been generated in the ocean, but would have had to accumulate on land; hence there must have been land present where the Atlantic is now situated.

FIGURE 7.1
Outline of the Atlantic Ocean showing the position of the Mid-Atlantic Ridge. Heavy lines indicate the true, submerged, margins of the continents, light lines those parts of the continents that are dry land.

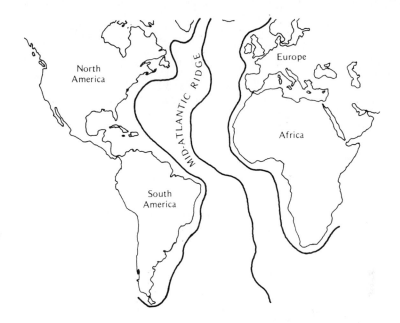

In addition to this physical evidence, there was also much paleontological evidence to support drift. Above the Permian glacial deposits in each of these areas there were terrestrial rocks that contained many fossil plants, in some places forming coal beds. The most distinctive of these plants were two kinds of extinct **gymnosperms**, related to modern conifers. These were seed bearing, and commonly called tongue ferns (Figure 7.3). Two genera, *Glossopteris* and *Gangamopteris* were found on all the southern hemisphere continents and on the peninsula of India, now in the northern hemisphere but considered by "drifters" to have migrated north after the Permian. These plants were not found anywhere in the northern

FIGURE 7.2
Reconstruction of Gondwanaland as it appeared about 250 million years ago. Arrows indicate directions of ice movement of ancient continental glaciers that covered most of the area. The South Pole (SP) is thought to have been situated on the coast of Antarctica.

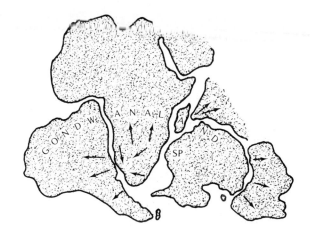

FIGURE 7.3
Leaves of the tongue fern *Glossopteris* from Permian rocks in New South Wales, Australia. The rock slab is 15 cm across. *Gangomopteris* is similar in aspect but lacks a distinct mid-vein in the leaves. (Photo by George R. Ringer, Indiana University.)

hemisphere, except in India, and were interpreted as a cool-climate flora because of their close association with ice deposits.

In addition to land plants, Permian rocks in these areas contained a unique marine fauna that was characterized especially by distinctive kinds of brachiopods and clams. This fauna was named the *Eurydesma* fauna after one of the large, thick-shelled bivalves that was found closely associated with glacial deposits (Figures 7.4 and 7.5). Some of the beds that contain this clam also have large boulders of granite, quartzite, and other rocks that are judged to have been dropped onto the Permian sea floor by melting icebergs.

Finally, Permian rocks in South America and Africa contain a small fossil reptile called *Mesosaurus*. The beds that yield this fossil are freshwater in origin, and the reptile is very distinctive, quite unlike any other reptiles known of this age in the northern hemisphere (Figure 7.6). Those who opposed drift had to explain how this freshwater animal could have migrated across the Atlantic Ocean

FIGURE 7.4
A large, thick-shelled marine bivalve, *Eurydesma*, found in close association with Permian marine glacial deposits of the southern hemisphere and India. The specimen is from Tasmania and is about 10 cm across. (Photo by George R. Ringer, Indiana University.)

○ — — — *Eurydesma*: marine fauna
□ — — — *Glossopteris–Gangamopteris*:
                terrestrial flora
△ — — — *Mesosaurus*: fresh-water
                terrestrial fauna

FIGURE 7.5
Distribution of Permian marine and terrestrial fossils associated with late Paleozoic continental glaciation in the southern hemisphere and India—areas that once constituted a single supercontinent, Gondwanaland, which later broke up and drifted apart.

FIGURE 7.6
A skeleton of the Permian reptile *Mesosaurus* from Brazil. This animal was important as early scientific evidence for the former connection of South America and Africa. (Courtesy of Takeo Susuki, U.C.L.A.)

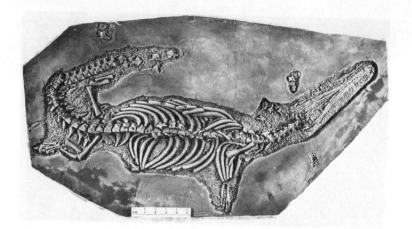

or else migrated from Africa through Europe and North America to South America without leaving any trace of it behind in intermediate areas.

These lines of evidence—the Permian glaciations, the *Glossopteris* flora, the *Eurydesma* fauna, and *Mesosaurus*—were used by the advocates of continental drift to strongly support the theory. The evidence indicated that the southern hemisphere land areas had once been joined together in one giant supercontinent. This supercontinent was named **Gondwana** by Edward Seuss, a Swiss geologist who wrote a worldwide synthesis of geology in 1900. The name is taken from an area in India where some of the critical rocks are exposed. Later, in 1912, Alfred Wegener not only accepted Seuss's name for this predrift continent to the south but further proposed that the northern hemisphere continents had also been united to each other at one time, such that all of the continental areas had once been together. He called this single continental area **Pangaea,** which means single land. He thought that Pangaea had broken up into two continents, Gondwana and a northern one that he called **Laurasia.** Each of these two continents, he believed, had divided to form our present continental configuration.

How were these arguments in favor of drift countered by the many geologists who did not accept the theory? The basic contention was that there was no adequate driving force known that could have caused the continents to have moved. The continental areas of the crust are composed of quite thick rocks that are lighter, or less dense, than the lower, thinner parts of the crust that underlie the ocean floors and the continents. To move, it seemed that the continents would had to have been pushed horizontally through this dense, lower part of the crust. The friction involved in such movements would have been enormous, and no forces were known that could have created such lateral pressures.

The glacial deposits cited as evidence of drift were thought by many geologists not to have been formed by ice. They were explained as having been formed by dense mud flows or turbidity currents carrying pebbles and boulders. Even if the deposits were accepted as glacial in origin, it was argued that there could have been several small ice caps, one on each continent, rather than a single large ice sheet. The faunal and floral evidence was dismissed by some with the argument that plants and animals have so many ways of dispersal that they may very well have migrated through northern areas without leaving behind any traces. Alternatively, some postulated that there had been a land bridge across the South Atlantic connecting South America and Africa. They contended that *Mesosaurus*, and perhaps *Glossopteris*, had migrated across this land bridge, which later sank into the ocean. Such a land bridge would had to have consisted of lighter continental rocks, and we now know that there is no evidence for the sinking of such a land mass in the southern oceans. It was also argued that if North America and Europe had indeed once been close together, then we should find much good fossil evidence in these areas where fossils have been most intensively studied. These and other arguments convinced many geologists that drift had not occurred and that the theory was mainly an intriguing idea that was entertained largely by a few geologists working in the southern hemisphere.

The discussion so far brings us up to the 1950s. By this time it had been discovered that many igneous and some sedimentary rocks preserve faint traces of the earth's magnetic field. When a lava is erupted and cools, tiny needlelike crystals of iron minerals become oriented in the earth's magnetic field, just as if they were many small compass needles. As the rock solidifies, the crystals become frozen into position and so record where the earth's north and south magnetic poles were situated at the time the lava cooled to a solid rock. Other small crystals of iron minerals that settle out of water, to form part of a sandstone bed, for instance, also become aligned in the magnetic field. Study of this preserved magnetism, called **remnant magnetism** because it remains in the rock, from many different areas and many different ages of rocks revealed that either the magnetic poles had not always been in their present positions, close to the north and south poles of rotation of the earth, or else they had always been close to their present positions and the continents themselves had changed position (Figure 7.7). **Magnetic reversal** was taken as strong evidence for continental drift by some scientists. Others argued that the magnetic field may very well have shifted position with time. The major difficulty in accepting or rejecting either hypothesis is that the underlying reasons for the generation of the earth's magnetic field are still not fully understood. Therefore, whether it could shift position or not was not entirely clear.

SEA-FLOOR SPREADING

Magnetic studies of rocks reveal not only that the positions of the magnetic poles, or the continents, have shifted through time but also that the poles have reversed their positions. The north and south magnetic poles have flip-flopped back and forth through time on the average of about every 400,000 years (Figure 7.8). The cause of these magnetic reversals is still debated, but the rocks that record such reversals can be dated radiometrically, and a time scale has gradually developed that gives the timing for each reversal.

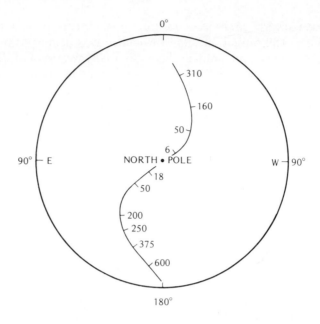

FIGURE 7.7

Wandering of the magnetic poles. The circle is at the equator and the North Pole is in the center. The upper curved line represents the position of the north magnetic pole as recorded in rocks for Australia from over 310 million years ago to 6 million years ago. The lower curve is similar for Europe from 600 to 18 million years ago. Because the curves do not coincide, they indicate that Europe and Australia have had different relative movements over the past several hundred million years.

Studies of the natural, or remnant, magnetism of rocks was extended from the continents to the ocean floors. It was found that the igneous rocks that make up the floor of the ocean exhibit both normal polarity (North Pole in its present position) and reverse polarity (North Pole near the position of the present south magnetic pole). Reverse and normally polarized rocks occur on either side of the Mid-Atlantic Ridge and other large ridges on the floors of other oceans in a candy-stripe pattern. Those rocks closest to the ridge on either side have normal polarity, followed by elongate areas with reverse polarity, and so on in alternation on both sides of the ridge (Figure 7.9). These rocks could also be dated and the patterns of polarity matched up on either side of a ridge. Such study led to the conclusion that the age of the sea floor increases with distance from an oceanic ridge. The only explanation for this is that new ocean crust is formed at a ridge and is then progressively pushed further and further away from the ridge as younger and younger crust is formed. Thus, we now believe that new oceanic crust is created at ridges and spreads out away from the ridges through a process called **sea-floor spreading.**

FIGURE 7.8
A sequence of volcanic rocks for which the polarity of the magnetic poles has been determined. Shaded intervals at the side indicate reverse polarity, white areas normal polarity. The time scale at the right indicates a duration of 10 million years.

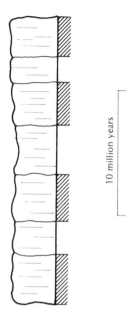

Much additional evidence for sea-floor spreading has accumulated. The rocks composing volcanic islands in the oceans are found to be increasingly younger the closer they are situated to an oceanic ridge. Dating of the oceanic crust in the Atlantic has revealed that none of these rocks is older than about 180 million years (Jurassic Period), and that therefore the entire floor of the present Atlantic Ocean has formed comparatively recently. The skeletons of various kinds of floating

FIGURE 7.9
Diagrammatic map of a portion of the sea floor showing an oceanic ridge in the center and bands of rocks of normal polarity (white) and reverse (diagonal rule) polarity on either side of the ridge. The similar pattern on either side of the ridge indicates that new crust is formed at the ridge and is pushed out equally to either side.

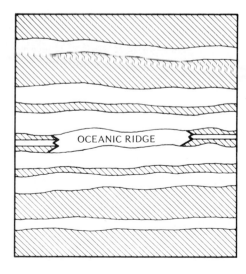

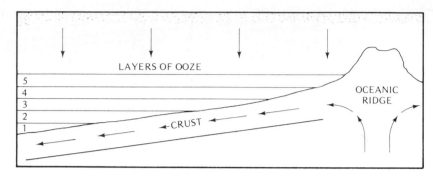

FIGURE 7.10
As new sea floor is formed at the oceanic ridge and moves away from it to the left, successively younger layers of fossil ooze (1–5) are deposited on top of the moving crust. The skeletons of tiny organisms living in the surface waters sink to the bottom, forming the ooze layers. Deep-sea coring of these sediments reveals the age differences at different sites.

protists and algae accumulate on the sea floor as the organisms die and the skeletons sink. It has been found that the further one moves away from a ridge the older such fossils become. These fossil deposits, called **oozes**, that overlay the crust are quite young and thin close to a ridge (Figure 7.10).

The demonstration of sea-floor spreading provides an adequate and sufficient mechanism for movement of the continents. It was shown that a continent did not have to plough its way through dense lower parts of the crust, like a ship through viscous tar. Instead, the continents could have floated on top of a continuously moving conveyor belt or moving sidewalk of oceanic crust that gradually moved them about. The concept of sea-floor spreading has been documented now in so many different ways, that all (save for a very few) geologists have come to accept this idea. However, we are still left with the question of what causes spreading from the oceanic ridges.

In the debate over the ultimate cause of spreading, several different hypotheses have been put forward. The process obviously takes place deep within the earth, where we cannot make direct observations of what is going on. The most generally accepted idea is that the material within the **mantle** is in constant motion, being heated near the core, moving upward under the crust, flowing sideways, and then, after cooling and becoming more dense, sinking again towards the core of the earth. This process produces convection currents, much like heating soup in a pot, so that it rises to the top, cools, and sinks. The ridges mark linear areas on the earth's surface where convection cells approach the surface. Hot mantle material melts and is forced upward as volcanos and molten lava produce new oceanic crust that is progressively pushed aside as more new crust is formed. The ridges where this happens are uplifted by these underlying forces and are termed **spreading centers.**

PLATE TECTONICS

In the preceding discussion we did not consider several questions that come to mind in light of the reality of sea-floor spreading. For instance, What happens to all of the old crust if only relatively young ocean floors are known? Does it just pile up somewhere, or is it destroyed? The answer is the latter case—the old crust is destroyed. The earth's crust is fairly rigid and is not easily deformed. The theory of **plate tectonics** states that the structure of the earth's crust consists of rigid bodies that move horizontally. These plates interact at their boundaries. Spreading centers form boundaries that separate large areas of the earth's crust from each other. These areas are called **plates**, and most plates have along one edge a spreading center, either active or inactive (Figures 7.11, 7.12). On the opposite side of a plate from the ridge there is commonly a deep ocean trench. These trenches are very deep, the deepest known places in the oceans. They are often the sites for many earthquakes, and curved chains of volcanic islands may be situated close to the trenches. Trenches are known now to be the sites at which the leading edges of plates, consisting of the oldest parts of oceanic crust, furthest from spreading centers, are turned back down into the mantle. These trenches are now called **subduction zones**, where the old crust is destroyed and melted as it is subducted, or turned down, into the earth. The lateral edges of plates, between subduction zones and spreading centers, are areas where one rigid plate slides past an adjacent plate. These edges are marked by huge elongate systems of fractures in the earth's crust.

What happens when a plate of crust with a continent riding on it is subducted near the edge of the continent? Apparently the continental areas, because they are composed of thick, light rocks, are not easily turned down. They tend to float over the subduction zone and may seal it off. They may collide with continental areas on an adjacent plate, the tremendous compression resulting in a squeezing up of large mountain ranges. When the continental mass now making up the peninsula of India moved northward and collided with the mainland of Asia, the rocks caught between were so squeezed and formed the world's loftiest mountains, the Himalayas.

The entire surface of the earth can now be divided into about eighteen plates. Some of these are very large, one occupying virtually all of the Pacific Ocean and one covering North America and the northwestern Atlantic. Others are quite small, the Arabian peninsula being one. Not all spreading centers are active, and the rate of spreading is very slow—just a few centimeters a year. Subduction goes on at the same slow rate. Putting all of this together into a coherent pattern of the past history of the continents and oceans is still going on. The further back in time we go, the less sure we are about past events and positions of continents. This is because much evidence has been obliterated with time and by subduction of old sea floor.

The position and activity of spreading centers has changed through time. The record for the past 150 to 200 million years of earth history is reasonably

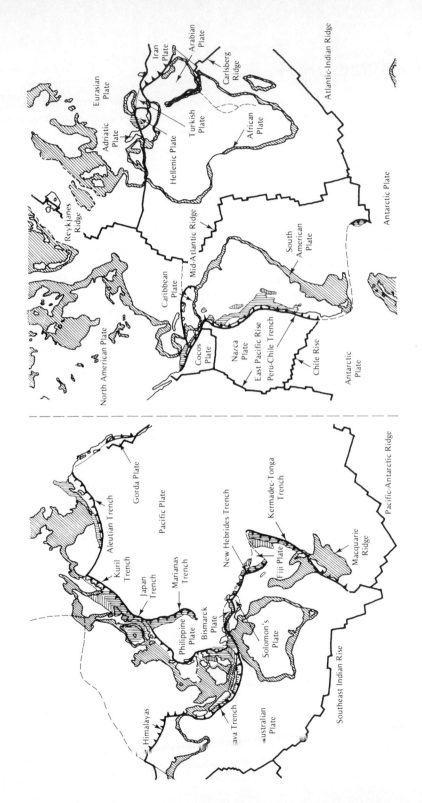

FIGURE 7.11

The present distribution of crustal plates, spreading centers, and subduction zones. Oceanic ridges and rises are active or inactive spreading centers. Oceanic trenches are subduction zones. Uncertain plate boundaries in Asia and southeast of South America are shown as dashed lines. (From J.F. Dewey, *Plate Tectonics*. Copyright © 1972, Scientific American, Inc.)

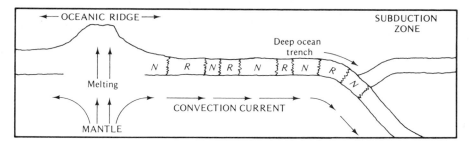

FIGURE 7.12

Cross section of an idealized ocean floor plate, between a spreading center (oceanic ridge) on the left and a subduction zone (oceanic trench) on the right. Convection in the mantle brings up deep rocks under the ridge that melt, cool, and are forced to each side by newly rising mantle. The *N* and *R* symbols in the crust indicate zones of rock that have normal and reversed magnetism.

straightforward and has been worked out in considerable detail (Table 7.1). Beyond about 200 million years we are still uncertain as to the exact sequence of events that took place. We do know that the present Atlantic Ocean is quite young. The ocean that was present before the closing together of Africa and North America is called Iapetus. The continents were all together, forming a single supercontinent, Pangaea, that began to break up about 220 million years ago, with spreading out of the crust from either side of the Mid-Atlantic Ridge. The motions of the plates were not uniform; Africa has moved north as well as east, closing up a large oceanic gap between it and Europe and resulting in a squeezing up of the Alpine mountain ranges and a reduction of the ocean area between to the present Mediterranean, Black, and Caspian seas. There is some evidence that early in the Paleozoic the continents were not together and that there was an older Atlantic that then was closed as Pangaea formed, only to open up as the continents again moved apart. What the configuration of land and sea may have been prior to the Paleozoic, in the Precambrian, is gradually becoming clearer.

THE EFFECT OF CONTINENTAL DRIFT ON LIFE

In this chapter we have devoted considerable space to discussion of continental drift and its underlying cause, plate tectonics. The position and arrangement of the continents surely has had a profound effect on life of different ages. Paleontologists are now busily investigating the relationship between these physical and biological events. We can specify several broad areas within which continental drift has been especially important in its influence on life.

As the continents shifted, they changed their relative position with respect to the poles of rotation, sometimes being close to the equator and at other times closer to a pole. For instance, the United States was close to the equator during

TABLE 7-1.

Sequence of events in sea-floor spreading and continental drift. Events prior to 200 million years are a subject of debate; they are not as well substantiated as are events after 200 million years.

Millions of Years Ago	Events
50	Separation of Antarctica and Australia.
100	India drifts northward.
150	Separation of Africa from India and Australia.
180	Separation of South America from Africa.
200	Beginning of separation of Europe and North America. Initiation of the Pangaea breakup.
300–400	Closing of Iapetus; formation of Pangaea; creation of the Appalachian mountains.
prior to 600	Formation of proto-Atlantic by breakup of an early Pangaea.

much of the late Paleozoic and has gradually drifted north, farther from the equator, since that time. This drift undoubtedly influenced the climate that was present over our country, which gradually changed from tropical to subtropical to temperate. Similar climatic changes occurred on all of the continents through time.

As the continents broke apart, came together, and then again separated, the distribution of both land and sea plants and animals was affected. When Pangaea was in effect, with a single large land mass, we would expect that life on land was quite cosmopolitan, the same species inhabiting large areas, subject, of course, to differences in latitude and climate. When the continents separated, with little or no direct land communication between them, land organisms evolved separately in each area, with little opportunity to cross from one continent to another. The result was distinctly different faunas in different areas with species circumscribed in their distribution. The same would have held for marine organisms. With a single world ocean and Pangaea we would expect many cosmopolitan marine species. Fragmentation of the ocean by separated continents would have resulted in local differences.

Virtually all of the marine fossils we know from continental areas lived in reasonably shallow water. The total area of shallow water available to such former life would have been greatly affected by whether or not the continents were apart or together. The total shoreline of several separate continents is conspicuously greater than is the shoreline of a single supercontinent. If the total number of species that can live in shallow water is proportional to the amount of area available for living sites, then the overall diversity of shallow-water marine life should have been low with Pangaea in effect and high when the continents were separate. Recent studies have shown that, indeed, this was the case.

The configuration of land areas, quite apart from their latitudinal position, has a profound effect on world climate. A single large land mass like Pangaea would have had an extreme continental climate, with wide fluctuation in temperature and rainfall, probably much more drastic than any such climates we see today. Breakup of the land mass into several continents would have resulted in each smaller land mass having much more climatic influence from the surrounding oceans, resulting in less drastic extremes. Widespread deserts and salt deposits during Permian and Triassic times, when Pangaea existed, are probable indicators of such climatic extremes.

It has also been suggested that movement of the continents apart from each other was also responsible for the widespread shallow seas that flooded the continents repeatedly during the past 600 million years. As the continents moved apart, the various midoceanic ridges built up. These displaced enormous volumes of ocean water, resulting in a significant raising of sea level relative to the land. As large areas of the continents were low, flat lands, just a slight elevation in sea level could have flooded immense areas of the continents. Thus, times of quickened volcanic activity along midoceanic ridges may have corresponded with times when several of the continents experienced widespread flooding by the sea.

We know that coral reefs today flourish in low-latitude, tropical environments. Yet, many ancient reefs are found in rocks that are at much higher latitudes today. Before continental drift, this apparent discrepancy was explained by the presence of widespread equitable climates. We now know that these areas were once located in low latitudes and that the continental area on which these fossil reefs were thriving has since drifted to higher latitudes. Thus, there are Silurian age reefs in southern Canada and just south of Hudson Bay. These reefs formed in tropical climates when North America was situated in low south latitudes.

The distribution of ancient coal beds can be explained in the same way. Coal swamps are thought to have required warm, humid climates. The occurrence of coals today at high latitudes is interpreted to indicate that these areas were in lower latitudes at the time the coal swamps flourished. This applies both to extensive Pennsylvanian age coals as well as to extensive Cretaceous and Tertiary coals.

KEY TERMS

continental drift	*Eurydesma*	Gondwana
core	*Gangamopteris*	Laurasia
crust	*Glossopteris*	magnetic reversal

magnetism   Pangaea   spreading center

mantle   plate tectonics   subduction zone

Mesosaurus   remnant magnetism

paleobiogeography   sea-floor spreading

READINGS

Kearey, P. 1990. *Global Tectonics.* Blackwell Scientific. 302 pages. An advanced-level textbook dealing with all aspects of plate tectonics.

The Living Planet: The Building of the Earth (videotape). 1988. BBC, with Time-Life; Ambrose Video. 55 minutes. The formation of the earth, moving continents, and volcanoes.

Smith, A.G.; Hurley, A.M.; and Briden, J.C. 1981. *Phanerozoic Paleocontinental World Maps.* Cambridge Univ. Press. 102 pages. A series of eighty-eight maps showing the distribution of land and sea from Cambrian to Recent times in two different map projections.

Wilson, J.T., ed. 1976. *Continents Adrift and Continents Aground.* Freeman. 219 pages. A selection of *Scientific American* articles by nineteen different authors on continental drift, sea-floor spreading, and plate tectonics.

8

Paleobiogeography

Fossils are distributed in rocks within both a time and a spatial framework. In this chapter the focus will be on the arrangement of fossils in space, the ancient geography of life, or **paleobiogeography**. We know that plants and animals are not evenly distributed across the face of the earth today, either in marine or terrestrial environments. We do not expect to find giraffes outside of central Africa or kangaroos outside of Australia, except, of course, in zoos. Coral reefs are confined today to near equatorial areas. Did the life of the past have similar distinctive patterns of distribution? Yes, it did, although the patterns were commonly very different from those of today. We will now look at some of the more important examples of the geographic differentiation of life in the fossil record.

Perhaps the single most important thing that one can say about ancient geography is that we live today at a time that is quite atypical. We have already seen that the fossil record for marine fossils is considerably better, more complete, and more enduring than that for land fossils. We have also noted that we have a much more complete knowledge of fossils found in rocks exposed on land than we do of fossils in rocks on the floors of the oceans. These two observations taken together must mean that many areas that are now land were once covered by the seas. Compared to the past, the present is a time when the continents stand relatively high above sea level; consequently, there are large areas of exposed land. In many past times this was not the case. Wide continental areas were flooded by the shallow seas that laid down marine rocks now exposed on the continents. The continents have been alternately flooded and dry, with seas transgressing over the

FIGURE 8.1
Distribution of land and sea over the United States during Mississippian time. Land areas are patterned; areas covered by seas are not. The position of the United States relative to the equator is shown.

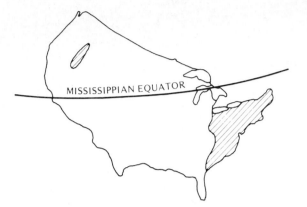

MISSISSIPPIAN EQUATOR

land and then later withdrawing. During times of regression, erosion took place, destroying some of the previously deposited sediments and resulting in the interrupted and incomplete rock record that we now have available for study.

Two of the greatest times of flooding were in the Ordovician and Mississippian periods of the Paleozoic (Figure 8.1). During the Mississippian, dry land in North America consisted of a group of large islands separated from each other by wide, shallow seaways. In addition, the climate was tropical or subtropical. How do we know this? There are two lines of evidence. During this time the supercontinent of Pangaea was situated well south of the present positions of northern hemisphere continents. This placed the equator right across the center of the United States. Later, breakup of Pangaea resulted in North America drifting west and north into higher latitudes and away from the equator. We can also estimate the position of the old equator through magnetic studies. If we know the position of the north magnetic pole during Mississippian time, and assume that it was close to the earth's pole of rotation, then the equator must have been about 90 degrees from the magnetic pole, which places it astraddle the United States.

During the Paleozoic, shallow seas covered parts of the United States during each of the periods of time from the Cambrian through the Permian. At the beginning of the Paleozoic, early and middle Cambrian seas were confined to the borders of the continent but flooded into the interior by late Cambrian time. Maximum withdrawal occurred at the close of the Permian when North America stood high relative to sea level. At this time marine rocks were deposited only along the southern and western edges of the United States. The last great inundation of North America occurred during the late Cretaceous, at the close of the Mesozoic Era. A seaway extended from Texas across the Great Plains region of the United States and Canada and far north into the Arctic. This seaway dried up at the end of the Cretaceous, from which time the continent has not had extensive flooding. Most of the wide seas that flooded the continents were quite shallow, probably less than 200 meters deep. This factor surely restricted currents and tidal action. If we wanted to study similar seas of the present time, we would be restricted to

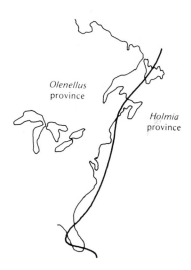

FIGURE 8.2
Map of eastern North America showing the boundary between the early Cambrian trilobite provinces—the *Olenellus* province to the west, the *Holmia* province to the east. The *Olenellus* province also occurs in Europe, in Norway, Scotland, and northern Ireland, as well as in western Africa.

bodies of water such as the Baltic Sea, between Scandinavia and Europe, or Hudson Bay. But since neither of these modern seas is in low-latitude, tropical or subtropical climates, they are not truly comparable to seas of the past.

FAUNAL AND FLORAL PROVINCES

We can recognize the distribution of any species of plant or animal as being **cosmopolitan** (worldwide), **endemic** (confined to a small area), or someplace in between. Actually the great majority of species fall in an intermediate category. Where a relatively large area is characterized by a suite of plants or animals that are restricted to that area and do not occur in neighboring areas, the region is spoken of as constituting a **faunal** or **floral province**, depending on whether one is studying plants or animals. Today we can recognize such large regions as the Boreal Province, constituting land and sea that surrounds the North Pole. Australia is both a faunal and floral province with many plants and animals that are unique to that continent.

We can also recognize such large areas in the fossil record, from the base of the Cambrian to very young fossils at the close of the Cenozoic Era. Some ancient provinces are based on marine fossils, others on land plants or animals. Prior to about fifteen years ago, reconstructions of ancient provinces were mostly based on the positions of land and sea with the continents and oceans fixed in their present positions. Now much of this earlier work is being reevaluated using the model of continental drift and very different arrangements of continents and oceans.

In Lower Cambrian rocks we find a distinctive assemblage of trilobites that occurs throughout western and eastern North America, except for a small region in the Maritime Provinces of Canada and in New England (Figure 8.2). In this latter area a quite different group of trilobites is found, also known from various localities in western Europe and in northwest Africa. The fossils predominant throughout North America are termed the *Olenellus* province, after one of the distinctive trilobites that occurs in this assemblage. The suite that dominates the small region in the Canadian Maritime Provinces and New England is called the *Holmia* province. A third trilobite assemblage occurs in Asia and Australia, named the *Redlichia* province. These distinctive assemblages of trilobites indicate that when we first get a good look at marine communities with hard parts in the Cambrian, life was already fragmented in different regions and clearly had a long prior history.

The small area of *Holmia* province trilobites in North America is judged to represent a piece of what was the European continent during Cambrian time, separated from North America by the Iapetus ocean, which separated *Holmia* and *Olenellus.* Later in the Paleozoic, when the continents came together to form Laurasia, this region was "welded" onto North America. When the continents drifted apart about 180 million years ago, this region in Canada and New England was left behind.

There are numerous other examples of marine faunal provinces. During the Silurian and Devonian periods marine faunas were marked by times of endemism; that is, many marine animals were confined to small areas. These intervals alternated with times of cosmopolitanism, when genera and species had a near worldwide distribution. What might have caused these differences in the size, number, and distinctiveness of marine faunal provinces? One answer is based on the fact that such differences are related to the ease with which marine animals could migrate or disperse from one area to another. If there had been numerous barriers to migration, then faunas with quite different aspects would have tended to evolve in different regions. An example of such a barrier might be land that surrounded a small seaway on a continent and prevented the animals living in that seaway from migrating to adjacent seas. Climate might also have been a controlling factor. Animals might not have been able to tolerate sharp differences in seawater temperature from one sea to another and so would not have migrated.

We have already seen from our discussion of continental drift (Chapter 7) that sharply defined faunal and floral provinces were in existence in the latter part of the Paleozoic Era, with a *Eurydesma* marine fauna and a *Glossopteris* terrestrial flora being confined to the **Gondwana** area. These provinces were surely induced by differences in climate, with Laurasian areas near the equator and Gondwana areas near the South Pole. The presence of continental ice sheets in the Gondwana region supports a climatic interpretation.

During Mesozoic time there is continuing fossil evidence for geographical differences in distribution of plant and animal life. In this instance there are distributions that show distinct changes with latitude in the northern hemisphere. We can recognize various groups that have either low-latitude distribution close to the equator or high-latitude distribution, closer to the pole. Mesozoic coral reefs, for example, are more common and larger in low-latitude areas. These reefs do extend further north, especially in Jurassic and Cretaceous time, than do coral reefs today, suggesting that the temperatures of this time may have been somewhat

FIGURE 8.3
Side view of a rudistid clam belonging to an extinct group of Mesozoic pelecypods that built small reefs and were confined to low latitudes. One valve, shown here, is elongate and coral-like in aspect. The other valve, not shown, forms a cap on top. Length of specimen, about 20 cm. (Courtesy of Field Museum of Natural History, Chicago.)

warmer than they are today. Some Mesozoic reefs are built of aberrant bivalves called **rudistids** (Figure 8.3) which had assumed a coral-like shape with a conical valve below and a cap-shaped valve on top. These clams became enormous, some up to 2 meters high, and built small patch reefs, especially in the Cretaceous. These reefs are common in Mexico and the Caribbean, but they barely penetrate northward into the United States where they are known mainly from Texas (Figure 8.4). Other Mesozoic fossils, such as the **belemnites**, a group of extinct cephalopods, have a high-latitude distribution, being much more common in the northern United States and Canada than they are further south (Figure 8.5). Belemnites had a solid, cigar shaped, internal skeleton and were squidlike animals.

These examples of latitudinal control on distribution of rudistids and belemnites are only part of a much more widespread differentiation with respect to latitude of various groups of marine fossils during Mesozoic time. The equatorial part of this latitudinal control goes under the general name of Tethyan. The **Tethys** was a great seaway of which the Mediterranean and Caspian seas are small, shrunken remnants. The seaway extended from the Caribbean area of the western hemisphere on the west, between Africa and Europe; across the Middle East, the present site of the Himalayas in India; and across southeastern Asia into the Indonesian archipelago. The life of this seaway had a distinctively tropical aspect, with many kinds of clams, snails, ammonites, and other fossils confined to this area. In contrast, shallow-water marine fossils further north in Europe, Asia, and North America lack this tropical aspect and are poorer in number and variety. Thus, many lower latitude Mesozoic marine faunas are said to have a Tethyan aspect, in reference to this major low-latitude feature of the earth. Clearly, the

FIGURE 8.4
Cretaceous marine faunal provinces in western North America. Rudistid pelecypods are most abundant in Mexico and Texas and become progressively rarer farther north. Jurassic and Cretaceous belemnites are most common in Canada but also occur in the northern United States.

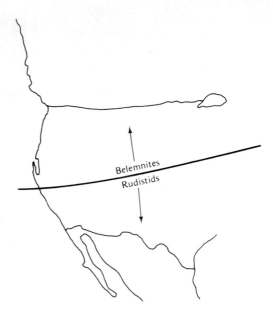

latitudinal control on fossil distribution in the Mesozoic reflects more fundamental controls of climate and temperature differences with latitude. Temperature was probably the ultimate control, limiting the distribution of many of these fossils.

Some of the best known ancient life provinces are recorded in terrestrial rocks of the Cenozoic Era. We find that both fossil mammals and fossil flowering plants, both of which were dominant elements of land life in the Cenozoic, have very distinctive patterns of distribution. Primitive groups of mammals were able to migrate to South America and to Australia near the beginning of the Cenozoic, when there were land areas available for such migration. When these land bridges were later submerged, mammals on these two continents evolved for millions of years in isolation, in the process becoming distinctive and unlike mammals evolving elsewhere. When South America later was connected to North America via the Isthmus of Panama, it was invaded by advanced mammals from the north, wiping out many of the distinctive South American mammals. Many South American

FIGURE 8.5
The internal skeleton of an extinct belemnite cephalopod from England. The specimen is Jurassic in age and about 19 cm long. (Courtesy of Field Museum of Natural History, Chicago.)

mammals moved north, such as the armadillo, which is still extending its range in the United States. This process is still going on in Australia, where the distinctive marsupial mammals (kangaroo, wombat, and the like) are faced with competition from man, cats, rats, and rabbits.

Floral provinces of the Cenozoic show radically different patterns of distribution early in the era than they do today. Subtropical plants extended far north into Alaska in the early Cenozoic and were gradually pushed southward with climatic cooling during this time. As climatic cooling took place, and an arid climate emerged, new assemblages of plants evolved in adjustment to these changed conditions. The widespread grasslands and prairies of the central part of the continent and the dry, desert flora of the southwestern United States evolved under this climatic influence, resulting in a more complex and fragmented series of floral provinces that is still in existence today. These floras were abruptly affected by four major continental glaciations that occurred in North America within the past 2 million years. The spreading and melting ice sheets caused floral provinces to alternately shrink and expand and to shift north and south as unusually sudden and extreme changes in climate took place.

MIGRATION, DISPERSAL, AND BARRIERS

In order for faunal and floral provinces to develop, there must be relatively easy migration and dispersal of plants and animals over some areas, so that they are mostly alike over the area that we call a province. There must also be restricted migration between such areas. A restriction to migration is called a **barrier.**

For land animals, the oceans and large freshwater lakes may constitute barriers. For marine life, land is obviously a barrier. Fresh water may or may not be a barrier to marine animals depending on the tolerance of an organism for changes in salinity.

G.G. Simpson, a well-known vertebrate paleontologist, has described three kinds of migration routes for land animals. The concepts he has formulated are also useful in thinking about migration and barriers for marine organisms. He recognizes corridors, filter bridges, and sweepstakes routes of migration (Figure 8.6). The first, **corridors**, provide for relatively easy movement back and forth for animals from both directions. The steppe region of Russia between Asia and Europe is an example of a corridor. In the case of **filter bridges**, migration is not so easy; generally some animals can utilize a bridge while others cannot. Thus, some are filtered out and not able to migrate. Two well-known examples of filter bridges for land animals are the Bering Strait and the **Isthmus of Panama.** At those times when sea level has stood lower than it does today, the Bering Strait was dry land connecting Asia and Alaska. Migration took place across this land bridge, but it was mainly restricted to animals that could tolerate climatic conditions at this high latitude. Animals that lived further south never had an opportunity to migrate, and so were denied access to the bridge.

FIGURE 8.6
Different kinds of dispersal pathways in North America. The broad area of the Great Plains provides a two-way corridor for migration. Filter bridges occur across the Isthmus of Panama and the Bering Strait, which at times in the past was dry land. Sweepstakes routes are found between the mainland and the Caribbean islands.

During much of the Cenozoic Era, the present land connection between North and South America did not exist. The Isthmus of Panama developed during late Pliocene time, not too many millions of years ago. About equal numbers of animals have migrated north and south across this bridge. The armadillos and ground sloths (later to become extinct) were northward migrants. Large cats, camels (llamas), and various small predators and rodents moved into South America, causing a near revolution among the native mammals and driving many of them to extinction.

In a **sweepstakes route**, migration is quite difficult. The name implies that a successful winning ticket is needed, perhaps one out of thousands, in order to migrate successfully across such a route. Most examples are islands, separated from a large mainland area by fairly wide expanses of water. The island fauna may be much less diverse than that of the neighboring mainland. It may also seem to us to be unbalanced, because some animals that one would think might have made the jump did not. Madagascar and many Islands in the Pacific are examples of sweepstakes routes.

Within the last 30 million years, since the Miocene Epoch of the Cenozoic, when the continents were approximately in their present positions, there have been several migration routes for land plants and animals. These include the ones we have already mentioned as well as a sweepstakes route made by island-hopping across Indonesia from mainland Asia to Australia; the land connection between the Middle East and Africa, now severed by the Suez Canal; and possibly a land connection, which existed at times, across the Straits of Gibraltar at the western end of the Mediterranean.

The idea of corridors and filter bridges can also be applied to marine invertebrates. After the Suez Canal was completed, a former barrier became a marine filter bridge for shallow-water organisms. Some animals were able to migrate through the canal; others were not.

The distribution of marine animals is also influenced by physical and chemical conditions in the oceans. Temperature provides a strong control on dispersal of shallow-water marine animals today and has done so in the past. On both the Atlantic and Pacific coasts of North America, there is a series of shallow-water life zones from south to north based on presence and absence of marine gastropods and pelecypods. Temperature (hence, latitude) seems to be the main control, although temperature is also greatly influenced by the distribution of cold- and warm-water currents. These molluscan life provinces can also be recognized in rocks of Cenozoic age along the two coasts, with shifting of provincial boundaries north and south providing evidence for shifting climates in the past.

REFUGES AND LIVING FOSSILS

When we survey the entire span of life from the Cambrian to the Recent, we find that many groups of plants and animals enjoyed periods when they were abundant, diversified, and widespread only to come upon hard times. In many instances these groups became extinct. In other cases we find that there have been survivors, sometimes called living fossils. Usually the surviving plant or animal is restricted to certain environments or to specific geographic areas where they have managed to hang on, perhaps because of reduced competition in the area. Such restricted areas or habitats are called **refuges**, although they have also been termed asylums or havens.

For marine animals, the most common instance is to find that primitive survivors persist in relatively deep water. These are animals that once lived in shallow water, as documented by their fossil record. One example is a very primitive group of mollusks, called **monoplacophorans**, that have a single shell that looks like that of a limpet, an intertidal rock dweller with a simple Chinese-hat-shaped shell. Monoplacophorans are found as fossils from the Cambrian through the Devonian period, when they were originally thought to have become extinct. In 1957, however, a living monoplacophoran, *Neopilina*, was dredged from deep-ocean trenches off Central and South America and in the Caribbean

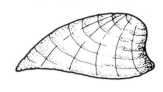

FIGURE 8.7
A living invertebrate that lives in a deep-water refuge. Top and side views of *Neopilina*, the only living monoplacophoran mollusk, fossil relatives of which are confined to Devonian and older rocks.

(Figure 8.7). Here, still alive, was an animal thought to have been extinct for over 350 million years!

Another example of a living fossil is a group of echinoderms (e.g., starfishes, sand dollars) called **crinoids**. These were exceedingly abundant animals during the Paleozoic, their remains forming thick beds of limestones. Most of them lived in quite shallow water and were raised up on a long stalk above the sea floor (Figure 8.8). Others later abandoned the stem or stalk, living directly on the sea floor. The stalked ones continued to live in shallow water into the Mesozoic Era, after which they became exceptionally rare. They still live today, but only in deep water generally greater than 200 meters in depth.

FIGURE 8.8
A slab of Mississippian limestone with examples of complete crinoids. Numerous long stems that supported the flowerlike head and arms are preserved. The slab is from LeGrand, Iowa. The paper clip at right indicates size. (Courtesy of National Museum of Natural History.)

FIGURE 8.9
The only living coelacanth fish, *Latimeria*. Note the thick, fleshy fins and the symmetrical tail. (Courtesy of Field Museum of Natural History, Chicago.)

A third living fossil is the well-known fish *Latimeria* (Figure 8.9). This is a coelacanth fish, related to living lungfishes and to an extinct group of fishes that gave rise to land vertebrates (tetrapods) in the Devonian period. Coelocanths are found as fossils through the Cretaceous and were thought to have become extinct. In the 1940s a single live specimen of a coelocanth was caught by a commercial fisherman in water several hundred meters deep off the coast of Madagascar. Since then other specimens have been obtained in the same general area. Coelocanths today seem to be restricted to deep water off the southeast coast of Africa. Many of the fossil coelocanths known are found in rocks deposited in much shallower water.

These, and other, examples indicate that animals that once lived in shallow water have simply not been able to compete successfully in this environment. The shallow-water environment is an especially favorable one to many organisms, with abundant food supplies close at hand and needed nutrients contributed by rivers. The light to persist in such conditions is relatively more intense than it is in habitats in deeper water. Whether monoplacophorans, stalked crinoids, and *Latimeria* were originally present in both deep and shallow water and became exterminated in the latter or whether they migrated from shallow to deep water is not clear.

Other instances of geographic restriction of marine animals do not seem to be related to changes in the water depth at which they live, but rather to persistence in a small area or refuge. The bivalve *Trigonia* provides an example. During the Mesozoic, this clam was worldwide in distribution and a very common fossil (Figure 8.10). *Trigonia* has not been found as a fossil since the Cretaceous, except in eastern Australia, where it still lives in shallow water off the eastern coast. The area of the southwestern Pacific seems to have been a refuge area for other kinds of marine animals for many millions of years. Persistence of suitable climate and environments in this region could be an explanation for the phenomenon, but this is still not certain.

FIGURE 8.10
External and internal views of one valve of the bivalve *Trigonia*. This specimen from Tennessee is of Cretaceous age. The shells are about 6 cm wide. (Courtesy of National Museum of Natural History.)

We can find the same patterns of distribution exclusively in the fossil record. Some marine animals became much restricted in their area of distribution, usually shortly before they became extinct. During the Mississippian Period, a very distinctive kind of bryozoan evolved. It is called *Archimedes* because it had a twisted central shaft shaped like an archimedean spiral (Figure 8.11). This bryozoan was cosmopolitan during the Mississippian but persisted into the early part of the Pennsylvanian only in Arkansas and Oklahoma, insofar as we can tell. In Middle

FIGURE 8.11
A small slab of limestone showing two of the spiral axes of *Archimedes*. The lacy bryozoan fronds that were supported by the spiral axis are scattered over the surface of the rock, although not all of the fronds were so attached. The slab is about 15 cm across and is Mississippian in age. (Courtesy of National Museum of Natural History.)

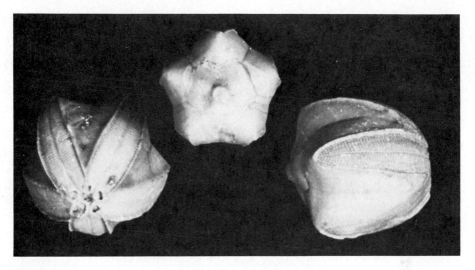

FIGURE 8.12
Three specimens of the blastoid *Pentremites*. Although this kind of blastoid is most common and diverse in Upper Mississippian rocks, it survives into the Lower Pennsylvanian in Oklahoma and Arkansas. The stem and food-gathering appendages are missing. Specimens are about 2 cm high. (Courtesy of Ward's Natural Science Establishment, Inc.)

Pennsylvanian rocks it is not found in the central United States but turns up in Utah. It has not been found anywhere in Upper Pennsylvanian rocks, but it has been reported in Alaska and Siberia in Permian strata, after which it became extinct.

An extinct group of echinoderms called **blastoids** were cosmopolitan during the Devonian and Mississippian (Figure 8.12). They, like *Archimedes*, are found in Lower Pennsylvanian rocks only in Arkansas and Oklahoma. After that they become exceedingly rare until the latter part of the Permian period. They are found in Permian strata, especially on the small island of Timor in Indonesia, where they are quite abundant and diverse. They have also been found in Permian strata in Russia and Australia, but still with a much more restricted distribution than they formerly enjoyed. They became extinct at the close of the Permian. Presumably both the blastoids and *Archimedes* were only able to survive under a specific set of conditions that became more and more restrictive, resulting in their confinement to certain small areas of the Paleozoic oceans. Final elimination of these habitats at the close of the Paleozoic resulted in the extinction of both groups.

So far we have confined our discussion of refuges to marine life, but there are also many examples among terrestrial plants and animals. During the Mesozoic Era there was a group of primitive coniferous trees, related to pines, called **araucarian pines**. These survive today strictly in the southern hemisphere, the best known example being the Norfolk Island pine, now commonly planted as an ornamental tree. During the Mesozoic, araucarians were widespread and abundant.

FIGURE 8.13
Leaf of the gymnosperm
Ginkgo from Paleocene rocks
in South Dakota, about 6 cm
wide. (Courtesy of David
Dilcher, University of
Florida.)

Many of the large fossil logs in the Petrified National Forest in Arizona, of Triassic age, are of these kinds of conifers. The relationship is established by the distinctive structure of the wood in both living and fossil logs.

Several other trees have a pattern similar to that of araucarians—widespread at a time in the past but now confined to a refuge (Figure 8.13). An example is a group of tree-sized gymnosperms related to conifers called **ginkgos**, or maidenhair trees. They have fan-shaped leaves with parallel veins. Ginkgos were very common elements of the terrestrial flora during the Mesozoic but disappeared from the fossil record in North America in the Miocene. They survived only in China, where they have been planted in Buddhist temple courtyards for centuries. No specimens are known in a wild state, apparently having been cut down for firewood centuries ago. Ginkgos once again have a worldwide distribution thanks to their dispersal by humans. They are useful street-side trees and are commonly grown as lawn specimens.

Another tree that survives in the wild only in China is *Metasequoia* (Figure 8.14). This is a close relative of *Sequoia* (the big redwood tree) of the coastal areas of western North America. *Metasequoia*, initially described from fossil remains in Korea, differs from *Sequoia* in the nature and arrangement of the needlelike leaves. In 1948 the tree was discovered alive in mountainous areas of China. It too has been spread by humans.

Examples of terrestrial animals that might be called living fossils include the tuatera lizard which is confined to a few small islands off the coast of New Zealand. This animal is not really a lizard but rather belongs to a group of reptiles, called **rhynchocephalians**, that were conspicuous in the Mesozoic. They flourished during that era but then disappeared from the fossil record. Now they have only this one small living representative.

Two mammals that live in Australia, the duck-billed platypus and the spiny echidna, or anteater, are unusual in that they are the only known egg-laying mammals. They have hair and mammary glands, and so qualify as mammals,

FIGURE 8.14
Metasequoia from the John Day Formation, Oligocene, Oregon. **A.** Leafy branchlet. **B.**
Branch. The scale on the right is in centimeters. (Courtesy of David Dilcher, University of
Florida.)

although they are seemingly very primitive. They may be survivors of a quite
archaic group of mammals, but unfortunately we are not sure to which group of
primltive mammals they belong. Most fossil mammals are identified on the basis
of their teeth, especially their molars. The teeth of these mammals are degenerate;
consequently, we have great difficulty trying to ally them with their possible
ancestors.

THERMOMETERS OF THE PAST

We know that present-day climate varies at different spots on the earth's surface.
At least some of these differences are caused by the angles at which the sun's rays
strike the earth's surface and the length of time that these areas are exposed to
daylight. These factors must have operated to produce climatic differences ever
since the earth was formed. Other important factors, such as the present distribution
of land and sea, and major oceanic currents have changed through time by
continental drift; thus, we could expect ancient climates to have been different
from what they are today. One of the most important aspects of climate is

temperature. We can make inferences about ancient climate if we know the average annual temperature for an area. Fortunately, we have a chemical thermometer that allows us to determine temperatures for the past 100 million years. For earlier periods of earth history, we must depend on the indirect evidence of rocks and fossils.

The method of determining temperature is called the **oxygen isotope** method. As we noted in discussing radioactive age dating (Chapter 1) many elements exist with different atomic weights called isotopes. Oxygen has several isotopes including O^{16}, the common isotope that accounts for 98 percent of all oxygen, as well as a heavier isotope, O^{18}, that accounts for about 0.2 percent of oxygen. The ratio of these two isotopes in seawater is a function of temperature. Thus, warmer seawater tends to have more O^{18} than colder water. This is because O^{16} is lighter and is preferentially given up in evaporation of CO_2 and H_2O in warm areas. Many marine organisms take up oxygen when they build their shells from calcium carbonate, $CaCO_3$. Although the organism uses the two oxygen isotopes in the precise abundances as they occur in the water, the organism cannot select whether or not it takes up an oxygen 16 or an oxygen 18. If the shell becomes fossilized and is relatively unaltered so that the oxygens in the limy shell are not disturbed, then we can collect that shell, find the ratio of the two isotopes, and determine the temperature at which the shell was formed. Shells that are preserved well enough to be studied using this technique are generally found in rocks as old as the Cretaceous. A few belemnite shells in the Jurassic have also been studied. They are so thick that they have undergone little alteration. Because many older fossils have been chemically altered, this method cannot be used on Paleozoic or older Mesozoic rocks. Nevertheless, the oxygen isotope method of studying temperature is a powerful tool for helping us understand the past history of the earth's climate.

The pattern of worldwide temperature fluctuations is now well understood. This history has unfolded by studying sediment cores from the major ocean basins during the deepsea drilling program (usually abbreviated as DSDP). From the Jurassic into the Cretaceous Period, there was a distinct warming trend. This was followed by a dip in temperatures in the very late Cretaceous Period and Early Paleocene Epoch. A second warming trend peaked during the Early Eocene Epoch, followed by a steep decline in temperature that bottomed out during the Oligocene (Figure 8.15). A second cooling trend occurred during the Miocene Epoch. The first big drop in the Oligocene probably led to the onset of what we think of as winter conditions. A cooler and drier climate resulted in the spread of prairie grasslands. The second cooling during the Miocene may have coincided with the formation of the Antarctic ice sheet. Certainly the sea level did drop as water was withdrawn from the oceans and became tied up as ice on land. The third drop in temperature began during the Pliocene Epoch and culminated in the ice ages of the Pleistocene. The earth's climate during the past 3 to 4 million years is known in great detail. Although we tend to think of four major advances and retreats of continental glaciers in the northern hemisphere, in reality there were many minor advances and retreats superimposed on the larger cycle.

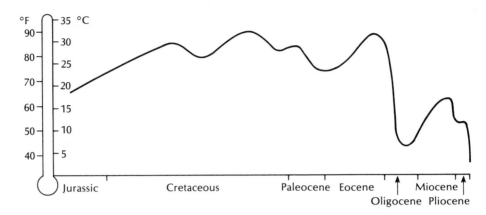

FIGURE 8.15
Average isotopic temperature of ocean waters since the Jurassic Period.

Many aspects of ancient life can be related to this pattern of changing climate. The spread of flowering plants and the development of major floras during the Cenozoic Era reflect these changes (see Chapter 14). The distribution of mammals is related to their adjustment to colder climates. And the pattern of coral reef development during the last 60 million years is tied to changes in world temperatures.

OLDER PALEOCLIMATES

In older rocks we cannot use the oxygen isotope method to measure temperature. Thus, we must depend on less precise methods to determine ancient climates. Some of these involve ancient life forms. For instance, we know that modern coral reefs do not develop where temperatures are much below 20°C or at higher than about 30° latitude. If ancient reefs had similar temperature constraints, then their widespread presence is an argument for tropical and subtropical conditions. The widespread development of reefs during the Silurian and Devonian periods indicates a generally warm climate and the restriction and rarity of reefs during the Permian and Triassic periods indicate restriction of tropical climes or extinction of reef-building organisms.

Various rock types such as evaporites (e.g., rock salt, gypsum) and red beds (shales and sandstones colored red by oxidized iron) are thought to reflect warm conditions and either arid or monsoonal rainfall conditions. The plants that produce coal beds require warm, humid climates. Many of the tree trunks in Pennsylvanian coals lack growth rings, which are typical of trees that live in areas with a distinct growing and dormancy season. Some Devonian woody tree trunks have such rings, as do Permian trunks in the southern hemisphere where glacial conditions existed.

Rocks that were clearly deposited by ice, rather than by wind or water, are also powerful climatic indicators. Such rocks, called **tillites**, are usually very poorly sorted and bedded. Tillites of various ages indicate that there has been a series of ice ages on earth, not just the current one that we are witnessing today, albeit from an interglacial viewpoint. We have discussed the Permian ice age in the southern hemisphere that played an important role in early debates about continental drift (Chapter 7). An earlier ice episode in the Ordovician period mainly affected Africa, which was over the South Pole at that time. Several ice ages during the late Precambrian peaked at about 670 million, 770 million, and 950 million years ago. The 670 million year glaciers were apparently worldwide. Evidence for them has been found on all continents, perhaps making this period the coldest time in the history of the earth.

KEY TERMS

Archimedes	filter bridge	oxygen isotope
barrier	ginkgos	refuge
belemnites	Isthmus of Panama	rudistids
corridor	*Latimeria*	sweepstakes route
cosmopolitan	*Metasequoia*	Tethys
endemic	*Neopilina*	tillites
faunal province	*Olenellus*	

READINGS

Briggs, J.C. *Biogeography and Plate Tectonics.* Elsevier. 204 pages. An up-to-date review of past plant and animal distribution related to continent and ocean positions.

Walliser, O.H., ed. 1986. *Global Bio-Events: A Critical Approach.* Springer-Verlag. 442 pages. This book includes technical chapters on various physical and biological events that affect the entire earth.

Whitmore, T.C., and Prance, G.T. 1986. *Biogeography and Quaternary History in Tropical America.* Oxford Monographs on Biogeography No. 3. Clarendon Press. 214 pages. Modern and ancient biogeography of Latin America, with emphasis on plants, birds, butterflies, and humans.

9

Origin and Evolution of Marine Communities

THE FOSSIL RECORD

Marine communities have a much longer fossil record than do plants and animals that lived on dry land. Our record of marine life goes back into the Precambrian, more than 600 million years ago. The oldest land plants and fossils are only about 400 million years old.

Precambrian Life

In the Precambrian there are two instances of fossils that are assuredly marine. These are the **Ediacaran fauna** and some recently reported microfossils from late Precambrian rocks in the Grand Canyon in Arizona. The Ediacaran fossils are all small and soft-bodied, with no traces of skeletons (Figure 9.1); see also Figure 5.6, page 73). They include a variety of fossils that seem to have relationships with the Coelenterata—the corals, jellyfishes, and sea anemones. These fossils point rather closely to a marine environment, probably in shallow water on a sandy bottom. Although there are freshwater coelenterates alive today, they are inconspicuous compared to their marine relatives, and the phylum certainly arose in the marine environment.

Newly discovered microfossils from the Grand Canyon belong to an enigmatic group called **chitinozoans** (Figure 9.2). These are exclusively late Precambrian and

FIGURE 9.1
Ediacaran fossils of late Precambrian age from southern
Australia. **A.** An exceptionally large jellyfish, about 8
cm across. **B.** Thought to be
the impression of a soft-bodied annelid worm, *Dickin-
sonia*, 2.5 cm long. **C.**
Thought to be a different
kind of annelid worm,
Spriggina, about 4 cm long.
(Courtesy of Mary Wade,
Queensland Museum, Bris-
bane, Australia.)

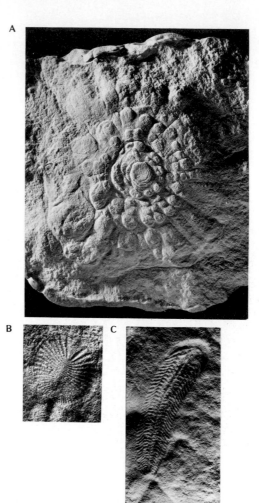

early Paleozoic fossils, the youngest known ones being Devonian in age. They
have a skeleton made of organic material, chitin, that is produced exclusively by
animals. These fossils may have been the housings for small animal-like protists,
or they may have been chitin-covered egg cases for larger animals. In any case,
they do provide evidence of animals living as early as 750 million years ago.

BASE OF THE CAMBRIAN

One of the most startling features of the fossil record is the sudden appearance of
shelled invertebrate fossils. This occurs at what we now call the beginning of the
Cambrian Period, the Tommotian stage, and is used to define the base of the
Cambrian and the base of the Paleozoic.

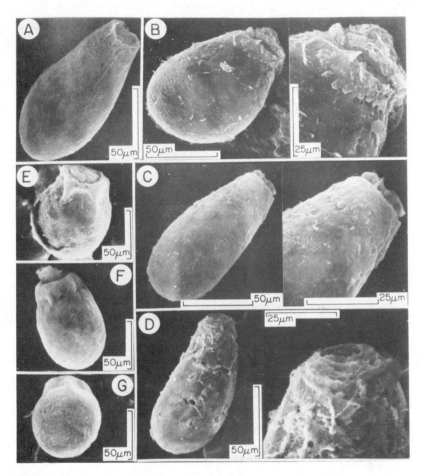

FIGURE 9.2
Precambrian microfossils called chitinozoans. These specimens from the Grand Canyon
are about 850 million years old. They were released from the rock with strong acids and
have been photographed using a scanning electron microscope. Their sizes are indicated
by the bars. (From B. Bloeser, et al., *Science* v. 195, fig. 2, 18 Feb., 1977; copyright ©
1977, Amer. Assoc. Adv. Science. Photo courtesy of J.W. Schopf, U.C.L.A.)

At one time it was thought that the oldest fossils were trilobites and that no
fossils appeared before them; that is, Precambrian rocks were believed to be entirely
unfossiliferous. We now know that the story is more complex than that, but
paleontologists still do not agree on the reasons why the fossil record appears as
it does. We now know that there are many microscopic fossils in Archean rocks.
The first indications we have of multicellular animal fossils is in the form of trace
fossils. These burrows and trails are sufficiently large that they must have been
made by metazoans. They appear in the latest Precambrian just before any metazoan
body fossils. The next indication is found in the fossils of soft-bodied organisms

FIGURE 9.3
An early Cambrian trilobite, *Olenellus*, from southern California. The large head and eyes and small tail are typical of many early Cambrian trilobites. The specimen is 12 cm long. (Courtesy of Takeo Susuki, U.C.L.A.)

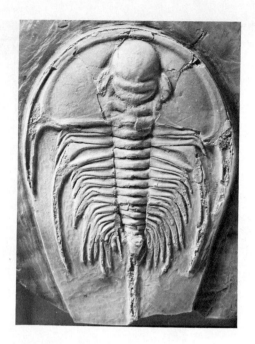

that lacked a skeleton, such as the Ediacaran fauna. This is followed by the first shelly fossils that are small, commonly phosphatic, and generally of uncertain affinities. These occur just before the oldest known trilobites.

After the trilobites appear, fossils with skeletons appear regularly in the fossil record until, by the close of the Ordovician Period, almost all major groups of marine invertebrates have appeared. The ammonites and various vertebrate fish groups come on the scene later, as do marine reptiles.

The base of the Cambrian, with the onset of preservable skeletons, marks one of the most important time intervals in geologic history. We will return to this feature a little later in the chapter.

Cambrian Life

Apart from the Ediacaran fauna and the microfossils in the Precambrian, our fossil record for marine communities really begins at the base of the Cambrian rocks and continues without major interruption to the present day. When we first get a good look at marine communities, they are very strange compared to those of modern oceans. Among the first fossils with hard skeletal parts that are likely to be preserved are **trilobites** (Figure 9.3). These are an extinct class of **arthropods**, or jointed-legged animals, related to crabs, lobsters, and shrimp. Arthropods are among the more advanced and complex of any of the various phyla or animals that we call invertebrates—animals without backbones—in contrast to the vertebrates, or animals with backbones. These first trilobites have large eyes and long

FIGURE 9.4
A slab of Middle Cambrian shale containing numerous specimens of the trilobite *Olenoides.* Carbon-film impressions of several other fossils are also shown on the slab. The specimens are about 7 cm long. (Courtesy of National Museum of Natural History.)

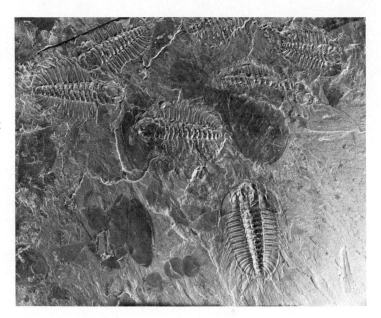

antennae (Figure 9.4). They clearly had a well-developed nervous system. They have many appendages that they used for swimming, walking, or feeding. Although trilobites are primitive in many respects as far as arthropods are concerned, they are quite advanced and complex in comparison with many other phyla of invertebrate animals. Simpler fossils appear in the Tommotian before the first trilobites.

Trilobites must have had a long evolutionary history before they suddenly appear as fossils, or else they evolved very rapidly, almost in an explosive manner. Paleontologists argue about which idea is correct—slow or fast evolution. It is certain that trilobites were around and active before they acquired a hard skeleton. The outer skeleton of trilobites is composed of chitin, a complex organic protein secreted by the animal, and the mineral apatite, composed of calcium phosphate. Thus we speak of them as having a chitinophosphatic skeleton.

During early Cambrian times, over 90 percent of all the fossil specimens known are of trilobites. The other animal groups include a phylum of invertebrates with two shells or valves, the **brachiopods**. Most brachiopods have a shell composed of calcium carbonate, in the form of the mineral calcite, but some (including all of the early Cambrian forms) have a chitinophosphatic shell like the trilobites. A few rare specimens of fossil **echinoderms** have also been found in Lower Cambrian rocks. These are the spiny-skinned invertebrates, whose living relatives include starfishes and sand dollars. The skeleton is calcite. The early Cambrian echinoderms include several types, some of which have very peculiar little football-shaped bodies and are called **helicoplacoids** (Figure 9.5). They could expand and contract in a spiral fashion, resembling little spiral concertinas. A

FIGURE 9.5
Specimen of a helicoplacoid from Lower Cambrian rocks of California, enlarged several times. The mouth was at the top and the spiral rows of small plates could be expanded and contracted when the animal was alive, thus changing the shape. (Courtesy of J.W. Durham, University of California, Berkeley.)

fourth group of early Cambrian animals are the **archaeocyathids**. These are extinct and are thought to be related either to sponges or corals. The skeleton looks much like a small porous ice-cream cone and is composed of calcite. These animals lived in dense aggregations and built up small moundlike reefs on the sea floor.

We see then that all of the major groups of animals that lived in early Cambrian time have been long extinct, except for a few relatives of the brachiopods. The great majority had a chitinophosphatic skeleton (see Figure 6.4, page 87) rather than a calcitic one, calcium carbonate becoming the principal skeletal component of most animals in later marine communities. It has been suggested that this difference in the chemical composition of major skeletal elements may reflect some difference in the chemical composition of ocean waters late in the Precambrian and early in the Paleozoic era. Calcium carbonate may have been more soluble and more difficult for the animal to precipitate as a solid mineral to build a skeleton. Yet Lower Cambrian rocks contain conspicuous beds of limestone that are composed of calcium carbonate, so we know it was being precipitated. These problems in the early history of marine animals are not yet fully resolved.

One additional aspect of these early marine faunas must be emphasized. In addition to fossils with preservable hard parts, Lower Cambrian rocks contain a wealth of what we call **trace fossils**. These are indirect evidences of the presence of life. Such traces consist of burrows, resting marks, fossilized dung and other such signs of activity. Trace fossils are abundant in many marine Cambrian rocks. Some of these traces are trackways, resting marks, or appendage impressions of trilobites. By studying these paleontologists have made meaningful interpretations of trilobite activities and lifestyles. Many other trace fossils cannot be certainly related to any of the fossils known from skeletons. These burrows and trackways are judged to have been formed by animals that lacked preservable hard parts. It seems clear that a large proportion of early marine communities consisted of animals that were entirely soft-bodied, for which we have little record, other than traces, in the early Cambrian.

The Burgess Shale

That there was a tremendous variety of soft-bodied animals living in Cambrian seas, of which we have only meager knowledge, is demonstrated by one of the most unusual and famous fossil localities in the world—the **Burgess Shale**. This locality is high on the side of Mount Field in the Canadian Rockies of British Columbia. The shale is black and platy, and on split surfaces are flattened impressions of many kinds of animals preserved as thin carbon films (Figure 9.6). All of these lacked hard skeletal materials and included jellyfishes, worms of several sorts, and a variety of arthropods that are only distantly related to trilobites or to living arthropods. Many of these animals are placed in classes and orders, or even phyla of their own. Some were pelagic, others benthonic, living both on and above the bottom. Apparently bottom waters were oxygen depleted, and when the organisms died and sank, decay-producing bacteria, most of which require oxygen, were lacking, so the bodies were buried in fine mud and compressed. They underwent destructive distillation in which light fractions of organic molecules were driven off, leaving behind very stable carbon. The carbon films preserve the body outline and details of structure in exquisite detail. Burgess fossils give us a hint that the fossil record is very incomplete with respect to soft-bodied organisms, and that there were many such kinds of animals around in the Cambrian. They also lend support to the idea just presented, based on trace fossils, that only a small part of what were probably quite complex Cambrian marine communities are represented by animals with skeletons.

The Acquisition of Preservable Skeletons

When we survey the record of marine fossils, we find several intriguing observations that have concerned paleontologists for many years. Among these observations are the following:

1. A preponderance of the earliest Cambrian fossils have chitinophosphatic shells.
2. There is considerable evidence from trace fossils and from the Burgess Shale fauna that there were many kinds of marine animals that did not have a preservable skeleton.
3. Various phyla of invertebrates appear in the fossil record abruptly and with the fully developed features of the phylum. Different groups ap-

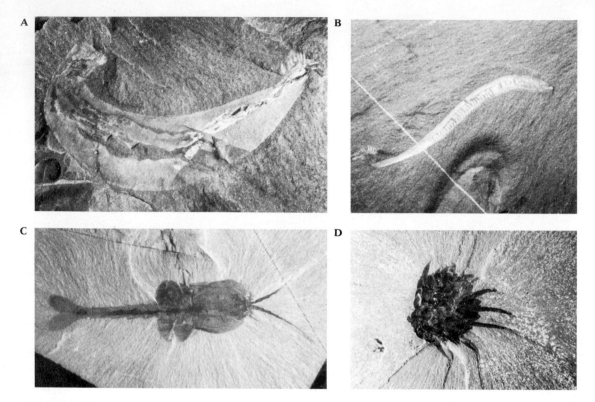

FIGURE 9.6
Burgess Shale fossils from British Columbia, Canada. **A.** *Ottoia*, a priapulid "worm" with small skeletons in the gut. **B.** *Pikaia*, the earliest known chordate. **C.** *Waptia*, a problematic crustacean arthropod. **D.** *Wiwaxia*, a problematic organism with skeleton composed of multiple small plates. (Courtesy of Derrick E.G. Briggs, University of Bristol.)

pear at different times in the Cambrian, and even some in the Ordovician (Figure 9.7). These appearances take place over several tens of millions of years.

These points have considerable bearing on how and when the different phyla of invertebrates evolved. It seems evident that most major groups of marine animals were already in existence long before they acquired a shell or hard parts. When they are found in the fossil record, they are already perfectly recognizable trilobites or brachiopods or gastropods. The time span over which these major groups appear seems to indicate that what we are seeing in the fossil record is different times of skeletonization. Each group must have evolved millions of years before first appearing as fossils, and each group surely had the capacity to live without a preservable skeleton. Therefore, the origins of the phyla of animals are lost to us, having taken place among soft-bodied animals that may never occur as fossils.

FIGURE 9.7
Times of first appearances of major groups of marine animals having a preservable skeleton. Records based on soft-bodied fossils and trace fossils are omitted.

Ordovician	Middle	Rugose and tabulate corals
	Lower	Bryozoans
Cambrian	Upper	Nautiloid cephalopods Jawless fishes (Agnatha)
	Middle	Bivalves
	Lower	Brachiopods, echinoderms, sponges, gastropods Trilobites Pre-trilobite shelled invertebrates
Proterozoic		Ediacaran fauna—oldest invertebrate body fossils Oldest metazoan trace fossils Decrease in stromatolite abundance

Why should major groups of animals have gradually acquired a skeleton over such a long period of time? Two general types of answers to this question have been proposed, neither of which is universally accepted. One answer has to do with the chemical composition of ocean water during late Precambrian and early Paleozoic time. It has been suggested that the oceans were deficient in calcium or bicarbonate ions, or both. Without sufficient concentrations of these materials, a shell is not easy to construct. The fact that such a large proportion of the earliest shelled fossils—trilobites and brachiopods—have a chitinophosphatic shell has been cited as evidence for such a hypothesis. Countering this argument is the fact that there are conspicuous limestone and dolomite rocks in the early Cambrian, the presence of which suggests that there was plenty of calcium and carbonate available.

A second kind of answer has to do not so much with the chemical nature of the oceans but rather with the biology of the organisms. It is suggested that there was abundant carbonate in the oceans and that during their metabolic processes animals accumulated surplus calcium that was difficult for them to get rid of. One way of disposing of such unwanted byproducts of metabolism is to secrete them as a solid. Thus, the external shell may have developed mainly as a way of disposing of such materials, creating a calcium sink for each individual. The shell may later have become useful in other ways, as protection against predators, as a firm site for attachment of muscles, to help prevent drying out of an animal in an intertidal situation, and so on.

Another variation on this kind of biological explanation is that the external skeleton had considerable survival value as marine predators became common and more adept at preying on various kinds of animals. The fossil record for predators is very poor in Cambrian rocks. The first important predators may have been the cephalopods, appearing first in the late Cambrian and not becoming common fossils until the Ordovician. There may very well have been soft-bodied predators for which we have no fossil record. The rise and increase of predation would have placed a premium on any features that would help an animal to survive and reproduce.

We have too little evidence to make a clear choice among these alternatives as possible explanations for the gradual acquisition of a hard skeleton by marine animals. As previously stated, it does seem clear that the various phyla of animals arose prior to the times that they began to acquire skeletons. They may have differentiated at the phylum level rather quickly once they had reached the stage of being multicellular. The Ediacaran fauna suggests that there were already several phyla of soft-bodied animals present in late Precambrian time. The time span between the Ediacaran fauna and the first appearance of shelly fossils at the base of the Cambrian is on the order of a hundred million years. This may seem like a long time, but it is a relatively short time for the evolution of major groups of animals that are distinctive enough to be classed as separate phyla. However, all animal phyla known as fossils had appeared by the end of the Ordovician Period.

THE STRUCTURE OF MARINE COMMUNITIES

Now that we are acquainted with the nature of the earliest fossil record for marine animals, we are ready to look at the gradual changes that have taken place in marine communities of plants and animals through time. In order to discuss the many complex interrelationships of life in the oceans, it is necessary to have a framework within which to place the major kinds of life. For this framework we will review the nature and structure of living marine communities, emphasizing certain aspects of the communities, living and fossil, that are most amenable to analysis in the fossil record.

The Relationship of Organisms to the Sediment-Water Interface and the Water-Air Interface

One group of marine organisms lives in the water column, above the bottom. Such organisms are called **pelagic**. Organisms of the pelagic realm are divided into two main groups: the swimmers, or **nekton**, and the floaters, or **plankton** (Figure 9.8). Fishes, squid, and whales are examples of nekton; the Portuguese man-of-war and many microscopic plants and protists are examples of plankton. The very upper

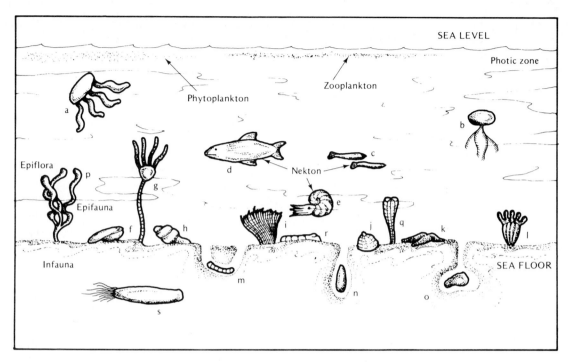

FIGURE 9.8
Relationship of different major groups of plants and animals to the surface of the sea and
to the sea floor. Nekton: (a) jellyfish, (c) conodont animals. Carnivorous nekton: (d) fish,
(e) cephalopod. Plankton: (b) graptolite. Epiflora: (p) seaweed. Epifaunal filter feeders:
(f) brachiopod, (g) crinoid, (i) bryozoan, (j) pelecypod, (q) sponge. Epifaunal detritus
feeders: (h) gastropod, (j) pelecypod, (r) trilobite. Epifaunal carnivore: (k) starfish, (l)
coral. Infaunal filter feeders: (n) pelecypod, (o) echinoid, (m) worm or crustacean (trace
fossil). Infaunal detritus feeder: (s) holothurian.

part of the water column, which is penetrated by sunlight, is called the **photic
zone.** Organisms living in this zone include virtually all autotrophs, or photosynthetic plants and algae. Those organisms living on the sea bottom are **benthos.**
Those that live directly *on* the bottom are called **epifauna** and **epiflora**, and those
living within bottom sediments are called **infauna.** There are no infaunal plants.

Pelagic animals can be divided into (a) those that are **motile** or **vagrant**—
capable of moving about over or in the sea floor—and (b) those that are **sessile**—
fixed in one place. Many gastropods (snails, worms, and starfishes) are examples
of epifaunal vagrants. Certain worms, pelecypods or clams, and some echinoderms
(echinoids and holothurians, or sea cucumbers, for example) mine their way
through soft bottom muds, some of them ingesting the mud to use tiny bits
of organic detritus for food. Sessile epifauna include most corals, bryozoans,
brachiopods, and many clams. These are all fixed, or attached to the bottom, in a

variety of ways and cannot move about. Most seaweeds in shallow water, in the photic zone, are sessile epiflora. Many infaunal animals are confined to stationary burrows and cannot or do not move about. Such animals are sessile infauna.

The Trophic or Feeding Levels of Different Organisms

In any community we can recognize several **trophic levels** of food production and consumption. At the first, and most important, level are the **primary producers**. These are the **autotrophs** which are able to manufacture their own food, mainly from carbon dioxide and water. These producers provide the food foundation for all of the consumers, that is, all of the animals or **heterotrophs**. We then recognize primary consumers. These feed directly on the autotrophs or primary producers. Over most of the oceans the primary producers are confined to shallow depths where sunlight penetrates. Except for near-shore larger seaweeds or floating weeds, such as *Sargassum*, virtually all marine primary producers are phytoplankton (microscopic floaters). Almost 90 percent of all organic production on earth takes place through the activity of phytoplankton. **Primary consumers** on this plankton are virtually all nekton or plankton, and most such consumers are quite small. **Secondary consumers** are those that feed on the primary consumers; thus they are predators, or **carnivores** (animal eaters). Many of the small nekton eat both small plants and animals indiscriminately; thus they are omnivores. Tertiary consumers are those that feed on secondary consumers. These are mainly large fishes such as sharks.

These relationships can be thought of in terms of a **food pyramid** or **food web**. The primary producers form the broad base of such a pyramid and the final consumers—tertiary, quaternary, or whatever—form the small peak. It takes many autotrophs to sustain a small crop of primary heterotrophs. Thus the building blocks of the pyramid, which represent the total mass of organic materials (biomass) at different levels, become progressively smaller and smaller at higher and higher trophic levels. A food web is a somewhat more realistic visualization of these feeding relations in that it displays some of the very complex interactions among the various consumers. Not all of the autotrophs get eaten up by primary consumers, just as not all green plants in terrestrial communities get digested by **herbivores**. If they do not get eaten, they die and sink to the bottom, forming part of the organic detritus of bottom sediments that is utilized by mining infauna. Thus there is a large class of detritus feeders that do not fit easily into a food pyramid or trophic level. Other animals are **scavengers**, eating the flesh of dead animals.

Feeding Strategies of Heterotrophic Organisms

In terms of feeding strategies, we can recognize **macrophagous and microphagous heterotrophs**, that feed on large and small organisms, respectively. The macrophagous types include mainly secondary and tertiary consumers. Many fishes, mammals, and even most corals are macrophagous. Considering that virtually all primary producers in the oceans are microscopic, and many primary consumers in the ocean are microscopic, it is not surprising that there are many kinds of microscopic feeders in the oceans. Many of these are what are called **filter feeders**, or **suspension feeders**. In a wide variety of ways, they strain microscopic plants and animals out of the water for food. They employ pumps, nets, screens, filters, tentacles, and a variety of other mechanisms to feed in this manner.

Ocean waters contain myriads of floating, drifting, and sinking bits of organic material—some still alive, some dead. This material is composed of the adult bodies of microscopic plants, protists, and animals and the early larval stages of larger forms of life. This rain, or snow, of organic particles constitutes the food supply for a great variety of filter feeders that are found in the oceans. These feeders intercept the organic rain on its way through the water. Obviously, not all of this food is going to be taken out of the water by filter feeders; some organic material does settle to the bottom, where it forms organic detritus.

In addition to filter feeders, there are many kinds of marine animals that feed on the detritus that settles on the sea floor. Some **detritus feeders** utilize material on the surface of the bottom; they are called epifaunal detritus feeders. Others are infaunal, eating their way through soft bottom muds and sands and utilizing organic material that has been buried under the floor of the sea.

In the next three chapters we will discuss the important changes that have taken place in marine communities through time, from the Cambrian to the present. In Chapter 10 we will concentrate on the primary producers, the base of the marine food chain. These are primarily planktonic organisms floating near the surface within the zone of water lighted by the sun, the photic zone. In this same chapter we will also discuss those animals that live in the water column—the animal nekton and plankton. These two groups of feeders live in the pelagic realm of the oceans. In Chapter 11 we will discuss the evolution of the filter feeders and detritus feeders. These are primary consumers, epifaunal or infaunal in their habitats, living off microscopic bits of organic detritus in the water or on the bottom. Finally, in Chapter 12, we will discuss secondary or tertiary consumers, the predators that live on a variety of smaller animals. These will include two main groups: the various kinds of fishes and the invertebrate cephalopods. Other groups of predators include starfishes, some advanced gastropods, corals, and marine reptiles and mammals.

CHANGES IN DOMINANCE OF MARINE COMMUNITIES

As evolution has taken place, new groups of plants and animals have appeared on the geologic scene at different times. Concurrently, other groups have become extinct. Thus, all assemblages or communities of plants and animals have changed their makeup over time. Some groups of fossils that were superabundant in older communities gradually dwindled away or declined abruptly, and their dominant position was taken by other groups. These changes in dominance do not need to be interpreted as a result of direct competition between the earlier and later

FIGURE 9.9
Generalized dominance in Cambrian marine bottom communities. **A.** U-shaped and vertical trace fossil burrows. **B.** Burrowing inarticulate brachiopods. **C.** Trilobites. **D.** Trilobite trail. **E.** Archaeocyathid. **F.** Siliceous sponge. **G.** Nektonic hyoliths.

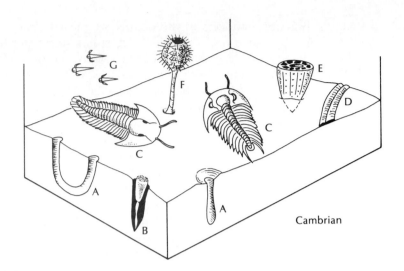

dominant groups. Rather, the later group may simply have had a life style similar to that of the older one and expanded into vacated habitats. In this section, we will briefly trace some of the major changes in dominance that have occurred in marine communities through time.

The earliest shelly marine communities are those of the Cambrian Period (Figure 9.9). The most conspicuous fossils are the trilobites, which make up about 50 percent of the species and 90 percent of the specimens in Cambrian marine rocks. These fossils are quite diverse, with many families, genera, and species. The next most common groups are the inarticulate and articulate brachiopods. Trace fossils are common and may be abundant in rocks that contain no skeletal fossils, indicating that completely soft-bodied communities were common. A variety of trilobite tracks and trails are especially conspicuous. Sponges, rare echinoderms, and enigmatic spongelike fossils called archaeocyathids are found in small reefs. Little is known about nektonic or planktonic life. Animals with small conoidal shells of uncertain affinity, called *Hyolithes*, may have been pelagic.

The contrast between Cambrian and Ordovician communities is striking (Figure 9.10). Ordovician rocks contain a much greater variety of shelly fossils than Cambrian rocks. Trilobites, on the wane, are no longer as conspicuous. A generalized community type consisting of bryozoans, brachiopods, stalked echinoderms, and corals was established during the Ordovician Period and persisted through the remainder of the Paleozoic Period. Obviously, within each of these periods there was considerable change in the dominant subgroups. Thus, Ordovician orthid brachiopods were mostly replaced by spiriferid brachiopods during the Devonian Period, and these were replaced by spiny productid brachiopods during the later Paleozoic Period. The same kinds of changes occur in stalked echinoderms, with early cystoids replaced by crinoids and blastoids. Shelled swimming predators became common in the nautiloid cephalopods. Many of the bryozoans were

FIGURE 9.10
Dominant fossils of the Ordovician Period. **A.** Spiral dwelling burrow. **B.** Brachiopods. **C.** Crinoid. **D.** Cystoid. **E.** Trilobite. **F.** Solitary horn coral. **G,H.** Branching and massive bryozoan colonies. **I.** Jawless fish. **J.** Nautiloid cephalopod.

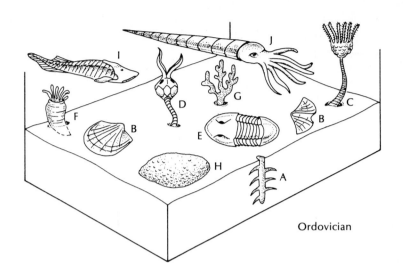

Ordovician

massive lumpy or branching types that were replaced later during the Paleozoic Period by more delicate branching or lacy forms.

By the Late Paleozoic Period, the same major groups of invertebrates were still in force, but within each group there had been conspicuous changes (Figure 9.11). The nautiloid cephalopods were largely replaced by a variety of sharks and bony fishes.

After the series of extinctions that took place in marine invertebrates at the close of the Permian Period, the previously dominant groups gradually assumed

FIGURE 9.11
Composite marine community dominance in the Late Paleozoic. **A.** Trace fossil burrow. **B,C.** Brachiopods (spiriferid, productid). **D.** Colonial coral. **E,F.** Fan-shaped and branching bryozoan colonies. **G.** Crinoid. **H.** Ammonoid with simple sutures. **I.** Primitive bony fish. **J.** Blastoid.

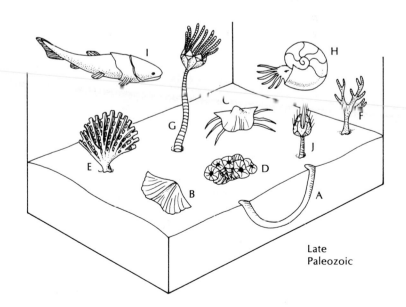

Late
Paleozoic

FIGURE 9.12
Generalized Mesozoic marine communities (not to scale).
A. Dwelling burrow. **B.** Burrowing bivalves. **C.** Burrowing echinoids. **D.** Partially burrowing oyster. **E.** Nestling bivalve. **F.** Reef-forming rudistid bivalves. **G.** Gastropods. **H.** Coiled and straight-shelled ammonoids with complex sutures. **I.** Bony fish.

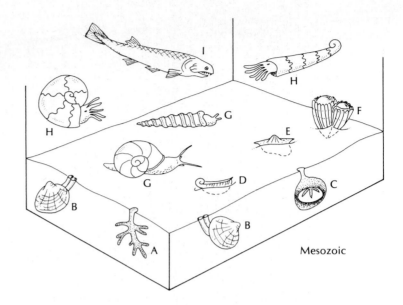

FIGURE 9.13
Generalized Cenozoic marine communities (not to scale).
A. Burrowing bivalves. **B.** Vertical arthropod burrow. **C.** Scaphopod mollusk. **D.** Oysters. **E.** Partially burrowing echinoid (sand dollar). **F.** Gastropod. **G.** Bony fish. **H.** Cartilaginous fish (shark).

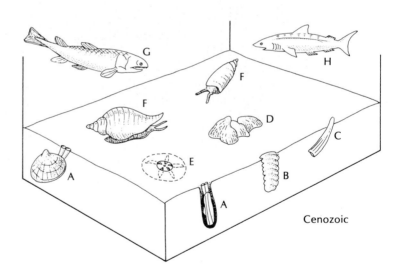

more modest roles. Although brachiopods and crinoids continued to be common elements in many Triassic and Jurassic marine communities, especially in western Europe, they were largely replaced by the mollusks. Gastropods, bivalves, and ammonoid cephalopods all became conspicuous elements in marine assemblages (Figure 9.12). Advanced types of bony fishes evolved. One of the conspicuous differences between Paleozoic and Mesozoic communities was the increase in variety and abundance of shelled burrowers. Many bivalves became burrowers as did the echinoid echinoderms. This change may have been a response to increasing predation pressure on epifauna by fishes and cephalopods.

Finally, the marine communities of the Cenozoic Era were much like modern communities, and direct comparisons of habitats and community roles can be made (Figure 9.13). The cephalopods dwindled with the extinction of the ammonoids, but the bivalves and gastropods became increasingly dominant. Coral reefs and oyster banks were similar to living ones. Fishes were conspicuous and both deep and shallow burrowing echinoids and bivalves were common. In the next three chapters we will look at the various components of these marine communities in somewhat greater detail.

KEY TERMS

arthropods	herbivore	primary producers
autotroph	heterotroph	scavenger
benthos	infauna	secondary consumers
brachiopods	nekton	sessile
Burgess Shale	pelagic	trace fossil
carnivore	photic zone	trilobites
epifauna	plankton	trophic level
filter feeder	primary consumers	

READINGS

For general textbooks dealing with various groups of invertebrate fossils, see the listings at the end of Chapters 11 and 12.

Conway-Morris, S., and Whittington, H.B. 1985. *Fossils of the Burgess Shale*. Geological Survey of Canada. Miscellaneous Report No. 43. 31 pages. This report features large black and white photographs of these important middle Cambrian fossils.

Conway-Morris, S. 1982. *Atlas of the Burgess Shale*. Palaeontological Association. 24 plates. Pictures of important fossils from this famous locality.

Gould, S.J. 1989. *Wonderful Life: The Burgess Shale and the Nature of History*. W.W. Norton. 347 pages. A readable account of the important Burgess Shale fauna with Gould's special view on the importance of these fossils.

Whittington, H.B. 1985. *The Burgess Shale*. Yale Univ. Press. 151 pages. The most complete discussion of the origin and significance of the Burgess Shale fauna.

10

The Fossil Record of
Plankton and Nekton

The foundation for the marine food web is the **phytoplankton**, the microscopic plants and photosynthetic protists that float near the surface of the oceans. Their conversion of simple inorganic substances into complex organic ones takes place only in surface waters where sunlight can penetrate, that is, in the photic or lighted zone. It has been estimated that marine phytoplankton account for about 90 percent of the total organic production on the earth, the other 10 percent being contributed by land plants, and a minuscule part by larger marine plants such as seaweed and calcareous algae.

Marine phytoplankton have undoubtedly been in existence for a very long time, well back into the Precambrian. As we will see in the next chapter, early Paleozoic faunas are dominated by animals that are filter or suspension feeders. They capture microscopic organic particles, alive or dead, from the water. In order for there to be a sufficient amount of such food for an extensive suspension-feeding fauna, there surely had to be a rich phytoplankton base.

Here we will consider these primary producers, the phytoplankton, and see how they have changed through time. Then we will take up the nonphotosynthetic plankton in light of the fossil record. Finally we will consider some of the evidence for fossil **nekton**, the swimming animals that live in the water column (Figure 10.1). We will consider swimmers that are also higher level consumers, the predators, in Chapter 12.

FIGURE 10.1
Relative prominence of major marine phytoplankton, zooplankton, and nekton in three geologic eras. Not to scale.

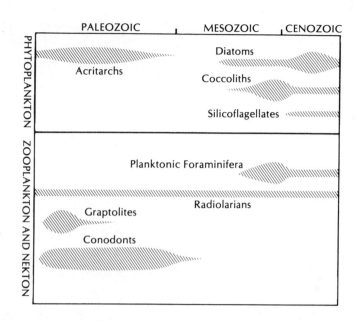

MARINE PHYTOPLANKTON

A majority of marine phytoplankton consists of microscopic single-celled organisms that do not have hard parts; thus, they are not at all likely to be preserved in the fossil record. Indeed, one may wonder how or why we have any fossil record at all for such delicate forms of life. Nevertheless, the phytoplankton do have a long fossil record, although it is admittedly incomplete and spotty. It also gives rise to some intriguing problems. The fossil record goes back into the Precambrian, where floras of cyanobacteria and green algae, such as in the Bitter Springs chert, provide us with evidence for phytoplankton.

Acritarchs

In the late Precambrian and continuing up into the Cambrian and younger Paleozoic rocks, the primary record for phytoplankton consists of a group of microscopic fossils called **acritarchs** (Figure 10.2). These are composed of highly resistant organic material, cellulosic in composition, that enables them to be preserved in such ancient rocks. Even so, they are not found everywhere, their absence being especially conspicuous in rocks that have undergone considerable temperature increase either by burial or by mountain-building processes. Thus, acritarchs are quite abundant in Lower and Middle Paleozoic rocks throughout much of the central United States, where the rocks have not been buried very deeply and have not been subjected to the heat and pressure of mountain building. In the western United States, where rocks have been deformed and uplifted, acritarchs are rarely found or are poorly preserved.

FIGURE 10.2
Photomicrographs of Devonian acritarchs, made with the scanning electron microscope, display variation in shape and ornament. These are from Devonian rocks of Ohio and are highly magnified. (Courtesy of E. Reed Wicander, Central Michigan University.)

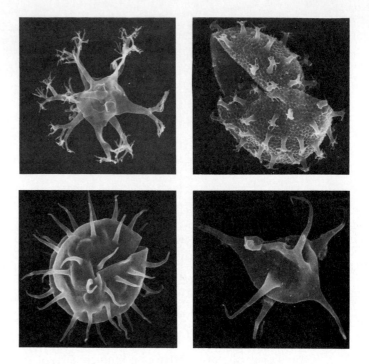

Although these tiny cellulosic bodies that we call acritarchs are sensitive to heat and pressure, they are very resistant to chemical changes. We can release these fossils by treating the rocks with a sequence of strong chemicals (mainly acids). This progressively dissolves the mineral components of the rocks, leaving behind a residue of highly resistant material, including the acritarchs.

The affinities of acritarchs are uncertain, and the name, meaning "old things that are confusing," was specifically coined to reflect that uncertainty. The acritarch body consists of a spherical or polygonal central sac or bag. This may have projecting from it a variety of different kinds of spines—some very long and fragile, others short and hairy. The spines may branch or have small secondary spines on them. The range of variation both in size and in character of the central sac and the spines is considerable; thus, a great many genera and species of acritarchs have been named.

These fossils are regarded as being the remains of reproductive cysts of one or several kinds of marine algae. Many marine algae have a complicated life cycle. After fertilization (union of two gametes) they may enter a dormant stage within which the zygote (first cell of the new generation) divides into a number of spores. In the dormant stage, the zygote is enclosed within a highly resistant organic-walled **cyst** that is comparable but not identical to many of the fossil acritarchs. The cyst ruptures, whereupon the developed spores are released and proceed to develop into new algal adult bodies. Presence of a cyst stage has considerable survival value in cases where adverse conditions may prevail for a period of time. One of the best known living groups of algae that have such organic-walled cysts are the **dinoflagellates** (officially named the Pyrrophyta). These have a very distinctive and complex arrangement of the organic material that makes up the cysts; that is, it is divided into numerous discrete areas. The patterns on the cysts of dinoflagellates do not match those on acritarchs. Although the two groups may be related, at least in part, we cannot firmly place acritarch fossils within this other category of living organisms.

In Early and Middle Paleozoic rock acritarchs are very common fossils. They undoubtedly record the presence of an abundant and diverse phytoplankton in oceanic waters. In some samples, tens of thousands of specimens can be obtained from just a few grams of rock. In the Cambrian, acritarchs are relatively simple spheres and are not especially large or ornamented with complicated spines. In the Ordovician and Silurian, they reach their peak of diversity, abundance, and complexity. Many genera and species are known, which has proven useful in biostratigraphy; fossil zones have been established based on acritarchs.

Although still common in the Devonian, acritarchs are not so diverse, and as one goes into younger and younger Devonian rocks, they become increasingly less common and show fewer types. By the beginning of the Mississippian, they have virtually vanished from the scene, leaving us with a very puzzling void for the remainder of the Paleozoic Era. What were the principal kinds of phytoplankton during the late Paleozoic from the Mississippian through the Permian? We cannot answer this question with certainty.

To the question, Where were the late Paleozoic primary producers? two kinds of answers have been formulated. One answer is that there was a real crisis in the oceans, that the decline in acritarchs records a serious dropoff in productivity, and that this had profound effects on all life in the oceans. Not only would a decline in phytoplankton erode the base of the food chains in the oceans, but it would also affect the oxygen balance in the atmosphere. Most of the photosynthesis that takes place on earth is by marine phytoplankton. A decrease in these organisms might very well have produced a decline in oxygen levels. A second effect would have been on those animals that depend directly on plankton and organic detritus for food—the filter feeders. At the close of the Paleozoic, toward the end of the Permian, many important groups of marine invertebrates became extinct. Virtually all of these—bryozoans, brachiopods, echinoderms, and others—were either filter feeders or detritus eaters.

The other answer is that no such crisis occurred in the oceans, that there was still an ample supply of phytoplankton, but the dominant groups did not have an encysted stage, and thus are not recorded as fossils. Such an answer is based on lack of evidence, making it very difficult to prove or disprove. One criticism of the hypothesis is that the acritarchs essentially disappeared at the beginning of the Mississippian, yet major extinctions of bottom-dwelling invertebrates did not occur until the close of the Permian—a time lag of almost 120 million years. Why should it have taken so long for a crucial decline in productivity to have taken effect?

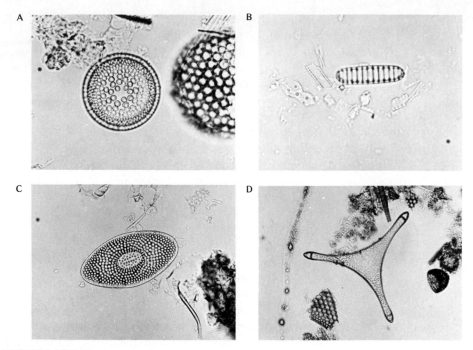

FIGURE 10.3
The siliceous skeletons of marine diatoms, highly magnified. **A** *(Actinocyclus)* and **B** *(Denticula)* are of Miocene age. **C** *(Coscinodiscus)* and **D** *(Trinacria)* are of Eocene age. (Courtesy of John A. Barron, U.S. Geological Survey.)

Other hypotheses have been put forward to explain the Permian extinctions, but these will be considered in the next chapter on filter feeders.

Diatoms and Coccoliths

Not only is our record for phytoplankton poor in the latter part of the Paleozoic, but the paucity of fossils continues into the Triassic at the beginning of the Mesozoic. By Jurassic time, fossil phytoplankton again appear. At first they are rare and poorly preserved, but they quickly become more abundant and diverse. When we get a good look at phytoplankton in the Jurassic and Cretaceous, we find that the same groups that are dominant today in the oceans are now found as fossils. Two of these groups, which together constitute the main producers in the oceans today, are the diatoms and the coccoliths. **Diatoms** belong to the algal phylum Bacillariophyta. They have a silica skeleton composed of two halves that fit together like the two parts of a pillbox. There are many different kinds important today in both marine and fresh waters (Figure 10.3). Diatoms reached a peak of abundance in the middle part of the Cenozoic, during the Miocene, where there

FIGURE 10.4
The surface of a small fragment of Cretaceous chalk
from western Kansas, viewed
at high magnification with
the scanning electron microscope. The horizontal bar indicates scale. Numerous individual coccolith platelets can
be seen scattered across the
surface, making up a significant portion of the chalk.
(Courtesy of Donald E. Hattin, Indiana University; photo
by George R. Ringer.)

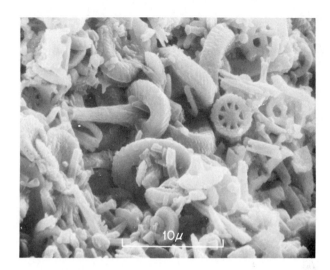

are thick deposits composed largely of their skeletons. These beds are called
diatomites, or diatomaceous earths, and are of commercial value. They are used
in insulating materials, abrasives, ceramics, and in filtering and filling materials.

Coccoliths belong to the algal phylum Haptophyta. They are very tiny
spherical organisms that secrete a series of small calcareous platelets (Figure 10.4).
These are produced throughout the life of an individual. As new plates are formed,
the old ones are forced off and drop to the ocean floor. An individual may have
all the platelets alike, or it may have two or three different kinds. Therefore, the
study of loose plates from the ocean floor is not a reliable index of how many
species produced the plates. Coccoliths are especially abundant and diverse in the
Cretaceous Period, in which their remains are the major constituent of **chalk**. One
cubic centimeter of chalk may contain several billions of these tiny plates, which
are so small that they cannot be readily resolved with an ordinary light microscope.
Most coccolith studies today are conducted using a scanning electron microscope,
an instrument that can raise magnification into the thousands or tens of thousands.
Chalks are not confined to the white cliffs of Dover, England. They are also present
in France and in other parts of Europe. In the United States, Cretaceous chalks are
widespread in western Kansas, Nebraska, Texas, and Alabama. After the Cretaceous,
coccoliths dwindled somewhat in variety, although they are still very important
primary producers in the oceans today.

In addition to these two groups, there are other, less important groups of
phytoplankton that have a fossil record. These include the **silicoflagellates** with
an open, latticework kind of skeleton composed of silica (Figure 10.5). These
belong to the yellow-green algal phylum Chrysophyta and are found in Cretaceous
rocks up to the present day.

In summary, the fossil record of phytoplankton is incomplete and not fully
understood. Acritarchs were apparently the dominant algal group(s) during the

FIGURE 10.5
Various fossil silicoflagellate skeletons, all of Tertiary age from California. The open meshwork of silica is typical of these microfossils which are here highly magnified. Specimen **A** is Pliocene; **B** is Miocene; **C** and **D** are of Eocene age. (Courtesy of John A. Barron, U.S. Geological Survey.)

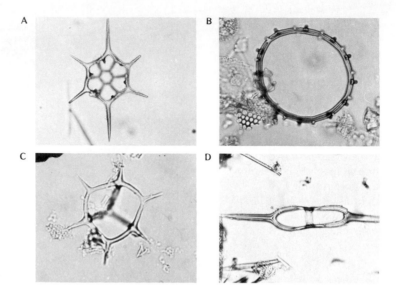

Early and Middle Paleozoic. We know very little about primary production from the Mississippian period through the Triassic. After the Triassic there is an essentially modern aspect to the phytoplankton, with peaks of diversity reached by the coccoliths in the Cretaceous and by diatoms in the Miocene.

MARINE ZOOPLANKTON

In addition to photosynthetic organisms that float in the water, there are many kinds of nonphotosynthetic protists and animals that have the same habit. Many of these organisms, called zooplankton, do not have a preservable skeleton and so do not appear in the fossil record. As far as paleontology is concerned, there are two major groups of zooplankton. One consists of microscopic protists called radiolarians and foraminifera. Both groups belong to the Phylum Sarcodina in the Kingdom Protista. The living ones are characterized by extensions of the protoplasm called **pseudopodia** (false feet) and are related to the amoeba. The second group consists of an extinct group of organisms called **graptolites**. We will consider the protists first and then the graptolites.

Radiolarians and Foraminifera

Radiolarians have a highly symmetrical, intricate skeleton composed of silica (Figure 10.6). The silica, like that of diatoms, sponges, and other silica-producing organisms, is opaline silica, which is not very stable under heat and pressure much greater than that at the surface of the earth. Nevertheless, radiolarians have a remarkably long fossil record, having been found without question in Cambrian

FIGURE 10.6
Several radiolarians of Miocene age from the southwest Pacific showing the intricate siliceous skeleton. Highly magnified. (Courtesy of Joyce R. Blueford, U.S. Geological Survey.)

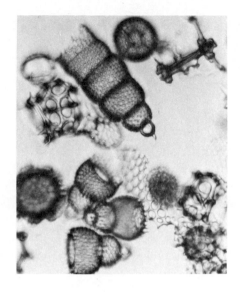

rocks. The record is spotty, however; they have been found in abundance in some formations but are entirely lacking in many. This may be explained by reference to their present-day distribution in which they tend to be concentrated in offshore waters where the water depths are from 3000 to 4000 meters. Thus, they occur far from shore and are rare in coastal waters. Since our fossil record in the main is contained in rocks laid down in quite shallow water over the continents, one might expect radiolarians to be relatively uncommon.

Foraminifera are much more common as fossils than are the radiolarians, but here we are concerned only with certain members of the "forams," those that were planktonic. Although all radiolarians were planktonic, foraminifera were for the greater part of their history benthonic (bottom-dwelling) protists (Figure 10.7). They are first found in Cambrian rocks, but planktonic representatives are not found until the Jurassic. The oldest forams built their skeletons out of foreign material; they gathered up bits of silt, sand, or other debris with their pseudopods and glued this all together with an organic cement that they secreted. Such a covering is said to be agglutinated (glued together) or arenaceous (composed of sandy particles). Some groups of forams soon abandoned this way of constructing a skeleton, although there are many living forams that still do so. A calcareous skeleton came to be secreted by forams beginning especially in Mississippian time, although a few calcareous forms are known from older rocks. Forams continued to be benthonic until the Jurassic, when some highly globular forms evolved (inflated or spherical) that could float. These planktonic foraminifera evolved rapidly and produced many new genera and species (Figure 10.8). They reached a peak of diversity in the Cretaceous, but at the close of that period many forms died out. Just as the close of the Paleozoic was a time of extinction for many marine animals, the close of the Mesozoic was also a time of extinction. One of

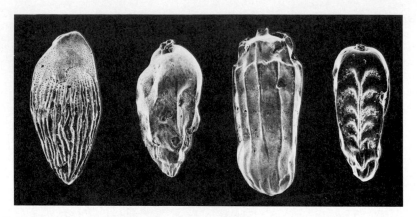

FIGURE 10.7
Four different kinds of calcareous benthonic (bottom-dwelling) foraminifera, highly magnified, from the Miocene of the Atlantic Coast. (Courtesy of Thomas M. Gibson, U.S. Geological Survey.)

the reasons the boundaries between eras of geologic time are placed where they are is because of the rapid changeovers in life that took place near these era boundaries. The extinction of many planktonic forams is one such event. The floating forams continue through the Cenozoic to the Recent, although reduced in number and variety. They have proven to be very useful index fossils for correlation and biostratigraphy because species were very widely distributed in ocean waters. They evolved rapidly, producing new types in relatively short periods of time,

FIGURE 10.8
Planktonic foraminifera of Miocene age from the Atlantic Coast, highly magnified with the scanning electron microscope. Highly inflated, globular chambers make up the skeleton. (Courtesy of Thomas M. Gibson, U.S. Geological Survey.)

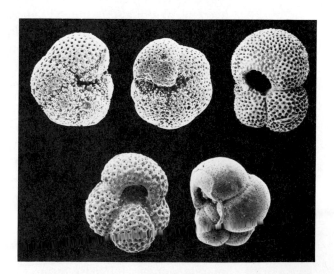

FIGURE 10.9
The two major kinds of graptolites. **A.** A primitive type,
living on the sea floor, with
many branches and crossbars.
Blow-up of a branch (in circle) shows polymorphic individuals. **B.** An advanced
graptolite that was planktonic, with eight branches
and no crossbars. Blow-up of
branch (in circle) shows that
all individuals are alike.

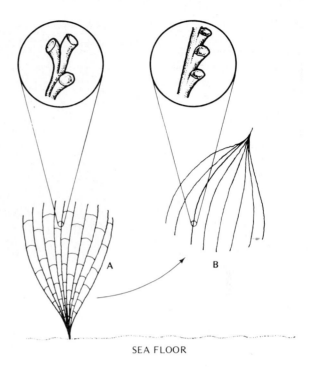

SEA FLOOR

geologically speaking. Along with coccoliths, planktonic forams are major components of chalks in the Cretaceous.

Graptolites

In addition to the microscopic zooplankton just discussed, another group of much
larger floating animals—the **graptolites**—flourished during Paleozoic time. These
were colonial animals that built skeletons of chitin, a highly resistant organic
material that is readily preserved upon burial. Graptolite colonies consist of a series
of branches, each branch built of a chain of thecae, or little houses. Each of the
thecae contained one small individual of the colony (Figure 10.9). The oldest
graptolites are found in the Upper Cambrian. They became extinct during the
Pennsylvanian, but had their heyday in the Ordovician and Silurian, when they
were very common and underwent conspicuous evolutionary changes. Three
aspects of graptolites are of special interest to us: (a) their change in habitat and
evidence for such change, (b) their relationship to living animals, and (c) the
nature of evolutionary change in the group.

  The oldest graptolites are thought to have been benthonic, living on the sea
floor. These colonies have many branches that are connected at intervals by little

FIGURE 10.10
The chitinous skeleton of a primitive graptolite colony, *Dictyonema*, from the Devonian of New York. The specimen shows the presence of many branches and crossbars connecting branches, magnified. (Courtesy of Ward's Natural Science Establishment, Inc.)

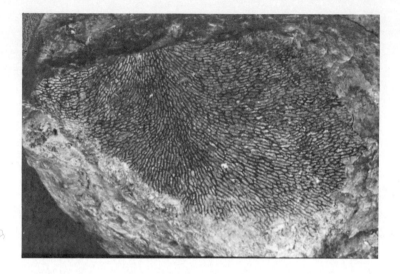

crossbars (Figure 10.10). Shrubby or mosslike in appearance, they have an attachment disc at the base. The individual thecae that make up a branch differ in size and shape; hence, they are said to be polymorphic (having many forms). Two kinds have a sequence down a branch of A-B-A-B. By Ordovician time a new group of graptolites had evolved that is thought to have been exclusively planktonic. These have no more than thirty-two branches in the colony, all of the thecae are alike down a branch (A-A-A), and the crossbars linking branches have disappeared (see Figure 10.9B). These more advanced types of graptolites evolved rapidly. Individual species and genera have wide geographic distribution, and many are found on several continents. Graptolites are most common in rocks that do not contain fossils of assuredly bottom-dwelling organisms. Dark shales and mudstones are the most common rock types that contain graptolites. Because of their wide distribution, appropriate for floating animals, and their occurrence in rocks that record conditions unsuitable for bottom life, graptolites are considered to have been floaters. In addition, one or two specimens have been found attached to a thin carbon film of circular outline that may record the presence of a gas-filled float.

During the Ordovician, graptolites underwent two kinds of changes: in the number of branches in the colonies and in the relative position of the branches (Figure 10.11). **Branch reduction** occurred over a period of 40 million years. Early Ordovician forms have thirty-two branches; then the number of branches is progressively halved, to sixteen, then to eight, four, and two. Finally the number is reduced to a single branch (Figure 10.12). During the Silurian, all of these graptolites are single branched, typified by a genus called *Monograptus*, meaning "one branch" (Figure 10.13). Floating graptolites continued into the early Devonian, when they became extinct. Their bottom-dwelling ancestors persisted even longer, without significant change, into the Mississippian. The other change the graptolites

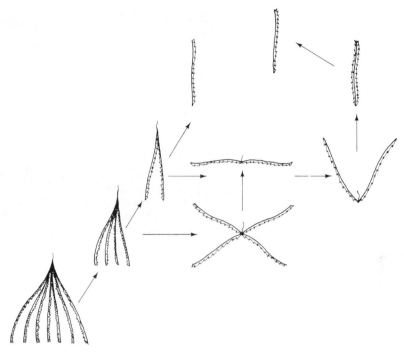

FIGURE 10.11
Evolution of planktonic graptolites during the Ordovician and Silurian, showing the two major evolutionary trends: reduction in number of branches and rotation of the branches.

FIGURE 10.12
A four-branched graptolite colony of Middle Ordovician age, somewhat enlarged, from Canada. The specimen is preserved as a black carbon film on the rock slab. (Courtesy of National Museum of Natural History.)

FIGURE 10.13
A single-branched graptolite colony, *Monograptus*, from the Silurian of New York. Each individual housing in the colony has a prominent projection. (Courtesy of Ward's Natural Science Establishment, Inc.)

underwent relates to **branch rotation**. Branches rotated from a drooping position to a flat position where they are 90 degrees from each other, to an in-line horizontal position, and finally to where they meet above the founding individual of the colony, so that they are back-to-back. Because these changes occurred more or less simultaneously in more than one lineage of graptolites, they have been called **programmed evolution**. The reduction in number of branches greatly reduces the number of individuals in a colony, from near a thousand in a thirty-two-branched form to as few as ten to seventeen individuals in *Monograptus*. Thus, the reduction may have reduced competition among the individuals of the colony and may have resulted in an overall reduction in the weight of the colony—an important advantage for plankton animals. As to why the branches rotated, there is no clear-cut explanation. The rotation does increasingly separate the individuals of one branch from those of another until finally they are back-to-back, rather than facing one another as they are in the drooping position. Thus, each branch may have been able to sample an increasingly different part of the surrounding water for food, although the thecae are actually closer together when they are back-to-back than they are when the branches are horizontal. Graptolites were interesting but puzzling animals. We do not yet have all of the answers as to the nature of the selection process that led to their conspicuous evolutionary changes.

Finally, there has been great debate concerning the affinities of the graptolites. For many years they were judged to be coelenterates, related to a group of this phylum called **hydrozoans**, and specifically to a group of hydrozoans called **siphonophores**. These include the Portuguese man-of-war, *Physalia*, a floating colony. One individual is highly modified into a gas-filled float; other polymorphic individuals occur in strings or branches below the float. Some capture food, others reproduce, and still others have powerful stinging cells for defense. That sounds like a pretty good comparison with the graptolites, but new evidence argues strongly against a graptolite-siphonophore tie. The great majority of graptolites, preserved in dark shales, are very thin, flattened carbon films in which details of the microstructure of the thecae are obliterated. However, some graptolites were discovered preserved in the round in chert (silica) nodules by a Polish paleontologist, Roman Kozlowski. He was able to release these fossils by dissolving the chert in hydrofluoric acid, thus revealing exquisite microscopical detail of the thecae. He found that these structures matched those of the living animal *Rhabdopleura*, which belongs to the phylum Chordata, the phylum that includes vertebrate animals. *Rhabdopleura* is a hemichordate, not a true vertebrate. It lacks a backbone but is colonial and wormlike with individual thecae for the members of the colony. The hypothetical relationship between hemichordates and graptolites has now been generally accepted by most paleontologists. If it is a true relationship, graptolites, along with the first fishes, provide us, in the late Cambrian, with the first occurrence of this important phylum of animals that was eventually to assume a dominant position both in the seas and on land.

MARINE NEKTON

Although many different marine animals live on the bottom, either sluggishly moving about or fixed in one position, and many protists, animals, and plants float in the water, relatively few animals actively swim about. The dominant swimmers are fishes, and these have been an important part of marine communities for hundreds of millions of years. There have also been marine reptiles, and there are still mammals today that actively move through ocean waters. Some reptiles—the sea turtles and snakes—still live in the ocean. All of these groups are primarily predaceous in habit, although there are some exceptions. Because they are higher level consumers (predators), these and other carnivores, along with the invertebrate cephalopods, will be considered in Chapter 12. Other swimmers include shell-less snails called nudibranchs, and various worms, but these have virtually no fossil record. For our purposes here we will consider only two groups of nekton: the jellyfishes, which surprisingly enough do have a fossil record despite their gelatinous body, and a group of extinct animals, called conodonts, that were important nekton during the Paleozoic.

Jellyfishes

Of all marine animals, one would expect **jellyfishes** to be among the least likely to be preserved as fossils. The body consists of over 90 percent water and there are no hard parts (Figure 10.14). Jellyfishes, however, have a remarkably long fossil record, the oldest ones being found in the Ediacaran fauna and in the Burgess Shale. Apparently, when a jellyfish dies it undergoes dehydration, so that as water is lost the tissues become more dense and compact. Most fossil jellyfish are preserved as either molds or impressions, showing the outline of the body and tentacles but without any organic material remaining. Some specimens from the

FIGURE 10.14
A fossil jellyfish preserved as an impression in a nodule from the Pennsylvanian of Illinois. The body, or umbrella part, is preserved at the top and several clusters of tentacles hang down from the sides and bottom of the umbrella. The specimen is about 10 cm high. (Courtesy of Field Museum of Natural History, Chicago.)

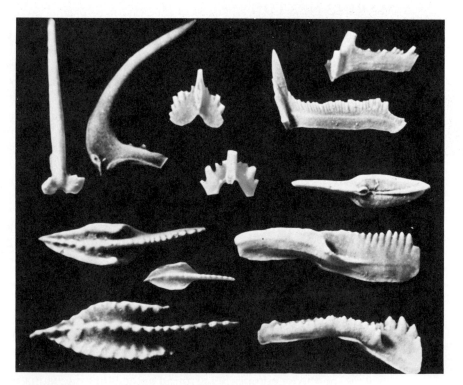

FIGURE 10.15
A variety of different kinds of conodont elements of early Mississippian age from Indiana. The pieces at the top are generally called bars or blades; those below are platform types. Individual fossils are about 0.5 mm long. (Courtesy of Carl B. Rexroad, Indiana Geological Survey.)

FIGURE 10.16
An assemblage of conodonts, highly magnified, showing the close association of several kinds of discrete elements. (Modified from Rhodes, 1954.)

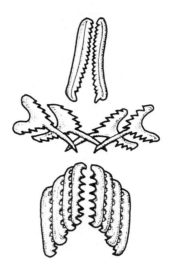

famous **Solenhofen beds** (of Jurassic age) in Germany do have traces of the original material as well as imprints of muscles preserved.

One important group of jellyfishes did have hard parts, in the form of a thin covering made of chitin. This group, the **conularids**, ranged from the Cambrian to the Triassic and had four-sided symmetry. Their systematic position was debated for many years, but they seem now to be allied to living jellyfishes which also have quadrangular symmetry. Conularids were unusual jellyfish in that they apparently were attached to the bottom, not swimming or floating as did most members of the group.

Conodonts

Conodonts are among the most enigmatic of fossils. They consist of tiny toothlike objects composed of apatite, the same mineral of which bone is made (Figure 10.15). Despite this chemical affinity, conodonts are thought not to be directly related to vertebrates; however, their true affinity is not known with certainty. These small fossils commonly occur in rocks singly, but rare specimens show that, in fact, each individual conodont was part of a complex apparatus of small hard parts that included several different forms (Figure 10.16). The conodonts grew from the center outward and are thus thought to have been embedded in soft tissue. Because they are found in all rock types, from black shales to conglomerates, they were probably parts of animals that could swim over any kind of bottom. The names of conodonts are complicated by the fact that most genera and species were named on the basis of individual elements before it was realized that several different kinds could occur together in one individual. Two or three genera and several species, based on isolated pieces, may occur together in one assemblage.

FIGURE 10.17
A carbon-film impression of a conodont-eating animal, about 6 cm long. (Courtesy of W.G. Melton, Jr., University of Montana.)

In recent years, a few specimens have been found with conodonts inside a carbon film outline of an animal that may have eaten the conodont animal (Figure 10.17). This fossil was found in Montana and is dated as being from the Mississippian age. The individual skeletal conodont pieces are situated in the gut, just above the circular dark patch in the lower center of the fossil.

The conodont animal has been discovered in Lower Carboniferous rocks of Scotland (Figure 10.18). The fossil is long and very slender and is divided into

FIGURE 10.18
The first complete conodont discovered from the Lower Carboniferous of Scotland. Inset shows the opposite side of the head with position and orientation of the skeletal apparatus. The animal is about 7 mm long. (Courtesy of R.J. Aldridge, Univ. of Leicester.)

segments with a conodont assemblage just behind the mouth. These fossil elements are thought to have served as teeth. The conodont animal has recently been classified as a chordate, and may have been a parallel development to the jawless agnathan vertebrates. Other researchers have placed the conodont in a phylum of their own.

Conodonts are very abundant in many marine rocks, which suggests that the animal was quite common during the Paleozoic Period. The oldest conodonts known lived during the Cambrian Period. The conodonts became extinct during the Triassic Period.

KEY TERMS

acritarchs	cyst	jellyfish
branch reduction	diatoms	nekton
branch rotation	diatomite	phytoplankton
chalk	dinoflagellates	radiolarians
coccolith	foraminifera	silicoflagellates
conodonts	graptolites	zooplankton

READINGS

Briggs, D.E.G., Clarkson, E.N.K., and Aldridge, R.J. 1983. "The Conodont Animal." *Lethaia* 16:1–14. This is the initial report on the animal now believed to have borne conodonts.

Haq, B.U., ed. 1983. *Calcareous Nannoplankton*. Benchmark Papers in Geology. Hutchinson Ross. 338 pages. A series of papers on very small marine plankton with calcareous skeletons, primarily discussing the coccoliths.

Loeblich, A.R., Jr., ed. 1970. *Ultramicroplankton*. Proceedings of the North American Paleontology Convention, Part G. Pages 705–1007. This volume includes nine technical articles on acritarchs, dinoflagellates, and other fossil marine phytoplankton.

Moore, R. C., ed. 1953–1981. *Treatise on Invertebrate Paleontology*. Univ. of Kansas, Geological Society of America. 27 volumes. This important series of monographs includes descriptions and illustrations of all known invertebrate and animal protist fossils. The twenty-seven volumes that have appeared so far include two revised volumes. Each book includes general chapters on evolution, ecology, and classification. Those volumes of interest in connection with this chapter include Part C on foraminifera; Part D on radiolarians; Part F on coelenterata, which includes fossil jellyfish; Part V on graptolites; and Part W, *Miscellanea*, which has four chapters on conodonts.

Ocean Drilling Program Proceedings. National Science Foundation. This series is a continuation of the reports of the Deep Sea Drilling Program. Over 120 volumes of results have been published over the past two decades. This includes many papers on all skeletal marine plankton, plant and animal. This is by far the most comprehensive and up-to-date series of fossil plankton and nekton.

11

History of the Filter Feeders and Detritus Feeders

The food obtained by the great majority of marine invertebrate animals is in the form of small particles. **Filter feeders** obtain these particles directly from the water; **detritus feeders** wait until the food falls to the sea floor and obtain it there. Many of these animals have hard parts and are preserved as fossils. Because they constitute the bulk of known fossils, most books on invertebrate paleontology place considerable emphasis on these feeding types. Several different phyla of invertebrates are included in the category of small-particle eaters. Most marine communities are dominated by either filter feeders or by detritus feeders, both types playing a significant part in the structure and complexity of these communities. The following generalizations can be made with regard to the fossil record of these animals:

1. Although many small-particle eaters have hard parts, many others (worms, crustacean arthropods, holothurians) lack, or almost lack, hard parts and are recorded mainly as trace fossils.
2. Filter feeders and detritus feeders may either be epifaunal or infaunal.
3. These animals are commonly stratified with respect to the sea floor, living different distances above or below the sediment-water interface. This adds to the complexity of many marine communities, past and present.
4. From the Cambrian to the Recent, there has been a constant turnover of major kinds of animals dominating these feeding roles.

5. At any given time in earth history, different kinds of filter feeders were dominant in different contemporaneous but geographically distinct communities.

DOMINANCE AMONG SMALL-PARTICLE FEEDERS

In the broadest possible sense, we can divide the last 600 million years into three categories with respect to the kinds of invertebrates that were most abundant and dominant. These three phases are (a) the Cambrian, (b) the period from the Ordovician through the Permian, and (c) post-Permian time.

The first period, the Cambrian, was dominated by **trilobites**. Brachiopods were present and second in abundance, but still neither very diverse nor common. Trilobites undoubtedly played a variety of food-gathering roles in the Cambrian; otherwise, their diversity is difficult to explain. Many of them were probably detritus feeders, moving across the sea floor. Others were probably shallow burrowers, obtaining food from the sediment. Still others may have been feeble swimmers, using their appendages to direct food-laden currents to their mouths, much like some living crustaceans such as fairy shrimp. Trilobites clearly occupied many roles in marine communities during the Cambrian. These communities were probably considerably simpler in structure than were the communities that replaced them in the Ordovician.

The abundance of trilobite fossils is partly attributable to the fact that they grow by casting off their exoskeleton in a series of **molt** stages. Thus, one animal might have produced ten or twelve potentially preservable skeletons during its lifetime. Trilobites had a considerable size range; adult specimens may be only a few millimeters long or from 20 to 30 centimeters in length. Some trilobites are characterized by large compound eyes, whereas others have no traces of eyes. The exterior may be very smooth, plain, and streamlined, or it may be highly ornamented with spines and nodes. Some of the highly spinose forms may have been active swimmers, the spines serving to decrease their volume-to-surface ratio, making it easier for them to move through the water. Still other trilobites could curl up so that the tip of the tail and the head met. This may have been a defensive posture; if so, it suggests that there were predators that fed on trilobites. Enrollment of some trilobites may also have been merely a death position. After the Cambrian, trilobites became a subordinate part of most marine communities. They continued to evolve but were limited to only a few groups. These gradually became extinct—many at the close of the Ordovician, others at the end of the Devonian (Figure 11.1). By the late Paleozoic only a few species remained, persisting until near the close of the Paleozoic, when they, too, became extinct.

Beginning in the Ordovician, a strikingly different kind of marine community came into existence and persisted until the close of the Permian—a time span of over 200 million years. The dominant members of these communities were all attached, sessile **epifauna** that were filter feeders. The three main groups were the brachiopods, bryozoans, and stalked echinoderms. Certainly, such other filter

FIGURE 11.1
A slab of Ordovician limestone containing many specimens of a large trilobite, *Ho-
motelus*, characteristic of that
period. The block is about 40
cm long. (Courtesy of Los
Angeles County Museum of
Natural History, photo by
Lawrence S. Reynold.)

feeders as the bivalves were present, but they were generally subordinate. Other important members of these communities were sessile carnivores (the corals), a variety of detritus feeders (such as gastropods), and predators (cephalopods and fishes).

Each of the three main groups—brachiopods, bryozoans, and stalked echino-derms—is characterized by two important attributes. First, within each group there is a definite succession of dominance. Those kinds of brachiopods that were predominant in the Ordovician gave way to other types in later periods. The same is true for the other two groups. Secondly, these groups provide the first evidence for conspicuous stratification of marine communities—the brachiopods living just above the sea floor, many bryozoan colonies being raised a few centimeters above the bottom, and stalked echinoderms being generally ten or more centimeters high. This stratification was largely lacking in Cambrian communities. Ordovician communities also differ from Cambrian ones in another respect; in the former, most of the common animals were sessile (fixed to the bottom), whereas in the latter, trilobites were mobile animals. Also, dominant Ordovician animals had a calcium carbonate shell, whereas Cambrian animals most often had a chitinophos-phatic shell.

We will now look at the succession from the Ordovician through the Permian of each of these three major groups of filter feeders in more detail.

Brachiopods

The earliest **brachiopods** have a shell that is chitinophosphatic in composition, and the two valves enclosing the soft parts are not hinged together (inarticulate condition). This kind of brachiopod predominates in Cambrian rocks but is replaced

FIGURE 11.2
A slab of Ordovician limestone containing many specimens of a wide-hinged
brachiopod that has been replaced by silica. Both interiors and exteriors of the
valves are shown. A bryo-
zoan colony is near the center of the picture. The specimen is somewhat reduced.
(Courtesy of National Mu-
seum of Natural History.)

abruptly in Ordovician rocks by calcareous brachiopods that have the two shells
hinged with a tooth-and-socket arrangement (articulate condition). Although the
inarticulate brachiopods persist to the Recent, they were never conspicuous fossils
after the Cambrian.

The calcareous brachiopods display several trends during the Paleozoic, when
they were among the most common marine filter feeders. These include a tendency
in some groups for the hingeline to change from wide to short, with a consequent
change in shape from a quadrangular to an oval outline. The mode of attachment
to the sea floor also was subject to several modifications. Ordovician brachiopods
typically have a wide hinge with coarse ribs (Figure 11.2). The shell is attached
by a horny tube called a **pedicle** that issues from the back of the shell through an
open triangular hole (Figure 11.3). As the hinge became shorter in some groups,

FIGURE 11.3
Cross section of a living
brachiopod showing major
anatomical features.

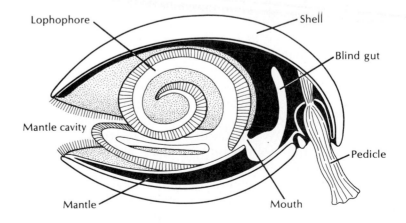

FIGURE 11.4
Two advanced brachiopods from Permian-age reefs in western Texas. The one on the right has a much reduced upper valve and lived an oysterlike existence attached to the reef. The one on the left is shown in both side and top views. This brachiopod adopted a coral shape and was attached by spines. The specimens are reduced in size and are silicified, having been etched from limestone. (Courtesy of National Museum of Natural History.)

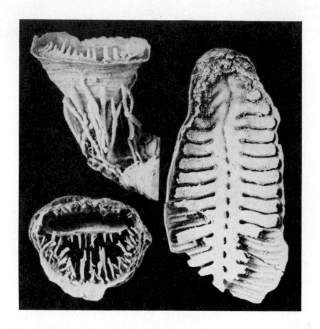

the hole for the pedicle became partly filled by calcareous plates, or it changed in shape from triangular to circular in outline. In still other lineages, a functional pedicle was lost altogether, the hole being completely closed. Such more advanced types either rested loose on the bottom or developed an alternate mode of fixation to the sea floor. Some had part of the shell buried in soft sediment, serving for anchorage. Others were cemented to other shells by the beak of the shell, much as oysters are attached (Figure 11.4). Still other groups developed spines, either along the hinge or over the valves, that penetrated into soft sediment and served for anchorage.

In addition to changes in shape and attachment, brachiopods also had progressive modification of the calcareous internal supports of their food-gathering apparatus, the **lophophore**. The lophophore consists of two coiled, fleshy arms covered with cilia that beat to direct food-laden water currents into and out of the food-gathering chamber within the shell. In early calcareous brachiopods the lophophore was supported only by two short straight rods at the base. In more advanced groups, the lophophore became more fully supported by internal structures. In one group, the **spiriferids**, the lophophore was supported by two helically coiled ribbons of calcite. In still others, for example, the **terebratulids**, the calcareous support formed a loop that was bent back on itself.

Water was drawn into the shell on both sides and expelled at the center of the shell opposite the hinge. Many features of the shell are related to this aspect of brachiopod biology. The central part of the shell is cut off from the two sides, resulting in a raised fold on one valve and a depressed sulcus on the other. This serves to separate incoming and outgoing water currents, helping to insure that

FIGURE 11.5
Major groups of brachiopods, all oriented with brachial valve in front, pedicle valve behind, and posterior up. **A.** Terebratulid. **B.** Orthid. **C.** Pentamerid. **D.** Productid. **E.** Spiriferid. **F.** Chonetid. **G.** Strophomenid. **H.** Rhynchonellid.

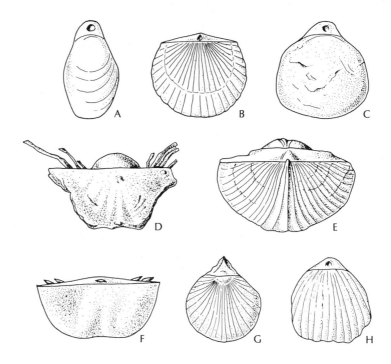

the brachiopod will not recycle already filtered water. The coarse ribbing on many shells results in a zigzag pattern near the opening between valves. This pattern produces an increased area through which water can pass so that the shells need not open wide enough to allow unwanted larger particles to get into the interior. Some brachiopods have small, close-spaced spines that project between the opened valves and act like a fence to keep out intruders.

The brachiopod shell is built by a thin layer of tissue called the mantle that lines the interiors of both valves. Growth proceeds along the edges of the shell and is called accretionary growth. Thus, the outer edges of the shell, where it opens, are the most recently formed part of the valves. Most of the space within the shell is called the mantle cavity, within which the lophophore is housed. The gut is reduced in size. In the **inarticulate** brachiopods the gut is complete, with a mouth and anus, but in the more advanced articulates the gut has a single opening that serves for both intake and ejection of food particles. The fecal material is called pseudofeces because there is no true anal opening. To expel accumulated pseudofeces from the mantle cavity, brachiopods rapidly clap the two valves together, causing water currents to wash the pseudofeces away.

During each period of the Paleozoic, one group or another of brachiopods tended to predominate in many marine communities. A primitive group, the **orthid brachiopods**, held sway in the Ordovician (Figure 11.5). These had a wide hinge, open triangular pedicle opening, and coarse to fine ribbing. The lophophore supports were stubby and primitive. Another group, the **pentamerids**, were especially common in many Silurian limestones, living on reefs. These large egg-shaped brachiopods had a short hinge and a smooth or faintly ribbed shell. Spiriferid brachiopods underwent a conspicuous adaptive radiation in the Devonian. Many had exceedingly long hinges, so that the shell appears to be winged. Others had short hinges and a round pedicle opening. Most were ornamented with ribs. The conspicuous aspect of these forms is the cone-shaped spiral ribbon of calcite that supported the lophophore. During the Mississippian and up through the Permian Period, three groups were common that had lost a functional pedicle. One of these, the **strophomenids**, either had the beak of the shell pushed down into sediment for anchorage or cemented to a hard substrate. The other two groups had spines, and the valves were concave and convex. The **chonetids** were small brachiopods with spines only along the hinge. The body cavity between the two valves was quite small. The other group, the **productids**, generally had spines over both valves or one valve. These spines served to anchor the brachiopod to the bottom. One valve was generally highly convex, whereas the other (uppermost in life position) was flat or gently concave, forming a lid over the body cavity.

By the close of the Paleozoic, most of these various groups of brachiopods (Figure 11.5), so conspicuous for millions of years, had become extinct. Brachiopods again were locally abundant in the Mesozoic but not nearly so diverse as before. One group, the **terebratulids**, with an oval shell, round pedicle opening, and a looped support for the lophophore, were most common and are still the most diverse group of brachiopods living today.

Bryozoans

Virtually any outcrop of Paleozoic marine rocks will yield fossil **bryozoans** if one makes a persistent search for them. In many such rocks, bryozoans are the single most common group to be found. Clearly, they were dominant animals in many Paleozoic marine communities (Figure 11.6). The individuals that make up each bryozoan colony are tiny; they usually cannot be readily seen without a hand lens. In order to identify a bryozoan as to the major group, genus, or species to which it belongs, it is usually necessary to examine the specimen under the microscope, preferably by using thin slices of the colony ground down on a glass slide so that light will pass through the rock. Such a **thin section**, as it is called, is essential for detailed research on these fossils.

Like brachiopods, bryozoans exhibit a succession of dominance in different groups through the Paleozoic. Cambrian bryozoans are very doubtfully known, but beginning in the Ordovician they suddenly become quite conspicuous. These early bryozoans are what we might call "stony" bryozoans. The colonies are typically massive, ranging in shape from an irregular lump to a large flat sheet to thick branching cylinders (Figure 11.7). They are important rock formers in many Ordovician limestones and shales. These bryozoans persist right through the Paleozoic, but they gradually become less and less prominent. By Devonian time, they have given way to a variety of other bryozoans that are less conspicuous,

FIGURE 11.6

A bryozoan-rich slab of lime-stone from the Rochester shale of Silurian age. Several different kinds of bryozoans (mostly twiggy, ramose types) are shown, as well as brach-iopods and other fossils. About natural size. (Courtesy of National Museum of Natural History.)

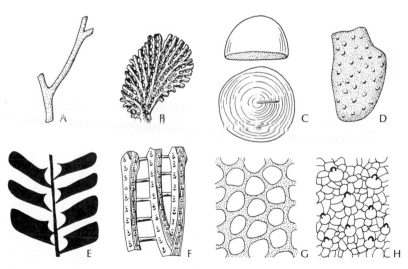

FIGURE 11.7

Colony shapes of representative Paleozoic bryozoans. **A.** Twiggy. **B.** Fan-shaped. **C.** Massive or stony. **D.** Bifoliate or laminar. **E.,F.** Cross sections of cryptostomes, magnified. **G.** Enlarged trepostome thin section. **H.** Enlarged cystoporate section.

FIGURE 11.8
Magnified cross section of an
individual bryozoan showing
the relationship between the
skeleton and major anatomical features. Highly mag-
nified.

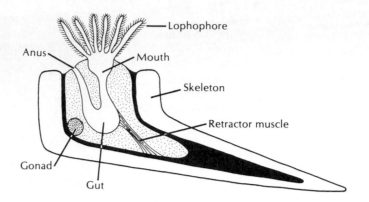

more delicate, and in many cases more abundant. The massive kinds are called
trepostome bryozoans and the more delicate forms, **cryptostomes.**

During the late Paleozoic, the cryptostomes are exceedingly varied and
abundant. These include a variety of "stick" bryozoans, delicate cylinders that may
range from one to several millimeters in diameter. These commonly branch several
times and are rarely found complete. Death and burial breaks up the delicate
branches. Other, bifoliate forms, are flat and ribbonlike with two sets of colonies
growing back-to-back. Still other bryozoans, called fenestrates, form colonies that
are an open meshwork, resulting in a lacy appearance. The skeleton is called a
frond, which may be cone shaped, opening upward, or in specialized types, such
as *Archimedes*, may be spirally arranged on a solid core of skeletal tissue.

All bryozoans are filter feeders as far as we know. They have a lophophore,
as do brachiopods, that is formed into a loop near the mouth (Figure 11.8). Soft
extensions of the lophophore and cilia on it capture microscopic food particles
from the water. In many instances bryozoans were clearly the dominant animals
in Paleozoic marine communities, although their abundance is easy to overlook.
Many of them are small and fragmented, making it difficult to get an accurate
estimate of the number of colonies present. The casual fossil collector will commonly
ignore them completely in favor of the more obvious and more interesting-looking
brachiopods, trilobites, or corals.

The most conspicuous Paleozoic groups became extinct at the end of the
Permian. One new type, the cheilostomes, that arose in the Mesozoic has persisted
to the present. These bryozoans are delicate and the colonies are usually small.
Many of them encrust on shells, seaweed, rocks, or on other hard objects. Although
they are quite diverse—there are many families and genera—they are rarely as
abundant as were their Paleozoic ancestors.

Stalked Echinoderms

An important component of many Paleozoic marine communities was the **stalked
echinoderms**. Like the brachiopods and bryozoans, this phylum is unusual in that

FIGURE 11.9
An unusually complete blastoid, *Orophocrinus*, showing a short stem, external growth lines on the thecal plates, and delicate, slender food-gathering appendages above. The specimen is Mississippian in age, from Iowa, and is about 4 cm tall. (Courtesy of D.B. Macurda, Jr., University of Michigan.)

it is represented in the fossil record by many more kinds of extinct forms, especially in the Lower and Middle Paleozoic, than it is by later representatives. A total of twenty-one classes have been established within this phylum. Of these, only five are still all alive today, leaving sixteen extinct major groups. Many of these extinct classes are confined to either Cambrian or Ordovician rocks. Several groups are known from only a handful of specimens that are assigned to a few genera and species. Yet these fossils are so distinct from each other that they have been separated at the class level. This separation seems to indicate that echinoderms underwent an extensive adaptive radiation early in the Paleozoic, developing many new and highly distinctive forms. Many of these new types were only successful for relatively short periods of time. Most did not evolve into new groups but became evolutionary blind alleys.

Some of these primitive echinoderms could move about. Others were attached by their body to the bottom. Still others had bodies on stalks or stems, above the sea floor, giving them a plantlike appearance.

The most common and diverse of these echinoderms were three stalked groups—the blastoids, cystoids, and crinoids. The **blastoids** are most common in Silurian through Mississippian rocks and became extinct in the Permian (Figure 11.9). The **cystoids** thrived from the Ordovician through the Devonian, when they became extinct. In each of these groups the stem is short, generally only a few centimeters long. The vital organs are encased within a theca composed of a few to many plates. From this theca there are five or more plated extensions that

FIGURE 11.10
Two stalked crinoids (large, *Taxocrinus*; small, *Dichocrinus*) from Mississippian rocks of Iowa. The stem is not complete on either specimen. Notice the array of food-gathering arms around the top of the head. The slab of rock is about 22 cm high. (Courtesy of Field Museum of Natural History, Chicago.)

project upward. These served as the primary food-gathering structures, presumably entrapping food particles from passing currents with cilia and mucus.

The **crinoids** were a much more successful group of echinoderms than were the blastoids or cystoids. They ranged from Ordovician to Recent and were very diverse; close to 6000 fossil species are known (Figure 11.10). Their basic structure is the same as the other two groups except for differences in the arrangement of thecal plates and in the structure of the food-gathering appendages, or arms. Crinoids generally have a longer and more robust stem than the other groups. The longest stems in Paleozoic forms are about one meter long, but some in the Jurassic are several meters long. Crinoids were able to arrange their arms into a feeding fan through which water currents could pass. Small side branches on the arms provided close-spaced meshwork within which food could be trapped. Crinoids reached their peak of abundance and diversity during the Mississippian Period, slowly declining in importance since that time.

The stalked echinoderms provide a distinctive component in many Paleozoic marine communities. Because of the variation in the distances they are elevated above the bottom, they cause the structure of many such communities to be tiered or stratified. If we now consider all the major components of many Paleozoic communities, we can note that the brachiopods are generally right on the sea floor, never more than a centimeter or two from the bottom. Bryozoans grow upward off the bottom, generally no more than 5 to 10 centimeters. Cystoids and blastoids have short stems that may be 10 to 15 centimeters long; thus they may be about the same distance above the bottom as bryozoans, or a little higher.

Crinoids form the highest part of such communities. Short-stemmed crinoids generally are raised above the bottom about the same distance as are blastoids, but there are many longer stemmed forms as well.

The primary impetus for this tiering of the communities was probably competition for the available food supply in the water. Each of the major groups was sampling a somewhat different part of the water, in a regime with horizontal water currents. This structure allowed many such communities to be more complex and more diverse than they would have been if all the animals had lived at the same level.

It should be noted that all of these animals were epifaunal, fixed to the bottom, and that they all had a skeleton. The predominance of sessile epifaunal filter feeders with a stratified community structure is characteristic of most Paleozoic marine communities.

Finally, we can note that those Paleozoic marine animals living as adults in the intermediate or higher levels of the community also had to be successful at lower levels. Bryozoans and echinoderms all start out growing up from the bottom. Thus, in their immature stages they must compete with brachiopods and other animals living next to the bottom. It is not yet clear whether the adaptive strategies of the young stages of such high-level adults changed as they grew upward and encountered progressively new and different competition.

Sponges

Sponges filter water through their bodies and feed on tiny food particles, such as detritus, zooplankton, and phytoplankton, that are carried into the body cavity with the water. Sponges have a low grade of organization and lack cells organized as discrete tissues. Cells with whiplike flagella beat in unison to force water currents through holes in the wall of the sponge (Figure 11.11A). The water then exits through a large hole in the top of the body. The sponge skeleton is composed of many small spicules, which are made up of silica, calcium carbonate, or organic material (spongin) as in the bath sponge (Figure 11.11B, C). The spicular skeleton commonly falls apart at death; therefore, individual spicules are common as fossils but whole sponges are rare. The siliceous or glass sponges (Figure 11.11D) were common Paleozoic fossils found in shallow water. Today, living glass sponges are found only in very deep water.

POST-PALEOZOIC FILTER FEEDERS

At the close of the Permian Period, one of the most conspicuous turnovers in dominance of marine animals occurred. Many of the groups that had been fundamentally important components of marine communities for millions of years became extinct. Included were several groups of brachiopods, bryozoans, and crinoids. The blastoids were wiped out, as were the trilobites. Other groups, such as the predaceous cephalopods, were affected as well.

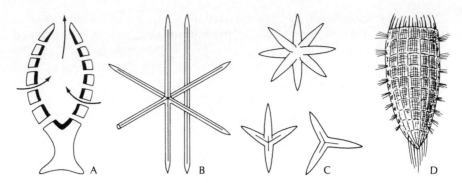

FIGURE 11.11
Fossil sponges. **A.** Cross section of a simple sponge, showing water currents entering the central cavity through pores and exiting through a large upper opening. The dark inner lining indicates position of food-capturing cells. **B.** Enlarged siliceous sponge spicule. **C.** Enlarged calcareous sponge spicule. **D.** Complete Devonian siliceous sponge with anchoring root tufts of spicules at the bottom. (Modified from Moore, 1952.)

Many hypotheses have been put forward to explain this crisis in marine life, which, by the way, did not affect terrestrial life to any important degree. One of the best inferences to date is that the extinctions were due primarily to progressive restriction of shallow seaways during the Permian and early Triassic. Most of our fossil record is from shallow seas that flooded the continents. If these seas withdrew, the amount of available living space was drastically reduced, resulting in increased competition and extermination of those groups that were not very well adapted to finding and holding on to a spot to live. After the Paleozoic, the brachiopods, bryozoans, and crinoids still survived but in greatly reduced numbers and diversity. The most conspicuous filter feeders now became the bivalve mollusks.

Bivalves

If you go shell collecting along any sandy beach today, you will find mostly gastropod and bivalve shells. In the shallow water of our time, **bivalves** are the dominant filter feeders and gastropods the principal detritus feeders, at least among those animals with a preservable skeleton. Bivalves have a long fossil history, the oldest ones being found in the Cambrian. During the Paleozoic, bivalves (also called pelecypods, clams, and in England lamellibranchs) were generally a minor component of most marine communities. Although it is possible to find one or two species at many fossil sites, only in certain situations were bivalves predominant. This was generally near shore, where the sediment was sandy or muddy. Dark shales, fine-grained sandstones, and other such kinds of rocks are the places to look for bivalves in Paleozoic rocks.

FIGURE 11.12
Gross anatomy of bivalves. Gills circulate water in the mantle cavity for food and oxygen. The foot can be withdrawn when the shell is closed or extended for loco-motion.

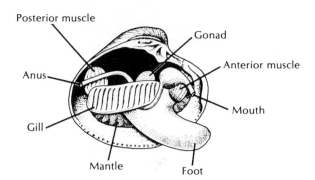

The great majority of the Paleozoic clams were epifaunal. Many of them attached themselves to the bottom by tough, horny threads like modern mussels attach to rocks. Others moved about, using their muscular foot, and fed from detritus on the sea floor (Figure 11.12). Most of them were not very good burrowers, generally having all or part of the shell exposed on the sea floor. Some can be called nestlers, hunkering down into the soft mud a short distance.

Beginning in the Mesozoic, bivalves become a conspicuous part of most marine communities (Figure 11.13). Suddenly they become one of the most common fossils to be found in many marine rocks (Figure 11.14). There are two apparent reasons for this. First, they surely successfully exploited some of the ways of life of the brachiopods that had become extinct. Certainly, this did not happen overnight. The extinction interval of the late Permian and the early Triassic was several millions of years in length. In a sense the bivalves did not outcompete the brachiopods; they outwaited them. Bivalves may also have been more efficient

FIGURE 11.13
A giant epifaunal, oysterlike bivalve, *Inoceramus*, from Cre-taceous chalks of western Kansas. The upper valve, with many small oysters attached to it, is almost 1 m high. This particular kind of clam became extinct at the end of the Cretaceous. (Courtesy of Sternberg Memorial Museum, Fort Hays State University, R.J. Zakrzewski, director.)

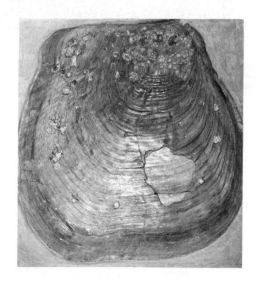

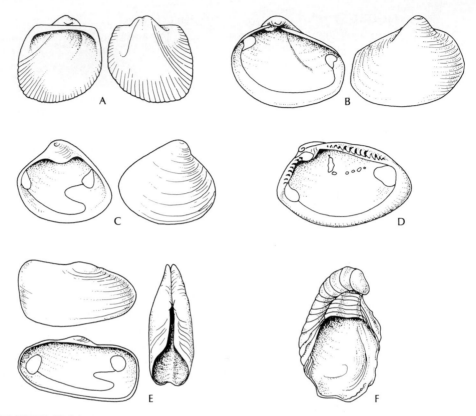

FIGURE 11.14
Major groups of fossil bivalves. **A.** Hinge with many small teeth and sockets (taxodont).
B. Hinge with single large tooth and socket. **C.** Advanced tooth and socket arrangement
(heterodont). **D.** Nuculid detritus-feeding bivalve. **E.** Burrowing bivalve with large gap
and internal notch along the mantle line where siphons are present. **F.** Oyster with large
single muscle scar. (Modified from Moore, 1952.)

epifaunal filter feeders than were the brachiopods; but, if so, it certainly took a
long time before this increased efficiency worked to their advantage. Perhaps they
gradually increased their food-gathering abilities. Direct evidence for this is lacking,
however; it would be found only in the soft parts, not in the shell.

A major factor that contributed to the increased diversity and dominance of
the bivalves beginning in the Mesozoic was that they became deep burrowers,
hiding the shell completely beneath the sea floor. A few of them were able to do
this in the late Paleozoic, but it was in the Mesozoic that burrowers became
progressively common and differentiated. Burrowing allowed them to escape
predators such as fishes which were undergoing extensive adaptive radiation during
the Mesozoic. This new adaptation involved some conspicuous changes in the soft

parts. Burrowers developed long, fleshy tubes, or **siphons**, that were rolled up parts of the **mantle**, the fleshy tissue that surrounded the body and secreted the shell. The edges of the mantle became fused to each other rather than remaining open, so that unwanted mud and sand could not easily get into the body cavity. At the front of the shell, where the muscular locomotor foot was situated, the shell developed a gap where the two valves did not close, so that the foot could be protruded without opening the shell (Figure 11.14E). A gap is also present at the other end where siphons protrude for pumping water into and out of the mantle.

One of the major differences between Paleozoic and Mesozoic marine communities was the trend toward burrowing by animals with hard parts in the Mesozoic. Virtually all burrowers in the Paleozoic were animals that mostly lacked hard parts. In addition to the bivalves, the echinoids were the other main group to show this trend. All Paleozoic echinoids were epifaunal, moving about over the bottom. During the Mesozoic, several different lines of echinoids became shallow burrowers, and one group, the heart urchins, became deep burrowers. Increased efficiency of predation by fishes may have been one factor that increased the survival value of burrowing.

DETRITUS FEEDERS

Although there are many groups of detritus feeders, all but a few have minor fossil records. Two groups with long fossil records are the gastropods and the echinoids (Figure 11.15).

Gastropods

Gastropods, or snails, have a long history and have been common fossils in some beds since the Ordovician (Figure 11.16). Many of the Paleozoic forms were probably herbivores, living in shallow water and grazing on calcareous algae and seaweed. Others moved over soft bottoms picking up bits of detritus. It is not completely correct to include all gastropods in this category. Although all or most Paleozoic snails were detritus feeders or herbivores, some of them became predaceous carnivores in the Mesozoic. They were able to drill holes in shells, especially those of pelecypods. They could then digest the soft parts through the hole. Shells will drilled holes in them are found in Paleozoic rocks as far back as the Ordovician. But such shells are quite rare, and it is not clear whether the holes were drilled by gastropods or by some other predator. Modern octopuses, for instance, can also drill holes in shells. At any rate, shells with drilled holes become increasingly conspicuous in the Mesozoic and continue to the present day.

Other Mollusks

In addition to the gastropods and bivalves, which are a very large and diverse group of mollusks, several other smaller and less diverse subgroups of this phylum

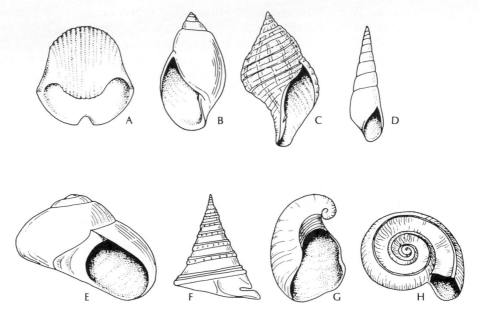

FIGURE 11.15
Various shapes of gastropod shells. **A.** Planispiral shell. **B.** Left-coiled (sinstral) shell with the aperture on the left; all other shells have an aperture on the right (dextral coiling). **C.** Advanced shell with elongated siphonal notch. **D–F.** Different styles of coiling. **G.** Gastropod shell normally attached over anus of Paleozoic crinoids. **H.** Flat coiled shell.

FIGURE 11.16
An aggregation of a high-spired gastropod, *Turitella*, from Eocene rocks in Virginia. Molds where the shells have been removed are on the upper left part of the block. These detritus-feeding snails commonly moved over the sea floor in herds and are commonly found in close-packed clusters as fossils. (Courtesy of National Museum of Natural History.)

FIGURE 11.17
Minor molluscan groups. **A.**
Scaphopod or tusk shell. **B.**
Polyplacophoran or chiton.
C. Monoplacophoran.

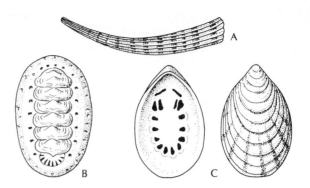

can be classified as filter feeders or detritus feeders. Three such groups are the monoplacophorans, the polyplacophorans, and the scaphopods (Figure 11.17). The first group is an ancient one, with a single living representative. The simple hat-shaped fossils of the monoplacophoran group are reasonably common in early Paleozoic rocks. The interior of the shell shows several pairs of muscle scars, indicating rudimentary remnants of segmentation of the body. This is interpreted to be evidence that the mollusks had a segmented ancestor similar to annelid worms.

The second group, the polyplacophorans or chitons, are mainly shallow-water (subtidal or intertidal) dwellers that live on hard rock surfaces. They scrape algae and detritus off the rocks for food. Eight plates are aligned along the top of the animal's body. When the animal dies and the flesh rots, these plates fall apart and are typically found as fossils.

The scaphopods or tusk shells have elongated, slightly curved conical skeletons. They burrow into soft sediment and leave the narrow tip projecting into the water. Seawater is drawn into the shell to obtain oxygen and microscopic food particles.

Echinoids

The mobile echinoderms called **echinoids** date back to the Ordovician. During the Paleozoic, they were exclusively epifaunal, moving about over the bottom with their spines. They picked up bits of detritus from the bottom, perhaps also feeding on marine vegetation and dead bodies. So they may have combined scavenging, detritus feeding, and a herbivorous habit. In the Mesozoic, echinoids underwent conspicuous change in their skeletal morphology that reflects changes in their habits. The epifaunal ones have very regular, pentagonal (five-sided) symmetry; they are shaped much like small pumpkins with spines (Figure 11.18). The mouth is directly in the center of the bottom surface, right next to the food supply, and the anal opening is in the center of the top surface. From this condition, informally called **regular**, echinoids evolved into what is called the **irregular condition**. Here

FIGURE 11.18
A regular echinoid viewed from the top. The prominent bosses on the test supported large spines that were used for movement and protection. The wavy tracts with two rows of small plates served for food gathering and uptake of oxygen. The mouth is down next to the sea floor and not visible in the photo. About natural size. (Courtesy of National Museum of Natural History.)

FIGURE 11.19
Top view, about natural size, of an irregular echinoid of Tertiary age from the island of Malta. The animal was partly buried in the sediment when alive. Both the mouth and the anus are on the underside. The five areas shaped like flower petals served for breathing. (Courtesy of National Museum of Natural History.)

the pentagonal symmetry becomes obscured, and superimposed on it is a strong bilateral symmetry. The spines become more numerous and smaller, giving a hairlike aspect to the outside of the shell. The anal opening shifts to the side and becomes posterior (rear) in position (Figure 11.19). All these changes, as well as many other more detailed ones, are a reflection of change from an epifaunal existence to one in which the echinoids lived partly or completely buried in the sediment. Still, they remained primarily detritus feeders and could plow their way through the soft mud or sand.

KEY TERMS

bivalves	detritus feeder	lophophore
blastoids	echinoids	molt
brachiopods	epifauna	pedicle
bryozoans	filter feeder	sponge
crinoids	gastropods	trilobites
cystoids	inarticulate	

READINGS

Boardman, R.C., and others. 1987. *Fossil Invertebrates*. Blackwell Scientific. 687 pages. A detailed account of the morphology and classification of all invertebrate fossils. Individual chapters vary somewhat in content and detail.

Brusca, R.C., and Brusca, G.J. 1990. *Invertebrates*. Sinauer. 922 pages. A modern textbook on living invertebrate animals.

McKinney, F.K., and Jackson, J.B.C. 1989. *Bryozoan Evolution*. Unwin Hyman. 238 pages. A comprehensive treatment of fossil and living bryozoans, their coloniality, biology, and evolution.

Willmer, P. 1990. *Invertebrate Relationships: Patterns in Animal Evolution*. Cambridge Univ. Press. 400 pages. Includes both biology and paleontology of invertebrates with emphasis on higher categories and patterns of phylogeny.

12
Marine Predators

In this chapter we will discuss three major groups of animals that were mainly predaceous, occupying second and third consumption levels within the trophic structure of marine communities. These groups are the corals, the cephalopods, and the fishes. We will also discuss, very briefly, two other groups of vertebrates that went from a terrestrial back to an aquatic existence: marine reptiles and marine mammals. These were important predators in the Mesozoic and in the Cenozoic, respectively. It is highly probable that not every species of each group was actively engaged in predation of larger animals. Earliest fishes were undoubtedly filter or detritus feeders, and other fishes were surely scavengers, feeding on dead organisms. Nevertheless, the great majority of species of each group was predaceous.

CORALS

Corals differ from all the other groups of carnivores considered in this chapter in that they have always been sessile animals, fixed to the bottom. They cannot move around in search of prey; they have to wait, instead, for a small motile animal to come within range of their tentacles. The oldest known corals are Ordovician. Just like calcareous brachiopods and bryozoans, which are rare or absent in Cambrian rocks, corals very quickly became a dominant aspect of marine communities beginning in the Ordovician and continuing right up to the present day. We presume that ancient corals played the same roles in many ancient marine communities as do their living counterparts in the marine communities of today.

Modern corals catch small crustaceans, fishes, and other nektonic animals with tentacles that are equipped with stinging cells to poison and quiet the prey. Sea anemones, corals that lack a stony exoskeleton, have the same feeding habits. Obviously the size of the coral partially determines the size of the prey that it can capture. Very small corals probably can capture animals only a millimeter or two in length, whereas larger individual corals, up to several centimeters across, can capture much larger swimming animals.

When we think about corals living today, images of coral reefs and tropical isles immediately come to mind. In shallow-water tropical areas, corals, along with calcareous algae and many other organisms, build up enormous structures that are composed of the skeletons of the organisms. These structures are called **reefs**, originally a nautical term that signified a hazard to navigation. The huge organic reefs are raised up above the sea floor. Most of the growth occurs in water that is quite shallow. The diversity of life on these reefs is amazing; they are the marine equivalent of tropical rain forests in terms of the number of species present. The corals that build reefs are called **hermatypic corals**. They contain in their soft tissues small unicellular protists called **zooxanthellae** which are photosynthetic. The relationship between the coral and the protists is probably mutually beneficial. Neither thrives without the other. It is because the zooxanthellae require sunlight that active reef growth is confined to shallow, sunlit waters. Living corals that lack zooxanthellae are called **ahermatypic**. Although they may live on reefs, they do not contribute to the rigid skeletal framework that causes the reef to grow upward. Organic reefs are common in the fossil record and have a long history. Not all of these were built primarily by corals and calcareous algae, as are modern reefs. Moreover, we have little direct evidence that fossil corals contained zooxanthellae, a relationship that seems vital to the formation of living modern reefs.

In the sense of a rigid organic framework raised above the sea floor, reefs extend back into the Precambrian when stromatolites built by algae qualify as small reefs. In the Cambrian there are small patch reefs built by **archaeocyathids**, an extinct group of animals probably related to sponges (Figure 12.1). Small algal reefs are present in the Ordovician. It is in the Silurian, however, that really large reefs became common. Some Silurian reefs in Illinois and Indiana are several hundred feet thick and over a square mile in area. The primary builders are algae; **stromatoporoids**, an extinct group of sponges (Figure 12.2); and an extinct group of corals, called **tabulate corals**. Tabulate corals are exclusively colonial. The tube for each individual in the colony is small, rarely more than a centimeter across (Figure 12.3). Each tube has a series of horizontal partitions in it, called tabulae (hence, the name *tabulate*), that serve as a floor for the soft coral polyp at successive growth stages (Figure 12.4). Tabulate corals were common participants in reef growth, especially in the Silurian and Devonian. Large oil-producing Devonian reefs are known in western Canada. Like the Silurian reefs, these are built mainly of algae, stromatoporoids, and tabulate corals.

In addition to tabulates, the other major group of Paleozoic corals is the **rugose corals**. These include both solitary individuals and colonies (Figure 12.5).

FIGURE 12.1
Several partial specimens (enlarged) of an extinct group of Cambrian fossils, archaeocyathids, probably related to sponges. The specimens are from Australia. They have been replaced by silica and partially etched from surrounding rock by acid. (Courtesy of Ward's Natural Science Establishment, Inc.)

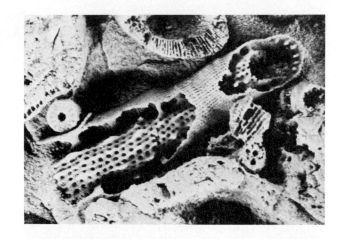

FIGURE 12.2
Skeleton representative of an extinct group of fossils, stromatoporoids, related to certain kinds of living sponges. These were important reef builders during the Silurian and Devonian. The specimen is several centimeters across and consists of concentric layers of silica, which has replaced the skeletal material. (Courtesy of Ward's Natural Science Establishment, Inc.)

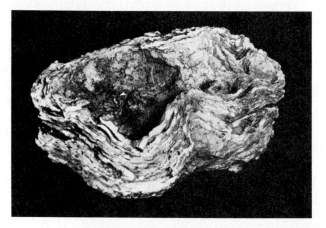

FIGURE 12.3
Two specimens of the tabulate coral *Favosites*, of Silurian age. Each specimen is a portion of a colony; each hole in the colony housed an individual coral polyp when the animals were alive. The specimens are about 4 cm in maximum dimension. (Courtesy of Field Museum of Natural History, Chicago.)

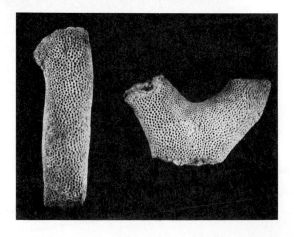

FIGURE 12.4
Magnified detail of the individual tubes in a Devonian colony of a tabulate coral, *Favosites*. Portions of the horizontal cross-partitions that served to support the soft polyp in life can be seen. (Courtesy of Ward's Natural Science Establishment, Inc.)

FIGURE 12.5
Solitary and colonial rugose corals. **A.** Two different solitary rugose corals shown from the side and somewhat enlarged. Notice the prominent radial septa showing at the top of each specimen. **B.** A colonial coral of Devonian age shown in both top and bottom views. The skeleton for each individual coral in the colony is tightly compressed against its neighbors. (Courtesy of National Museum of Natural History.)

A

B

FIGURE 12.6
Thin section through a colony of rugose corals showing the internal structure of the skeleton. The specimen is of Devonian age. Thin dark walls separate each individual from adjacent corals. Many radiating septa are present within each individual. Notice two small individuals along the left edge that have budded from adjacent ones. (Courtesy of National Museum of Natural History.)

The individuals are conspicuously larger than are those of tabulates. Some solitary rugosans may be up to a meter long, and colonies are up to a meter across. Rugose corals further differ from tabulates in having, in addition to tabulae, vertical partitions called **septa** (singular *septum*) that divide the soft polyp into a series of compartments (Figure 12.6). The internal structure may be quite complex and is used as a basis for classifying families, genera, and species.

Interestingly enough, rugose corals virtually never take part in forming organic reefs raised above the sea floor. They may live on such reefs, as do ahermatypic corals today, but they do not contribute to the rigid organic framework. Some beds in the Paleozoic are crowded with rugose corals, such as the famous Devonian beds exposed at the Falls of the Ohio near Louisville, Kentucky. But these corals form a horizontal bed; they do not bind together into an organic framework raised above the surrounding sea floor. Thus, it is at least possible that tabulate corals may have had zooxanthellae and that rugose corals may not have. We have little direct evidence to support such a contention, as there is no obvious morphological way to distinguish the skeleton of a modern hermatypic coral from one that lacks zooxanthellae.

During the Late Paleozoic, reefs were composed of a variety of organisms but neither tabulate nor rugose corals were conspicuous. Sponges, bryozoans, calcareous alage, and even some kinds of brachiopods contributed to the rigid framework.

Both rugose and tabulate corals became extinct at the close of the Permian. Tabulates persisted into the Mesozoic before they became extinct, but they are rare, insignificant, and do not take part in reef growth in the later part of their history. Beginning in the Triassic, another group of corals arose, perhaps from the rugose corals, or perhaps from soft corals acquiring hard parts. The evolutionary transition from one group to another is obscure. This new group is called **scleractinian corals**. They differ from rugosans in the symmetry of septal arrangement within an individual. The earliest scleractinians were small and solitary. They lived on Triassic reefs, but were not framework builders. Triassic reefs generally have a sponge-algal framework. By Jurassic time, larger colonial scleractinians had evolved, and they began to become important reef builders. They continued in that role through the Cenozoic to the present. Presumably, scleractinians did not inherit zooxanthellae from their rugose ancestors; neither rugosans nor the earliest scleractinians are reef builders. Therefore, the intimate relationship between these protists and the corals must have developed sometime in the first half of the Mesozoic Era.

Coral reefs today are restricted to warm-water areas generally between 20 degrees north and south latitude. Ancient reefs are found over very large areas of the continents and at quite high latitudes compared to the present. There are two factors that explain this disparity in distribution. During certain times, especially the Mesozoic, climates were not as severe as they are today, and the belt of tropical climates was clearly much broader than it is now. Secondly, we know that the continents have drifted extensively across latitudes, generally in a north-south direction. As different parts of a continent passed through equatorial belts, suitable climatic conditions were present for extensive reef growth. Silurian reefs in North America are primarily found in Wisconsin, Michigan, Illinois, Indiana, and as far north as the southern tip of Hudson Bay. We now know that this entire area was within 20 degrees north or south of the Silurian equator, as North America was situated much further south than it is today, having drifted northward to its present position during the Mesozoic Era.

CEPHALOPODS

Among the best known and most common shelled predators in the fossil record are the **cephalopods**, represented today by squids, octopuses, and one living representative with an external shell, the pearly nautilus. Although cephalopods are known from the Late Cambrian, they are very rare. Beginning in the Ordovician, cephalopods undergo an extensive adaptive radiation, evolving many different groups, some of which include forms that reach impressive sizes, up to 5 meters long. These earliest cephalopods are called **nautiloids** and are related to the living *Nautilus*. They mostly have a long, cone-shaped, tapering shell that is divided into a series of liquid- and gas-filled chambers. Others have a slightly curved shell or

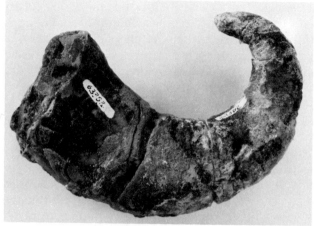

FIGURE 12.7
Two fossil nautiloid cephalopods of Paleozoic age. **A.** This specimen is from Ordovician rocks of Ohio and has a straight shell. The final living chamber is at the left, and watch-glass-shaped partitions dividing the shell into chambers can be seen to the right. **B.** This specimen, Silurian in age, is openly coiled rather than straight; the chambers cannot be readily seen. Both specimens are considerably reduced in size. (Courtesy of National Museum of Natural History.)

one with small, tapering early chambers and inflated adult chambers (Figure 12.7). Although very diverse and common, most of these early nautiloids probably were rather sluggish swimmers and lived near the bottom. They may have preyed on a variety of bottom-dwelling invertebrates, such as trilobites, brachiopods, and echinoderms.

By Devonian time, the nautiloids had already passed their peak of evolution. Many of the Ordovician and Silurian varieties had become extinct. The Devonian witnessed the origin of another group of cephalopods that evolved from the nautiloids, the **ammonoids**. These were rapidly to become the most common fossil cephalopods; they were so diverse in the Mesozoic that it is sometimes called the Era of Ammonoids. Ammonoids, or ammonites as they are informally called, differ mainly from nautiloids in the nature of the shelly partitions that divide the shell

FIGURE 12.8
A primitive ammonoid, *Tor-noceras*, from the Devonian period, has a few very simple wavy septa (goniatite) dividing the shell into chambers. The specimen is 4.4 cm high. (Courtesy of Takeo Susuki, U.C.L.A.)

into chambers. In nautiloids these partitions, or septa, are simple with straight edges. In ammonoids the edges of the septa become fluted and are arranged into a series of waves, so that the edge of the septum, as seen from the shell exterior, traces a series of folds across the shell. The functional significance of this evolutionary change has been much debated. The most generally accepted theory is that the shell edge helps to strengthen the shell, avoiding crushing or implosion of the shell if the animal changes its living depth rapidly. The chambers provide buoyancy, being partly gas filled, and external water pressure on them can be severe. Most ammonoids, and some nautiloids, especially advanced ones, have the chambers coiled so that they are above the living chamber. This helps keep the animal upright in the water, with the lighter gas-filled chambers located above the heavier animal in its living chamber.

Although ammonoids are small and not very common in the Devonian, they rapidly increase in size and abundance through the remainder of the Paleozoic. The most conspicuous change they undergo is increasing complication of the septal edges. The wavy septa develop small, secondary crinkles on them, first on every other fold and then on each fold. The primitive simple type is called a **goniatite septum** (Figure 12.8), the one with secondary crinkles on alternating folds is called a **ceratite** (Figure 12.9), and the most complicated forms are called **ammonite septa** (Figure 12.10). By Permian time, most ammonoids had evolved the complicated ammonite septum. If the prime impetus for this evolutionary change was to strengthen the shell, then some of the more advanced forms must have been living at increasingly greater depths, or else they were accustomed to changing their living depth quite rapidly. The ammonoids underwent a crisis at the close of the Permian. Most of the forms that had been common during that period of time became extinct, and only a few genera survived to provide the ancestral stock for Triassic ammonoids. All of the ones with complicated ammonitic septa became extinct; the survivors were ones with ceratite septa. There are over 300 genera of

FIGURE 12.9
A typical Triassic ammonoid, *Submeekoceras*, with ceratite septa. Alternate waves of the septa have secondary crenulations on them. The specimen is from the Lower Triassic of Idaho. (Courtesy of Takeo Susuki, U.C.L.A.)

FIGURE 12.10
An ammonoid with complex ammonite septa. This late Cretaceous specimen of *Placenticeras* is from South Dakota and is 22 cm high. (Courtesy of Takeo Susuki, U.C.L.A.)

ceratite-type cephalopods known from the Triassic. These, in turn, underwent a crisis at the close of the Triassic; all cephalopods with ceratite septa became extinct at that time. A few forms with the advanced ammonite septum had evolved during the Triassic. These persisted into the Jurassic, giving rise to another burst of ammonoid evolution. Several hundred genera are known from the Jurassic.

During the Jurassic and Cretaceous, ammonoids reached their peak of abundance, diversity, and rapidity of evolution. They are used for correlation of marine rocks of these ages on a worldwide basis. The Jurassic Period is divided into about twenty zones based on these fossils, and intervals of about 1.5 million years can be discerned with their use. In the Cretaceous, some of the ammonoids

FIGURE 12.11
One of the heteromorph ammonoids, *Hamites*, from the Cretaceous of England. The early growth stages are to the right (partly broken away on this specimen), and the mature part is to the left. Length along the outside of the curve is about 90 cm. (Courtesy of Field Museum of Natural History, Chicago.)

became exceptionally large, their coiled shells reaching 2 meters in diameter. Another group of Cretaceous ammonoids came unstuck, so to speak. The regularly coiled shell changed to one of several other varieties of form. Some of these ammonoids were coiled in very young stages and then had a straight shell. Others were coiled, then straight, then had a half-turn at the end, producing a canoe-shaped shell with the animal directed back toward the early part of the shell (Figure 12.11). Still others had a helically coiled shell like that of the gastropods, and still others had the shell in the form of an irregular knot. These types are especially common in Cretaceous rocks, although a few unrelated forms do occur in the Jurassic. They are called **heteromorphic** because they depart from the normal, coiled shell.

The ammonoids became extinct at the very end of the Cretaceous. Why they became extinct has been a long-continuing puzzle in paleontology. Up to the time they disappeared they were abundant and diverse, yet they vanished abruptly from marine rocks. Because some of the youngest ammonoids are highly ornamented, or heteromorphic, some paleontologists have suggested that these ornaments signal the racial senescence, or old age, of the ammonoids. Yet these heteromorphs were surely successful animals in their time; they were common and they survived for millions of years. Another, more plausible, explanation for ammonoid extinction is that first nautiloids and then ammonoids were unable to compete very successfully with fishes. It is certainly true that fishes underwent an extensive adaptive radiation in the Mesozoic, and they may very well have competed with ammonoids directly by preying on them and indirectly by competing for the same food supply. However, if such competition did adversely affect the ammonoids, it would have been more likely for them to slowly dwindle away instead of to suddenly disappear after having been so common. We do not really have an adequate answer as to why the ammonoids became extinct.

FIGURE 12.12
Bottom view of an articulated
specimen of a Devonian ostracoderm, *Cardipeltis*. The
heavy bony scales provide an
armor over the body and tail.
(Courtesy of Field Museum of
Natural History, Chicago.)

FISHES

An event that was to have enormous consequences for the nature of marine
communities was the appearance of the first vertebrate animals, the fishes, in the
Cambrian. The first fishes were small, only a few centimeters long, and were
harmless animals at best. They were not predators but rather filter or detritus
feeders, and so did not compete directly with the large cephalopods that were
about. Our knowledge of Ordovician fishes is scanty. They are known mostly from
a single formation, the **Harding sandstone**, in the Rocky Mountains. The fossils
consist of isolated plates and spines, no complete skeletons ever having been found.
Fishes are known from middle Silurian rocks in Asia and Europe.

The earliest fishes are jawless. The front end of the gut, the mouth, is simply
a round, immovable hole. These animals apparently sucked in water, mud, or both
into the pharynx and expelled the water through numerous gills that have separate
external openings. The gills presumably served as a filter or screen to strain out
small food particles from the water; they also absorbed oxygen from the water.
Thus gills were initially dual-purpose structures used for both feeding and
respiration. The little early fishes had a flattened body with a heavy external armor
of bony plates. They lacked an internal bony skeleton.

The early fishes are the oldest known vertebrate animals or chordates, that
phylum of animals characterized by a dorsal nerve cord and a backbone or spinal
column. This jawless group of water-dwelling forms, called **ostracoderms**, belong
to a group called the **Agnatha** (*A* for without, *gnatha* for jaws) (Figure 12.12). The
armored fishes found as fossils are known from the Ordovician through the
Devonian, at which latter time they became extinct. Before their extinction,
however, another group of jawless vertebrates evolved from them. This group lacks
any bony skeleton, having an internal skeleton of soft cartilage that does not

FIGURE 12.13
Bottom view of a model of a primitive ostracoderm, *Hemicyclaspis*. The body is covered with heavy, bony scales; the mouth is simply a hole, lacking jaws; and there are numerous separate gill openings around the margin of the head. The peculiar lateral fins are unlike those of more advanced fishes. Original specimens are about 20 cm long. (Courtesy of Field Museum of Natural History, Chicago.)

preserve as fossil. Two kinds of these soft jawless fishes are living: the lamprey eel and the hagfish, both of which are parasitic on other fishes. In addition, lampreys are known from carbon-film impressions in the Pennsylvanian.

Let us turn now to the fishes of the late Silurian and Devonian. Ostracoderms are an important part of these fish faunas, and several major groups of jawless fishes are represented. In addition, a new group of fishes is present, the **placoderms**, the most primitive of the armored, jawed fishes. These fishes were active predators, although still small in the early Devonian and no match in size for the cephalopods. The presence of movable, biting jaws is a clear indication of their predaceous habits. How did these jaws, which characterize all the rest of vertebrates both in water and on land, come about? In the ostracoderms, we mentioned that there are numerous gills, each with a separate opening (Figure 12.13). Each of the fleshy walls between gills contains a series of small bones that supports and strengthens the partitions between the gill chambers. These series of bones are called **gill arches**. Modern fishes have only five pairs of gills, but ostracoderms had seven or nine gill pairs. The bony arch of one of the front pairs of gills gradually moved forward until it was positioned in the side of the face just behind the mouth. The forward gill chamber became obsolete. This front set of gill bones ultimately became incorporated into the rim of the mouth, and the bones of each arch became articulated, or hinged, near the center, so that the lower half of each arch became one-half of the lower jaw and the upper half of each arch became one-half of the upper jaw. The mouth could now open and close on the hinge. Gradually, bony teeth developed on these bones that were now jaws. Thus, the typical vertebrate bony jaws, teeth, and mouth evolved in the transition from ostracoderms to placoderms.

The evolution of jaws signaled other changes in the placoderm skeleton that are related to changes in feeding methods. The ostracoderms were all heavily armored and were undoubtedly slow, sluggish swimmers; they did not need to be active to obtain their diet of microscopic food particles. The placoderms took up an active life of predation, and the filter-feeding ostracoderms gradually declined in importance. In placoderms we see the change from a flattened body shape suitable for slow movement over the sea floor to a more streamlined, fusiform, or torpedolike, body shape. The heavy armor becomes progressively reduced, allowing the fish to swim faster. There are also changes in the structure of the tail, the main propulsive organ of the fishes, that permit more efficient swimming motions. The fins on the sides, top, and bottom undergo a series of changes that allow for increased efficiency in balancing, braking, and making sharp turns in the water. All of these changes were advantageous to active predators such as the placoderms.

The Devonian Period was a time of rapid evolution of fishes and is sometimes called the Age of Fishes. In addition to placoderms, two other major groups are found in Devonian rocks. These are the sharks and the advanced bony fishes, which together comprise virtually all modern fishes. **Sharks** are the cartilaginous fishes; their internal skeleton is composed exclusively of cartilage, not bone. The only bone is in their teeth, in bony spines that support and strengthen the leading edges of some of their fins, and in tiny bony plates embedded in the skin, called **dermal denticles**. These give the sandpaperlike feel to the skin of living sharks. When a shark died, only these small parts of the skeleton were likely to be preserved. Our record of Devonian sharks is based almost exclusively on these isolated teeth and spines. The **bony fishes** differ from placoderms in a variety of ways. They lack the heavy external armor, having replaced it with series of small, lighter weight scales. In addition they have an internal skeleton of bone that is lacking in most placoderms. The structure of the tail is now stabilized, and the number and arrangement of lateral, top (dorsal), and bottom (ventral) fins is fixed.

The fishes added a new dimension to the complexity of marine communities. They rapidly took over as the dominant carnivores of the sea—a role they still play today. Apparently, the nautiloid cephalopods were not able to stand up to this direct competition, so that by the end of the Devonian they had started to dwindle in variety and abundance. Eventually the ammonoid cephalopods also suffered from this competition. The success of Devonian fishes is epitomized by the giant placoderm of the Middle Devonian, *Dunkleosteus*, which reached a length of 10 meters (Figure 12.14).

Among the bony fishes we can recognize two major groups: the **ray-finned fishes** and the **lobe-finned fishes**. The former group has many thin parallel bones supporting the fins; the latter has a thick, fleshy fin with a central axis of larger bones as the main fin support. Both groups lived in both salt and fresh water.

The lobe-finned fishes were never very diverse or common. They are especially significant because they include one group, the **crossopterygian fishes**, that evolved directly into the four-legged (tetrapod) vertebrate animals that first lived on land. They include living lungfishes and the **coelocanths**, the latter of which for years were thought to be extinct. The coelocanths appeared first in the Devonian and were thought to have become extinct in the Cretaceous, when the youngest fossil ones are found. A single living representative is now known—*Latimeria*—a

FIGURE 12.14
Reconstruction of the skull and neck region of the giant Devonian placoderm *Dunkleosteus*. Complete specimens reached 10 m in length. The peculiar jaws, with tusks and shearing edges, are quite unlike the jaws of other fishes. (Courtesy of Field Museum of Natural History, Chicago.)

large marine fish caught off the southeast coast of Africa in the 1940s (See Figure 8.9, page 125). We will discuss lobe-fins again when we come to the transition that animals made from water to a land-dwelling life.

The ray-finned fishes constitute virtually all of the common living fishes with which you are familiar, except for sharks and rays. They have been and are today enormously successful animals (Figure 12.15). They underwent a series of evolutionary changes during the Late Paleozoic and especially in the Mesozoic. The tail became shorter and reduced in size. The bones that supported the lateral fins were reduced in number, and the scales became thinner, with fewer bony layers. The most conspicuous changes occurred in the nature of the mouth and in the position of the lateral fins. Older bony fishes had a mouth that was long and slitlike, parallel to the main axis of the body. The mouth was progressively shortened, with the jaw hinge shifting downward, so that advanced types always

FIGURE 12.15
A giant bony fish, *Xiphactinus*, about 3 m long, from Cretaceous chalk of western Kansas. A smaller bony fish is preserved in the rib cage; overindulgence may have caused the death of the larger fish. (Courtesy of Sternberg Memorial Museum, Fort Hays State University, R.J. Zakrzewski, director; photo by E.C. Almquist.)

FIGURE 12.16
A specimen of an ichthyosaur of Jurassic age from Germany, much reduced. Notice the large number of bones that support the lateral fins and the way the vertebral column extends into the lower lobe of the tail. (Courtesy of National Museum of Natural History.)

look a little "down in the mouth." The front pair of lateral fins shifted upward in position, closer to the backbone, and the hind pair of fins moved forward until they were directly under the front fins or even slightly in advance of them. By the close of the Cretaceous, most of the modern groups of fishes appear in the fossil record. Holdovers from the more primitive early Mesozoic grade of fishes include living sturgeons (the source of caviar) and paddlefish. Intermediate grades include freshwater garpikes and the bowfin, both of which are rapidly approaching extinction in major rivers of the central United States.

LATER PREDATORS

Before we conclude our discussion of marine predators, we must mention two groups that came on the scene quite late in the fossil record. These are the marine reptiles of the Mesozoic and the marine mammals (whales, porpoises, and seals) of the Cenozoic Era. Not all of these animals were active predators—marine turtles, for example, are herbivores—but the great majority of them were.

Marine Reptiles

Marine reptiles evolved from terrestrial reptiles in the Mesozoic and took up an aquatic existence as a secondary adaptation. The giants of the Mesozoic seas are included in the reptiles, as are the land giants of the time, the dinosaurs. Several groups are represented. Large marine turtles are found, especially in Cretaceous chalk beds of western Kansas and Nebraska. Some of these turtles reach 4 meters in length. Another conspicuous group is the ichthyosaurs (Figure 12.16). These were dolphinlike reptiles (or perhaps we should say that the dolphins are ichthyosaurlike mammals). They were streamlined, and probably highly active

FIGURE 12.17
A short-necked plesiosaur about 3 m long from Cretaceous chalk of western Kansas. A large number of bones make up the oarlike paddles. (Courtesy of Sternberg Memorial Museum, Fort Hays State University, R.J. Zakrzewski, director.)

predaceous animals. Mostly small ichthyosaurs are found in the Triassic, although some in Nevada are 10 meters long. The peak of diversity and abundance is in the Jurassic, at which time they reached lengths of 7 meters. Other reptiles include large marine lizards, called **mosasaurs**, which reached 10 meters, and the giants of the Cretaceous seas, the **plesiosaurs**, some of which had very long necks and tails and reached 16 meters (Figure 12.17). These latter reptiles had the limbs modified into large, paddlelike fins in which the finger bones, the phalanges, were multiplied over and over again, so that there might be as many as fifty bones in a paddle. Reptiles in the Mesozoic underwent an extensive radiation in the marine environment; several groups independently took up a fully aquatic or semiaquatic existence. Many of these reptiles must have been at or very near the top of the marine food chains of the Mesozoic. They probably fed mainly on fishes and ammonites

Marine Mammals

Several groups of marine mammals that are active predators today have a fossil record in the Cenozoic Era. Seals actively feed on fish and squid. Some sea otters eat sea urchins as a major part of their diet. The most diverse group of predatory marine mammals are the whales and their relatives. Dolphins and porpoises are active swimmers that feed mostly on fishes and squid. Among the whales, the sperm and bottlenose whales feed mostly on squid, and killer whales feed on large squid. These are all toothed whales. In contrast, the whalebone whales mainly strain seawater through the long strands of baleen in their mouths and feed on zooplankton, although small fishes and crustaceans up to 4 centimeters long are also eaten. The right whale simply opens its mouth and lets water flow through the baleen as it swims forward. The other whalebone whale, the rorqual, feeds in just the opposite way; it takes in a mouthful of seawater and then expels it, trapping plankton in its baleen.

KEY TERMS

Agnatha	gill arch	plesiosaurs
ahermatypic corals	hermatypic corals	ray-finned fish
ammonite septum	heteromorphic	reef
ammonoids	ichthyosaurs	rugose corals
archaeocyathids	*Latimeria*	scleractinian corals
bony fishes	lobe-finned fish	septum
cephalopods	mosasaurs	sharks
ceratite septum	nautiloids	stromatoporoids
coelacanth	ostracoderms	tabulate corals
goniatite septum	placoderms	zooxanthellae

READINGS

Beerbower, J.R. 1960. *Search for the Past*. Prentice-Hall. 533 pages. This text includes general principles of paleontology as well as coverage of both invertebrate and vertebrate animals.

Kuhlmann, D.H.H. 1985. *Living Coral Reefs of the World*. Arco. 185 pages. Primarily a picture book with color pictures.

Roessler, C. 1986. *Coral Kingdom*. Abrams. 216 pages. A book with many pictures of corals and coral reefs.

Schultze, H.P., ed. 1986–. *Handbook of Paleoichthyology*. Gustaf Fischer. A planned series of 10 volumes, each on a major group of fossil fishes. Very well illustrated. Four volumes have been published to date.

Tasch, P. 1973. *Paleobiology of the Invertebrates*. Wiley. 923 pages. This large book covers protists and invertebrates, with emphasis on morphology and classification.

Wiedmann, J., and Kullman, J. 1988. *Cephalopods Present and Past*. E. Schweizerbart. 765 pages. A series of papers by various authors on living cephalopods, especially *Nautilus*, and reconstruction of soft-part anatomy of fossil cephalopods.

13

Origin and Early Evolution of Terrestrial Communities

$\mathbf{W}$e have now finished discussing the major components of marine communities through time. We have seen the tremendous differences between Cambrian communities dominated by trilobites and soft-bodied organisms that left traces, and the mollusk-fish–dominated communities of today. In tracing this record through 600 million years, one aspect is outstanding: the marine record, especially for benthonic life, is virtually continuous. For almost every time interval, we have sedimentary rocks someplace on earth that reveal what sea-floor life was like during that time. The greatest gap in knowledge of marine life occurs close to the Paleozoic-Mesozoic boundary.

In sharp contrast to the marine record, the fossil record for terrestrial life is spotty, discontinuous, and full of gaps, even in younger rocks. There are a number of reasons for this. In the first place, the areas in which sediments are deposited on land are much smaller than marine areas, many of the former being areas of erosion, not deposition. Once sediments are deposited on land—on a flood plain, in a lake, or elsewhere—they are much more likely to be eroded than are marine sediments. Even after nonmarine rocks are deposited and buried with their fossil content, they are generally situated at a somewhat higher elevation than are marine rocks of the same age, above sea level rather than below. Later in earth history, if there is a period of uplift and erosion, the terrestrial rocks are eroded first and may be eroded selectively, leaving behind marine rocks at a lower level. For these reasons, our fossil record of terrestrial rocks leaves much to be desired, certainly in comparison to the marine record.

We have only to look at the record of terrestrial rocks during the early part of the Paleozoic Era to discover the dearth of terrestrial fossils. No definitely terrestrial rocks that contain fossils are known before the latter part of the Silurian Period. This is a time span of almost 200 million years. One could argue that there was no terrestrial life during this early time, that all life still lived in the sea, where it undoubtedly originated. This argument is based on negative evidence. The easiest way to certainly identify terrestrial rocks is to find fossils in them. If no such fossils are found, the question usually remains as to whether the rocks are terrestrial or marine in origin. Thus, there may be terrestrial rocks from the Early Paleozoic, but we cannot identify them as such because they contain no fossils. On the other hand, the record seems to show there were widespread shallow seas over the continents during much of this time, certainly during the latter part of the Cambrian and much of the Ordovician. With land areas so restricted, there was little opportunity for terrestrial sediments to be deposited. Those that were, were subsequently removed by erosion, so no record is left. We are not really sure which of these hypotheses is correct, or whether there is some truth in each of them. The fact remains that our discussions of land life must necessarily begin with fossils of Silurian age because these are the first firm records we have of terrestrial life.

We will begin our discussion of ancient terrestrial life by considering the origin of land plants and the origin of land animals separately. These events are rather widely separated in time; furthermore, each deserves separate treatment because of the complex evolutionary factors that led to profound changes with the transition of life from submersion in an aqueous, watery environment to existence in an atmospheric, air-based environment. We will first set the stage for origins of these communities; then, in later discussion of communities, we will consider first the primary producers (plants), then primary consumers (herbivores), and finally secondary consumers (carnivores).

ORIGIN AND EARLY EVOLUTION OF LAND PLANTS

The Transition from Water to Land

The problems that are involved in making the transition from water to land are formidable for any organism. There are two possible pathways by which such a change can take place. One is for an organism to make a direct transition from marine waters to dry land—an approach that might be called "over the beaches." The other is for an organism to adjust first from salt to fresh water and then, from there, make the step onto dry land. The latter approach is generally accepted as the one plants and animals have followed. It obviates at least one major difficulty having to do with differences in the osmotic pressure of cells and tissues in fresh and salt water. If that difficulty can be handled while the organism is still immersed in water, then adjustment to rain water as a source of moisture on land is more easily made.

Let us now look at some of the other problems involved in making the transition. A plant living in water is buoyed up by relatively dense water. It does not need to fully support itself; the water helps do this. Air, however, is much less dense than water; thus, in order to lift itself up off the ground, a plant needs some kind of stiff, supportive tissues. Imagine yourself on a beach with a long piece of seaweed and a long branch from a land-dwelling tree. You can stick the branch in the sand and it will remain erect, but if you try to put the seaweed upright, as it was when it was buoyed up in the water, it will flop over and become sand covered. So, the problem of support was one that had to be solved.

Not only did a land plant need some support, but it also needed some kind of anchor. The stick you put in the sand would have blown down in a stiff breeze because it had no means of support, that is, no roots. Seaweeds also need an anchor, and many have a rooting type of structure, especially huge plants, such as giant brown kelp, and shallow-water seaweeds exposed to strong tidal currents and waves.

Water-dwelling plants do not need any special way to obtain water for photosynthesis or to maintain water balance in their tissues; they can simply take it up from their immediate environment. For land plants, however, living in air, the water problem is much more difficult. Air does not contain enough moisture for many plants to survive on as their sole water source. The humidity of air is also likely to fluctuate in a way that could be fatal to a plant. The most reliable source of moisture is from water trapped in the ground, in soil. Thus, a land plant needs some kind of special structure to take up water from the ground. Since the roots or rootlike structures are already needed to provide support, they are a likely candidate to provide a water-uptake system. After the water is taken up, it must be distributed to those parts of the plant that are not buried in the ground. These above-ground parts, exposed to the sun, are obviously those in which photosynthesis is going to take place. Thus, we have water needed for photosynthesis being taken up at one end of the plant which is buried, and photosynthesis taking place at the other end, up in the sunlight. Clearly land plants also need some kind of distribution or circulation system in order to move the water and dissolved minerals and salts throughout the plant and to distribute the manufactured carbohydrates of photosynthesis, somewhat similar in function to the circulatory system of animals. Land plants have evolved a special system of conductive tissues to perform these services called the **vascular system**, or system of vessels. This consists of many very narrow, elongate, hollow cells through which water and food can circulate within the plant. Virtually all land plants are also vascular plants; the terms are almost synonymous, but not quite. The vascular tissues also fulfill an important function we have already mentioned; they provide support for the plant body. The walls of the vascular tissues contain organic compounds called **lignin** and **cellulose**, which are rigid and sturdy, thus strengthening the plant body. The vascular tissues are typically situated in the center of the plant body. They are called **xylem** and **phloem**, the former tissue providing for upward movement of water from the soil and the latter for downward movement of manufactured food.

Since air is relatively dry, a land plant must also have some way of preventing its tissues from drying out—water is lost to the air. This is typically provided by an outer layer of cells that are impervious to water, which help maintain the inner environment of the plant.

A final adaptation that is necessary for a plant to live successfully on land has to do with reproduction. Algae, lower plants that live in water, reproduce exclusively by means of **spores**. These spores are distributed by water currents. An early land plant, in order to reproduce effectively, had to have spores that could be transported by wind rather than by water; and the spore had to be able to resist drying out—a situation that is not a problem for a water plant. Thus, spores of land plants were light, with a resistant waxy outer cuticle for protection.

The Fossil Record

Now let us see how the earliest known fossil land plants conform to this series of requirements for successful life on land. The oldest known land plants are of early Devonian age and have been found mainly in Australia. This earliest flora is named after one of the genera in the flora, *Baragwanathia*, an unusual name taken from an aboriginal place name. The next younger floras, in the middle part of the Devonian, are also named after the most common and characteristic plants found, the *Psilophyton* and *Rhynia* floras (Figure 13.1). Actually, all of these plants share a number of common features. Thus, they all can be considered together as typical of the earliest known vascular plants to inhabit land.

The plants of these floras were all quite small. The stem was less than a meter high (commonly only a few centimeters). The main axis or stem of the plant had a central solid core of vascular tissue, the xylem forming the center of the stem, surrounded by a cylinder of phloem tissue. There were no leaves differentiated specially for photosynthesis, but the entire outer walls of the plant contained photosynthetic pigments, so that the entire above-ground part of the plant was green. There was no true root system, but anchorage and water uptake were provided by an underground, undifferentiated part of the stem, called a **rhizome**. The spore-producing structures, the **sporangia**, are situated at the terminal tips of the plant branches, which, compared to more advanced plants, is judged to be a primitive feature. The plants have **dichotomous branching**; that is, each branch has two equal halves, like a tuning fork. This again is thought to be a primitive feature inherited from algal ancestors. A more advanced mode of branching is **monopodial branching**, which is characteristic of younger land plants. These have a central, larger main axis, like the trunk of a tree, with smaller, more slender side branches.

FIGURE 13.1
A model of the Devonian psi-
lopsid *Rhynia.* Note the lack
of leaves and true roots; also
note the dichotomous
branching, terminal sporangia, and small size, about 20
cm high. (Courtesy of Field
Museum of Natural History,
Chicago.)

The earliest land plants show many very primitive features. The sediments in which they are found indicate that they lived in low, wet, marshy bogs of fresh water. They still had their feet in the water, so to speak, but had managed to raise their stem into an upright position with vascular tissues; they had solved the internal conduction problems of living on land.

As previously mentioned, a few plants that live on land are not vascular. These are primarily the mosses. These plants are all quite small and can live only in low, moist places because they have not solved all of the problems of land living. They do not have an adequate conduction system or root structures to allow them to become larger and move out into drier habitats. Mosses may well have preceded vascular plants onto the land, considering their primitive features and lack of adaptation for land life. If so, we have no fossil record of them during the middle part of the Paleozoic when vascular plants first appear.

An interesting question is, What was the landscape like during the Early Paleozoic, the Cambrian, and Ordovician? We have no record of land plants. Does this mean that dry land was bare rock? Does it mean that there were primitive vascular plants for which we have no fossil record? Or, could the earth have been clothed with mosses, lichens, and other lower forms of land plants? The first vascular plants were so simple, and they evolved so rapidly after they first appeared, that it seems unlikely that they had taken up this mode of life for very long before they first appear in the fossil record. We are left with two possibilities. Land areas may have had only bare and weathered rocks. In the absence of land plants there would have been no soil, or humus, in the modern sense, although rock fragments may have been broken down by bacteria. Or plants, if they existed, such as moss ancestors, would have been confined to low, wet areas. Certainly the landscape would have been stark and very different from that of today.

Where did the vascular plants come from? We have noted that they probably arose from some kind of freshwater algae. The most likely candidates are found among the **green algae**. This is because these water-dwelling plants have a variety of photosynthetic pigments, including particular kinds of chlorophyll, that are identical to those in vascular plants. In addition, green algae include many types that have relatively large and multicellular plant bodies. Other groups of algae are mostly small and single celled or confined largely to marine water or have a complement of pigments that is different from that of land plants.

Changes in the Reproductive Cycle

We will see shortly how land plants evolved leaves, roots, and other important structures early in their evolutionary history (Chapter 14). Before turning to that, however, we must consider one of the most important aspects of early plant evolution that has to do with the reproductive cycle. Many of the major events in plant evolution, such as the changes from spore-producing to seed-producing plants and the evolution of flowering plants, involve changes in reproduction.

The earliest vascular plants were spore producers. What does this mean? A spore is produced by the conspicuous plant body with which we are all familiar. A spore, a single cell, is dispersed, and when it lands in a suitable habitat it germinates and begins to grow by a series of cell divisions. But it does not grow into a new plant body like the one that produced it. Instead, it becomes a new and different plant body that remains very small; it is confined to low, moist habitats and has no vascular tissues. This new plant body may even be microscopic and subterranean, living under the ground. The large, conspicuous plant that produces spores is called a **sporophyte**. The small, inconspicuous plant body that is produced by the spores is called a **gametophyte**, because it produces gametes, that is, sexual products. Male and female sexual organs develop on the gametophyte and produce sperms and eggs. When released, these unite to form a **zygote**, which then grows into a new sporophyte plant body. In other words, in these primitive vascular plants two separate and distinct individual plants take part in a single reproductive cycle.

Not only do these two phases—the sporophyte and the gametophyte—differ in size, appearance, and function, but they also differ genetically. When the spores are produced by the sporophyte, reduction division, or **meiosis**, takes place, so that each spore cell has one-half the number of chromosomes as the cells of the sporophyte plant (see Figure 14.4, page 222). The spores are thus **haploid** (N number of chromosomes), whereas the cells of the sporophyte are **diploid** ($2N$ number of chromosomes). When the spores germinate, they do so without change in ploidy, so that all the cells of the gametophyte are also haploid and the gametes, or sex cells, are produced by mitosis, or ordinary cell division. A diploid condition again results in the cycle, with union of two gametes to form a zygote. This then grows into a new diploid sporophyte. Such a life cycle is typical of the earliest land plants, as well as of many of their aquatic algal ancestors. It is also found in many of the somewhat more advanced vascular land plants, such as the ferns.

The vascular plant reproductive life cycle contrasts sharply with the ordinary reproductive cycle of higher animals. The major difference between the reproductive cycle of an animal and a plant is that in animals, meiosis, or reduction from diploid to haploid cells, directly produces sexual products that lead, with a zygote, to a new diploid individual. In spore-bearing plants, meiosis and diploidy are separated from each other by an entirely haploid individual, the gametophyte. What are the advantages or disadvantages of this kind of life cycle? One obvious answer is that a great many spores can be produced and released. If they find a suitable habitat, they can then germinate into a new individual without the need, that animals have, for close contact with another of their species. Thus, wide distribution is possible without dependence on chance encounters. Because plants are stationary and cannot move about to find a sexual partner, some dispersal system such as we have described is necessary. In the sexual phase, the gametophyte lives in a wet situation, mainly because moisture is necessary in order for the sperm to swim to an egg for fertilization. In a sense, the gametophyte is still almost an aquatic plant, not being removed nearly as far from the ancestral home of vascular plants, the water, as the sporophyte phase.

ORIGIN AND EARLY EVOLUTION OF ANIMALS ON LAND

Terrestrial Invertebrates

The oldest known land animals, including animals that live both on dry land (terrestrial) and in freshwater streams and lakes, are invertebrates. The oldest known land animals are arthropods, which have jointed legs and are related to trilobites, crabs, shrimps, and other animals of the sea. The oldest land arthropod fossils are found in Upper Silurian rocks and consist of three kinds of scorpions and a fossil millipede, or "thousand-legged" arthropod. One variety of scorpion is thought to have lived in fresh water because of the sediments in which fossils of these arthropods were found. The others are thought to have been dry-land animals. One variety of scorpion was infested with fossil nematode worms. Other land arthropods are found in Devonian rocks.

Fossils of insects and spiders exhibit many primitive features. Early insects were wingless and related to living silverfish. Once wings evolved, they initially could not be folded back along the body but, instead, projected out to the sides, like the wings of modern dragonflies. Various cockroaches, beetles, and relatives of crickets and grasshoppers were especially common. Hundreds of fossil insect species date back to the Pennsylvanian Period. Certain advanced groups of insects that interact with flowering plants, such as bees, wasps, flies, and ants, do not appear in the fossil record until the Cretaceous Period, when flowering plants evolved. One must remember that the fossil record of insects is very spotty and that some groups may have evolved long before they appear as fossils.

During the Devonian Period other freshwater invertebrates evolved. Freshwater bivalves have been found in the Old Red Sandstone in Great Britain and similar types of rocks in North America. Freshwater and land gastropods appear in the fossil record during the Pennsylvanian Period along with two additional groups of bivalves, worms with limy tubes, and freshwater arthropods with bivalved shells called clam shrimp and ostracodes. This freshwater and land community persisted through the remainder of the Paleozoic Era and continued into the Mesozoic Era when additional species evolved, increasing the diversity of terrestrial invertebrate groups.

Terrestrial Vertebrates

One of the most significant evolutionary events that occurred on earth was the transition from water-dwelling fishes to land-going tetrapods. As we have seen, fishes probably originated in the oceans and our first records of them are in marine rocks. By the Devonian, however, they had radiated into almost all available aquatic habitats, including fresh water. One of the groups that is especially common in rocks deposited in fresh water is that of the lobe-finned fishes, even though the only surviving lobe-fin, *Latimeria*, lives in the sea.

The freshwater Devonian fishes of interest here are the **crossopterygians**, or "crossops," as they are known. These fishes lived in ponds and streams on large deltas. The deltaic rocks in which the fossils are found are commonly red, due to oxidized iron minerals, indicating that the deltas formed in a climate that had alternate wet and dry periods. If there were periods of drought, any adaptations allowing the fishes to survive them would have been advantageous. In these crossops fishes we can see several such adaptations. First, we know that they had lungs as well as gills for breathing. Cross sections cut through some of the fossils reveal that the mud filling the interior of the carcass was of different consistency and texture in different parts of the fish. These differences reveal a saclike cavity below the front end of the gut that can only be interpreted as a lung. The gills were undoubtedly the main source of oxygen, the lungs serving as an auxiliary breathing device for gulping air only when the water became oxygen depleted, as it does during periods of drought. So, these fishes had already evolved one of the prime requisites for living on land, that is, the ability to use air instead of water as a source of oxygen. A second adaptation of these fishes was in the structure of the lobe fins (Figure 13.2). The fins were thick, fleshy, and quite sturdy, with a median axis of bone down the center (Figure 13.3). They could have been used as feeble locomotor devices on land, perhaps good enough to allow a fish to flop its way from one pool of water that was almost dry to an adjacent pond that had enough water for survival. These fins gradually changed into short stubby legs. The bones of the fins of a Devonian crossops fish exactly match in number and position the limb bones of the earliest known tetrapods, the amphibians. It should

FIGURE 13.2
A specimen of the Devonian lobe-finned fish *Eusthenopteron*. Notice the thick, fleshy fin at about the middle of the body and the peculiar tail with prominent middle section. Considerably reduced. (Courtesy of National Museum of Natural History.)

be emphasized that the evolution of lungs and limbs was in no sense an anticipation of future life on land. These adaptations helped the fish to remain a successful fish. It was serendipitous that these developments were also, later, useful for life on land.

What ecological pressures might have caused these fishes to gradually abandon their watery habitat and become increasingly land-dwelling creatures? Changes in climate during the Devonian may have had something to do with this if freshwater areas became progressively more restricted. Another impetus may have been new sources of food. The edges of ponds and streams surely had many

FIGURE 13.3
A. Idealized bone arrangement in the front (pectoral) fin of a primitive crossopterygian fish. **B.** Such an arrangement in the front limb of an amphibian. Bones labeled *a* (humerus), *b* (radius), and *c* (ulna) are the same in each instance.

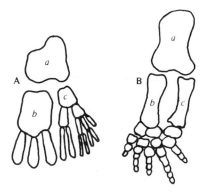

FIGURE 13.4
Side views of skulls. **A.** The earliest amphibian *Ichthyostega*. **B.** Its ancestor, a crossopterygian fish. Note the otic or ear notch at the back of the amphibian skull and reduction in size of the opercular bones (stippled).

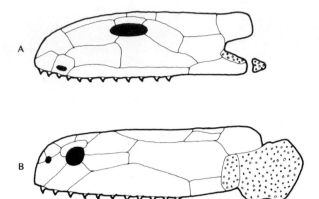

dead fishes and other water-dwelling animals on them, making them a rich if rather rank source of protein that could be exploited by an animal that could easily climb out of the water. Adjacent land areas may have had a rich supply of dead or slow-moving arthropods available as food. We have already seen that several kinds were already living on land. There is no evidence that these earliest tetrapods utilized land plants as foods; they were presumably all carnivorous fish eaters and had not developed the ability to feed on plant life.

The earliest known land animals are **amphibians**. They are related to the living amphibians—toads, frogs, and salamanders. These first land vertebrates are found in very late Devonian rocks in Greenland. They belong to genera like *Ichthyostega*, a name that refers to their fishlike skull (Figure 13.4). Actually, many of the skeletal parts of ichthyostegids, especially the skull, are very similar to their fishy crossops ancestors. The tail is long and fishlike, and the limbs are short, stubby, and sprawled to the side of the body, barely long enough to keep the trunk from dragging in the mud. The internal bony supports for the limbs, and the girdles (pelvic behind and pectoral in front) are weak and small. The backbone, too, is not strongly constructed. The skull, especially in the back part, has many bones that are identical to the parallel bones of a crossops. The exterior had a bony armor of scales, inherited from fish, that probably aided in retaining body moisture, so the animal would not dry out on land. All amphibians have to return to the water to lay their eggs; the eggs lack a hard outer shell to keep them from drying out on land. So, these first amphibians undoubtedly spent a large part of their life in the water, certainly during the breeding season. They were surely slow and awkward moving about on land.

Another feature these amphibians exhibit has to do with sensing their new environment. Fishes sense water movements largely through a special series of fluid-filled tubes in the skin on the sides of the body and on the head. These constitute the lateral line system. Water is incompressible, and these tubes were affected by water movement next to the fish. Because this system was not useful on land, where air is highly compressible, amphibians developed an ear to sense

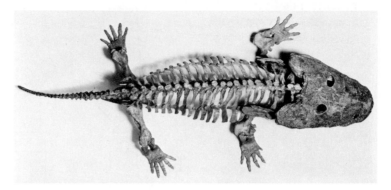

FIGURE 13.5
A mounted skeleton of the Permian amphibian *Eryops* from Texas. The otic notches in the back of the skull are on either side of the neck, the skull is low and broad, and the short limbs sprawl to the side. The skull is about 45 cm long. (Courtesy of Field Museum of Natural History, Chicago.)

changes in air pressure. The ear was situated low on the back part of the skull where a tympanic membrane, or eardrum, stretched across the **otic notch**, or ear notch, in the bone at the back of the skull (Figure 13.5). Movements of the eardrum had to be transmitted into the interior of the skull where they were recorded by nerves. This transmission was effected by a single earbone, called the **stapes**. You will recall from our discussion of the evolution of jaws that we mentioned the bones supporting the gill just behind the jaws. This arch of bones is called the **hyomandibular arch**. It moved forward during the evolution of fishes and came to support and strengthen the hinge area of the jaws and to prop the upper jaw against the braincase. The gill opening caught between the jaw arch and the hyomandibular was reduced in size, migrated upward, and became a respiratory tube (spiracle) in more advanced fishes. In amphibians, the old fish hyomandibular bone became the earbone or stapes. It was present in the area where the ear developed and was converted to this new function. The fish spiracle became the eustachian tube of amphibians.

During their early stages of evolution, amphibians underwent a number of evolutionary trends that improved their fitness for living on land. The bones of the pectoral and pelvic girdles, which supported the limbs, became larger, stronger, and more firmly attached to the backbone for support. Bones at the back of the skull became reduced in size and some were lost. A major change occurred in the nature and construction of the bones comprising the vertebral column. In fishes these are weak and disc-shaped, much of the weight of the fish being buoyed up by water. In amphibians these bones become larger, with bony crests for muscle attachment and with processes that articulate and support one bone in relation to another, front to back. Several different lineages of amphibians evolved from their ichthyostegid ancestors, each of which has as one diagnostic character in the

FIGURE 13.6
Side views of arrangements
of bones in single vertebrae
of fossil amphibians. The two
or three bony elements in
each vertebra have a different
arrangement and prominence
in each of the four major
groups of fossil amphibia represented.

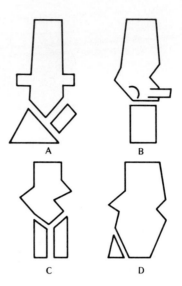

construction of the backbone (Figure 13.6). Some of these early amphibians
probably spent much of their life in water. They still have relatively weak limbs.
Others developed very massive limbs that are lengthened; these animals probably
returned to the water only to breed.

KEY TERMS

amphibian	hyomandibular arch	spore
Baragwanathia	*Ichthyostega*	sporophyte
crossopterygians	meiosis	stapes
dichotomous branching	mitosis	vascular system
diploid	monopodial branching	xylem
gametophyte	otic notch	zygote
haploid	phloem	

READINGS

Alexander, R.M. 1975. *The Chordates*. Cambridge Univ. Press. 480 pages. The most
comprehensive book yet written on the biology of living vertebrates. Includes a minor
discussion of fossils.

Banks, H.P. 1970. *Evolution and Plants of the Past*. Fundamentals of Botany Series. Wadsworth.
166 pages. An excellent short paperback that emphasizes the early history of land
plants by one of the authorities on Devonian-age plants.

Benton, M.J., ed. 1988. *The Phylogeny and Classification of the Tetrapods.* Systematics Association Special Vol. No. 35. 2 vols. Clarendon Press. Detailed discussion by a series of authors on the evolution and classification of all land-dwelling vertebrates.

Carroll, R.L. 1988. *Vertebrate Paleontology and Evolution.* Freeman. 698 pages. The modern-day version of the classic Romer textbook on vertebrate fossils. Detailed discussion of all fossil vertebrates with a classification listing all valid fossil genera.

Gee, H. 1988. "Oldest Known Reptile Found in Scotland." *Nature* 336:427. A short note describing the purported earliest reptile.

Gensel, P.G. 1986. "Diversification of Land Plants in the Early and Middle Devonian." In R.A. Gastaldo, ed., *Land Plants: Notes for a Short Course.* Studies in Geology 15. Univ. of Tennessee, Dept. of Geological Sciences. Pages 64–80. A good up-to-date discussion of early land plant evolution and radiation.

Knoll, A.H., and others. 1986. "The Early Evolution of Land Plants." In R.A. Gastaldo, ed., *Land Plants: Notes for a Short Course.* Studies in Geology 15. Univ. of Tennessee, Dept. of Geological Sciences. Pages 45–62. A technical discussion of physiological aspects of early land plant evolution.

14

Terrestrial Primary Producers: The Land Plants

In the preceding chapter we discussed the basic requisites for making the transition from water-dwelling algae to land plants. The oldest and most primitive land plants display characters indicating that they have made some but not all of the steps along this path. The first known land plants are all quite small, rarely more than a few centimeters high. Many seem to have lived in bogs and other wet places, so that while they were truly land plants they still had their "feet" in the water.

These simple plants, called **psilopsids**, lack true roots and leaves (see Figure 13.1). They were anchored by an underground extension of the upright stem, called a **rhizome**. This structure also furnished water and dissolved nutrients. The entire above-ground part of the plant was green, contained chlorophyll, and carried on photosynthesis. This plant had two other primitive features: (a) the plant had dichotomous branching, and (b) the reproductive structures, or **sporangia**, where spores were produced, were on the tips of the branches. In dichotomous branching, each branch is equal. This type of branch proliferation can be contrasted with the more advanced monopodial branching in which there is a main central stem, generally the largest and thickest part of the plant, with small side branches scattered along the sides. In more advanced land plants, the terminal sporangia are shifted down the plant body and become situated at the junctures of leaves and stems, or they are present on the undersides of leaves.

These earliest land plants are vascular plants. That is, they have the specialized conductive tissues typical of land plants. These tissues also serve to keep the plant

FIGURE 14.1
Stem and leaves of *Baragwa-nathia,* a primitive early Devonian lycopod from Australia. Long, slender, simple leaves cover the stem. Slightly reduced. (Courtesy of F.M. Hueber, National Museum of Natural History.)

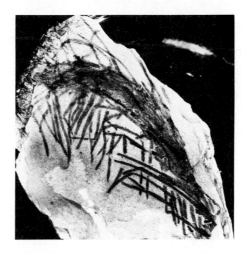

body erect above the ground. The oldest land plants known are late Silurian and early Devonian in age. The *Baragwanathia* flora of Australia has a variety of primitive plants, including psilopsids, but also somewhat more advanced types of plants (Figure 14.1). The flora is named after one of the genera that characterize it. Some of these plants already have true roots and leaves. The roots are differentiated with respect to the structure of their internal vascular tissues. The leaves are small, elongate, and very simple, with a single midvein and no lateral veins.

EVOLUTION OF LEAVES

Early in the history of land plants, two quite different kinds of leaf structures are present. We have just described one of them, which has small, simple leaves issuing directly from the stem and a simple midvein (Figure 14.2A). Leaves of this type are thought to have evolved from small spines on early plant stems. Several of the leafless fossil psilopsids have abundant, hairlike spines preserved on the stems. These spines are thought to have enlarged and, eventually, to have had a branch of vascular tissue in them that conducted water to the leaves and moved manufactured food to the rest of the plant body (Figure 14.3). Chlorophyll eventually became concentrated in these structures, making them the principal food-manufacturing sites of the plant.

Other early plants had leaves that were quite large and much divided, similar to the leaves of ferns (Figure 14.2B). They are thought to have evolved not from spines, but from the flattened and increasingly subdivided tips of branches. That is, the ends of the branches became thin and flattened, and then they were richly divided into small leaflets, hundreds of which might make up a single leaf. Chlorophyll became concentrated in these modified branch tips, which contained vascular tissue. This kind of foliage was developed in the early ferns, which also

FIGURE 14.2
Two contrasting types of primitive leaves. **A.** The leaf of a lycopod—long, slender, and with a midvein. **B.** A leaflet of a much-branched, fernlike foliage with a complex network of veins. (Courtesy of Field Museum of Natural History, Chicago.)

FIGURE 14.3
The two modes of evolution of leaves in vascular plants. Small, simple leaves (above) evolved from spines on the stem. Large, complex leaves (below) evolved from flattened tips of branches.

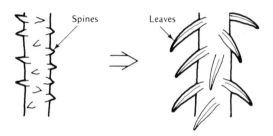

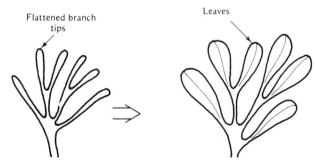

are distinctive in having the reproductive sporangia on the undersides of the leaflets.

The first type of leaf structure, small leaves clothing the stems, was characteristic of two other major groups of primitive plants, the lycopods and the sphenopsids. The **lycopods** include the giant scale-trees of the Pennsylvanian coal swamps. These were up to 30 meters high with secondary woody tissue in the trunk. The **sphenopsids** had a jointed stem with leaves concentrated at the nodes between joints. Some of these also reached tree-sized proportions.

EVOLUTION OF VASCULAR TISSUE

The xylem and phloem tissues that together constitute the vascular system of a land plant initially formed a solid cylinder in the center of a primitive plant stem. The **xylem** was situated in the center of the stem surrounded by the **phloem**. As long as plants were small, as with the early psilopsids, this structural arrangement was satisfactory. But there was an early trend, beginning in the Devonian, for plants to become increasingly larger, with more stress put on the stem by winds. Plants evolved two features of the vascular tissue that allowed the plant body to increase in size. The primary vascular tissues evolved from a solid central core to a hollow cylinder, with the center filled with soft, pithy tissue. A hollow cylinder has considerably more strength to resist bending and breaking than does a solid cylinder; thus, this arrangement of vascular tissues helped increase the strength of the stem. The primary vascular tissues are generally not woody. Woody tissue is stiff and resistant because it is strengthened by lignin. Wood, so characteristic of the larger plants that we call trees and shrubs, evolved from the xylem tissues. The softer, conductive xylem is called **primary xylem**. It produces **secondary xylem**, or **wood**, which is nonconductive but supports and strengthens the stem and branches of the plant. Woody tissues are found in Devonian plants and are characteristic of many later plants, especially large ones. Wood is among the most commonly found fossil plant material because of its resistance to decay and destruction.

EVOLUTION OF REPRODUCTIVE STRUCTURES

The most complex and important evolutionary changes that took place in land plants had to do with their reproductive structures—their conformation, complexity, and position on the plant. One can almost say that the fossil history of plants is principally a history of plant reproduction. In order to understand this aspect of plants, we must first briefly summarize the generalized life cycle of land plants.

We will begin with the adult plant body that is typically large and conspicuous; it is this that you usually think of as the plant. The cells of this plant body are all diploid; that is, they have the 2N number of chromosomes per cell (Figure 14.4). The plant has specialized reproductive structures, called sporangia, where spores are produced. This takes place by reduction division (meiosis), so that each spore has a haploid (N) number of chromosomes. In early land plants the sporangia were small and simple, but in later ones they became larger and arranged in clusters. Upon release, a small spore is carried by the wind to a new living site, where it germinates and grows into a new, different, and usually quite small plant body, unlike the other, larger plant body that produced the spores. The spore-producing body is called a **sporophyte** because it makes the spores. The new plant body is called a **gametophyte** because it produces sex cells or gametes. Notice that the spores do not undergo combination or fertilization before they produce the gametophyte. The gametophyte has all haploid cells from the spore. It develops

FIGURE 14.4
Generalized life cycle of a
vascular plant with two separate individuals—one haploid
(gametophyte) and one diploid (sporophyte).

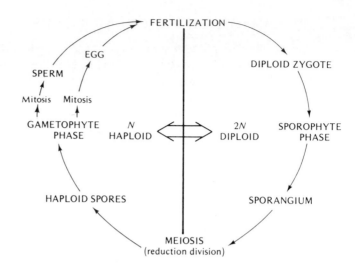

male and female sex organs in which the sex cells are produced by mitosis, or
ordinary cell division. The male and female gametes that are produced unite to
form a new diploid (2N) cell, the zygote, which then develops into a new diploid
sporophyte plant body.

This generalized life cycle of land plants differs essentially from that of most
higher animals in that an additional adult body is interposed between meiotic
division, which takes place in spore formation, and the union of haploid cells in
fertilization to form a diploid adult. The interposed adult phase is, of course, the
gametophyte. In all of the land plants discussed here, the sporophyte is the dominant
phase. In mosses and liverworts (not important as fossils), the gametophyte is
dominant and the sporophyte inconspicuous. In lower land plants, such as ferns,
the gametophyte is a separate body, although it may be microscopic or subterranean.
In more advanced land plants, that is, the seed-producing plants, the spores are
retained on the sporophyte and develop into a tiny gametophyte consisting of only
a few cells. This gametophyte may remain attached to the sporophyte. Sex cells
are produced, fertilized, and develop into an embryo sporophyte encased in a
protective coat and food supply. This is the **seed** which contains the developing
sporophyte of the next generation (Figure 14.5). Pollen is the male gametophyte
of seed plants and may be dispersed by wind, water, or insects; the female
gametophyte remains attached to the sporophyte plant and develops eggs. These
are fertilized by sperm formed by the pollen grains.

In the most primitive land plants, reproductive evolution involved the nature
of the sporangia. We know almost nothing about the nature of tiny gametophyte
phases which are not preserved as fossils. The sporangia are initially small, simple,
and all of the same size. They are located on the tips of the branches. The sporangia
then shift to a position where they are protected in the angle between a leaf and
a stem. Two different things then happen to the sporangia, not necessarily both in

FIGURE 14.5

Two large seeds from Pennsylvanian-age seed ferns. Each seed was contained within an ironstone nodule that has been split in half, so that two views of each seed can be seen. Approximately 2/3 natural size. (Courtesy of Field Museum of Natural History, Chicago.)

the same plant. First they arrange themselves in clusters, forming a large conspicuous structure called a **cone** (Figure 14.6). Secondly, two different kinds of sporangia evolve, one that produces a female gametophyte and another that produces a male gametophyte (Figure 14.7). Rather than have both sexes present on one gametophyte body, there are thus two separate gametophytes. Combining these two features, plants come to have separate male and female cones. In seed plants the female spores are retained within the cone, pollen is produced by the male cone, and seeds ultimately develop in the female cone. This is the familiar condition in pines and other seed-producing conifers: the pine cone produces the female spores and the seeds. The male pollen-producing cones are much smaller and inconspicuous. Both male- and female-producing reproductive structures need not be on the same plant. An entire sporophyte may produce only male or female cones. Even if both sexes are present on the same plant, they need not be self-

FIGURE 14.6

A large compressed cone, *Lepidostrobus,* of a Pennsylvanian lycopod, several centimeters in length. Notice the modified leaves (called sporophylls) that make up the cone, preserved around the edges and base of the cone. The specimen is of Pennsylvanian age from Illinois. (Courtesy of Field Museum of Natural History, Chicago.)

FIGURE 14.7
Generalized pattern of evolution in the reproductive
structures of vascular plants.
A. Terminal small sporangium. **B.** Single sporangium
in angle between leaf and
stem. **C.** Cluster of sporangia
(all with the same size
spores) and associated leaves,
forming a cone. **D.** Separation of male cones with small
spores and female cones with
large spores.

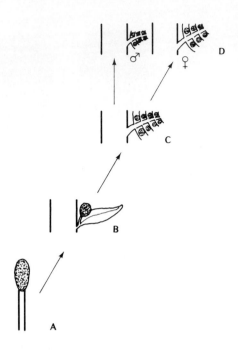

fertile but may require the pollen from another individual to fertilize the eggs from
the female gametophyte.

Silurian and Devonian Plants

We know that land surfaces were devoid of vascular land plants during the
Precambrian and earliest part of the Paleozoic Era. More primitive plants such as
lichens and mosses may have existed but left no fossil record. The oldest fossil land
plants are so primitive that they could not have been in existence for very long,
geologically speaking. There may have been small low plants in moist situations,
such as mosses, lichens, and fungi, but certainly most of the continents were bare
rock. The earth certainly was not green as it is today. The scarcity of plants would
have had a profound effect on various weathering processes, both chemical and
physical, and on soil-forming processes, especially because humic acids produced
by land plants, which play such an important, even definitive, role in modern soil
formation, would be lacking. Even the weather and climate would have been
affected because land plants play an important role in recycling water from the
soil back into the air. On a summer day an apple orchard of forty trees can move
sixteen tons of water from the soil back into the atmosphere.

The oldest land plants are found in upper Silurian rocks. Tissue scraps and
spores of even older plants have been reported from Ordovician rocks, but there
is uncertainty as to whether these belong to true land plants or to aquatic green
algae that were beginning to evolve into land plants. Once established on land,

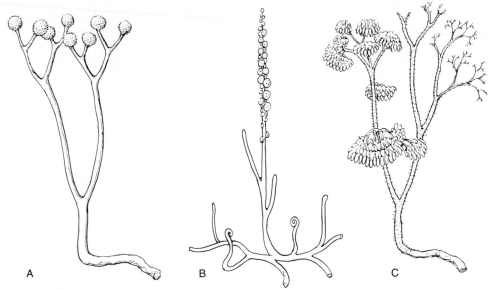

A   B   C

FIGURE 14.8
Typical Devonian land plants, all about 30 to 45 cm tall. **A.** One of the earliest known
land plants, *Cooksonia,* with dichotomous branching and simple terminal sporangia. **B.** A
zosterophyle vascular plant with smooth stem and rhizome, dichotomous branching, and
sporangia arranged along sides of a fertile branch. **C.** A psilophyte with clusters of terminal sporangia and hairlike spines on the stem. (Modified from Gensel, 1986.)

plants diversified very rapidly; the Devonian Period was a time of very rapid land
plant evolution (Figure 14.8). By the end of the period most of the major Paleozoic
land plant groups had evolved.

The simplest and oldest of these early land plants are typified by the Late
Silurian genus *Cooksonia*. These plants are small, less than 50 centimeters high,
lack woody tissues, lack leaves or true roots, and have small, single sporangia at
the branch tips. The stem branches dichotomously. These plants seem to have lived
in permanently wet habitats, such as shallow ponds and bogs.

Somewhat more advanced land plants occur in Lower Devonian rocks. These
are somewhat larger, up to 1.5 meters tall, with more advanced monopodial
branching of the stem. They still lack roots but have small scales on the stem that
may be green and precursors of leaves. The sporangia are globose and arranged in
rows on the stem. These plants, typified by *Zosterophyllum*, are thought to have
been the direct ancestors of the lycopods, important late Paleozoic plants prominent
in the coal swamps of the Pennsylvanian.

A third primitive group of land plants consists again of quite small plants,
about 45 centimeters tall, that have very complex branching patterns which may
be a precursor to leaf development. Some have hairy or spiny stems, and the

FIGURE 14.9
Portion of a restoration of the Pennsylvanian scale tree *Lepidodendron*. Long, simple, straplike leaves are arranged in spirals that clothe the stem. A large female cone is also visible. (Courtesy of Field Museum of Natural History, Chicago.)

sporangia are complex, arranged in drooping clusters. The genus *Psilophyton* is typical.

From these groups came more advanced spore-bearing plants—the lycopods, sphenopsids, ferns, and forerunners of the gymnosperms. These groups developed true leaves—some large and complex, others small and simple—and true roots. They developed woody tissues in the stem and hence had a much stronger stem and could grow much taller. Ultimately, some of these advanced spore-bearing plants developed seeds.

The **lycopods** are represented today by two living genera; both are small plants that seldom grow more than a few inches tall. The common names for these include club moss and ground pine. Their extinct relatives were large trees up to 30 meters tall. The lycopods are characterized by long, slender, very simple leaves that issue directly from the trunk and that are arranged in spirals. The bark of the trunk has distinctive leaf pad scars that are diamond shaped and reflect the spiral leaf pattern. Lycopods are more advanced in their reproductive products than any of the Devonian plants discussed above in that the sporangia may be arranged into tight clusters called cones, and both smaller male spores (microspores) and larger female spores (megaspores) may be present in separate cones. These result in separate male and female gametophytes. By far the best known genus of this group is the coal swamp tree *Lepidodendron* (Figure 14.9).

FIGURE 14.10
A branch of a sphenopsid, *Annularia,* preserved in an ironstone nodule of Pennsylvanian age from Illinois. The stem is divided into segments, with a whorl of leaves present at the junctures of segments. (Courtesy of Field Museum of Natural History, Chicago.)

The **sphenopsids** are sometimes called the joint-stemmed plants (Figure 14.10). There is a single living genus, *Equisetum,* which is quite common in roadside ditches and other wet places and has about twenty species. Common names include scouring rush and horsetail. All sphenospids are characterized by a hollow stem that is jointed. Leaves and sporangia occur at the nodes where two joints meet. In *Equisetum* the leaves are reduced to tiny scraps, but fossil forms have larger leaves with spiral or fan-shaped leaflet patterns. The sporangia in *Equisetum* are terminal, and generally only a few individuals are fertile in a stand. The name scouring rush comes from the fact that the stem tissues of this living form are full of microscopic crystals of silica, creating a very hard and harsh tissue. Pioneers used bundles of these stems to scour skillets and pots. The silica crystals, however, cause the stems to be poisonous to livestock. Fossil sphenopsids may be large trees, like the coal swamp *Calamites,* or nonwoody trailing vines or llianas.

The third group of advanced spore-bearers are the **ferns**. These have large, complex leaves. The spores are often in rows of small, dotlike sporangia on the undersides of leaflets. Some ferns are nonwoody, but other living and fossil ferns are woody and called tree ferns. The origin of the ferns is poorly understood. They were very common Late Paleozoic plants and are common fossils in the Mesozoic, but the relationship of these fossils to living ferns is not very clear. They are the most common and diverse spore-bearing land plants today, with over 10,000 species.

Finally, there is a small group of spore-bearers that are the direct ancestors of the seed-bearing **gymnosperms** that seem to represent the spore-to-seed evolutionary transition. These are called the Progymnosperms. These plants have leaves that look pretty much like those of ferns. However, the woody stems have vascular tissues that are essentially similar to those of slightly younger gymnosperms. They may be either homosporous (all spores the same size) or heterosporous (large

FIGURE 14.11
Impression of the surface of the trunk of *Lepidodendron*, showing the diamond-shaped leaf bases; a simple leaf was attached to each base. Specimen is of Pennsylvanian age from Illinois. (Courtesy of Field Museum of Natural History, Chicago.)

female spores and small male spores), and the spores may be in cones. These plants are confined to Devonian and Mississippian rocks. During the Pennsylvanian the first true gymnosperms appear, with true seeds. Many of these advanced spore-bearing plant groups became fossilized in coal swamps and are evident in the resulting coal deposits of the Late Paleozoic.

Coal-Swamp Plants

Rocks of Pennsylvanian age are the major source of the world's coal supply. Coal beds are composed of the altered and degraded remnants of plants that grew in very large, low, swampy areas. We know these swamps were mostly coastal, bordering the sea, because marine rocks and fossils are commonly found next above coal beds, deposited when the sea flooded over the swamp and killed all of the terrestrial life that had built the swamp. The areas most like these swamps we have today are some of the low swamps along the southeastern coast of the United States, such as the Okefenokee and Dismal swamps.

These coal swamps had a rich and varied flora and fauna. The plants that dominated the swamps were diverse and probably highly structured in their community relationships. Large trees were present that stood 30 meters tall and were a meter across at the base. These were mainly lycopods, represented especially by *Lepidodendron* (Figures 14.11 and 14.12), and sphenopsids, represented by *Calamites* (Figure 14.13). These were probably the two most common plants with large, woody stems composed of secondary xylem tissue. Another common tree was *Cordaites*, which belongs to an extinct group of gymnospermous plants, the Cordaitales (Figure 14.14). These had long, narrow, straplike leaves and fruiting bodies of loosely constructed cones, or **catkins**, with the female catkins producing large, rounded seeds. These three trees, each belonging to a different group of plants, towered over the remainder of the coal-swamp flora. A middle story was formed by **tree ferns**. These are ferns and seed ferns that had a woody trunk (Figures 14.15 and 14.16). They attained heights of 2 to 4 meters and produced an abundance of fern-type foliage. The understory of these forests consisted

FIGURE 14.12
Typical Pennsylvanian coal swamp lycopods, all parts of *Lepidodendron*. **A.** Trunk or stem with bark pattern showing diamond-shaped leaf scars. **B.** Female cone. **C.** Root, with large pith cavity and scars where rootlets attached. **D.** Long slender leaf with simple midvein.

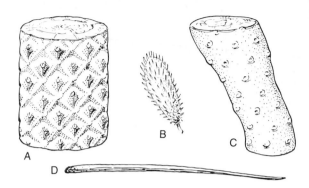

FIGURE 14.13
Pennsylvanian coal swamp sphenopsids. **A.** Jointed stem of *Calamites* with large central pith. **B.** Vinelike whorled leaves of *Annularia*. **C.** Stem and small leaves of *Calamostachys*.

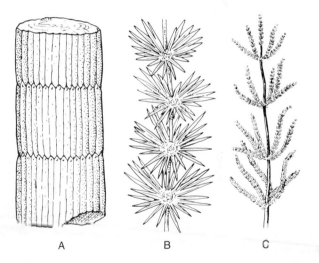

predominantly of ferns, of which there was a wide variety. These are identified as fossils mainly on the basis of the shape of the final little leaflets. Each leaf is immense, as much as a meter across, with numerous subdivisions, each of which has many small leaflets.

Some of our best evidence for the nature of plants in Pennsylvanian swamps comes from **coal balls.** These are rounded concretions found in coal that formed early in the history of the swamp, before the plant material had time to be buried, compressed, and coalified, the last process obliterating all fine detail of the plant material. The plants that are found in coal balls are puzzling in terms of their

FIGURE 14.14
Restoration of a Pennsylvanian *Cordaites* tree, a typical coal-swamp gymnosperm. The long, slender leaves are quite unlike the small, needlelike leaves of most living gymnosperms. (Courtesy of Field Museum of Natural History, Chicago.)

FIGURE 14.15
Leaflets from a seed fern, *Neuropteris,* of Pennsylvanian age from Illinois, preserved in an ironstone nodule. (Courtesy of Field Museum of Natural History, Chicago.)

abundance and diversity. At some localities, virtually every coal ball will contain remnants of *Lepidodendron,* but no other plants. At other localities, an amazing diversity of plants has been found. This seems to indicate that coal swamps were not uniform in their floral content. Imagine one edge of a swamp at the margin of a sea, with the landward plants becoming progressively higher and drier until dryland borders of the swamp are reached. It seems reasonable that different plants

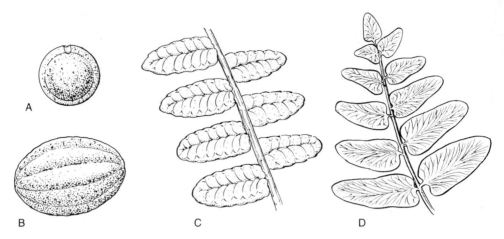

FIGURE 14.16
Pennsylvanian coal swamp fern and seed-fern fossils. **A, B.** Typical seeds. **C, D.** Portions of two different large leaves, showing shape of leaflets or pinnules. All slightly larger than natural size.

would respond differently according to whether or not the water was fresh, brackish, or salty and to the extent of wetting and drying. There is some evidence that *Cordaites* tended to dominate around the dryland edges of the swamp and that *Lepidodendron* occupied the seaward edges. It is quite clear that there were immense pure stands of plants such as dense stands of *Lepidodendron* and broad clearings occupied by low-growing ferns. In this respect these ancient forests may have resembled modern coastal swamps.

Permian Plants

The close of the Paleozoic Era was a time of tremendous change in terrestrial communities. Coal swamps had virtually disappeared in the northern hemisphere, although they still persisted in cooler climates of the southern hemisphere where *Glossopteris, Gangomopteris*, and other plants formed thick layers of coal. With increasing continentality in the Permian, broad alluvial plains with low relief formed over large areas of the southwestern United States. Seasonal rainfall oxidized iron in freshly deposited sands and muds, giving them a red color. These alluvial areas spread from Kansas down into Oklahoma and northern Texas, across into New Mexico, Arizona, and Colorado. In this area we have one of the best fossil records of terrestrial life, especially during the early part of Permian time.

The plants of the Permian reflect changes in climate. The dominants of the Pennsylvanian coal swamps, such as *Lepidodendron* and *Calamites*, were still present, but they were much less abundant. *Cordaites* and its relatives, which occupied edges of swamps, were one of the most conspicuous plants in the Permian, living in drier, more upland habitats. These gymnosperms were joined in the Permian by the first conifers, which are rarely present in the Pennsylvanian. The oldest known conifer is *Walchia*. The small needlelike leaves that clothe the branches are in sharp contrast to the fern foliage of the seed ferns and the long, straplike leaves of the cordaiteans. *Walchia* and related forms became one of the most conspicuous groups in the Permian. Seed ferns were present, but declining in number; true ferns were still conspicuous. The overall aspect of Permian floras is one of plants growing in floodplains and more upland areas, with various groups of gymnosperms dominating the floras. The evolution of seeds had by now been clearly demonstrated as a feature of great adaptive significance. There was no longer a need for a low, wet habitat in which a separate gametophyte plant could germinate and grow. A seed was able to persist in the soil and germinate only when growing conditions, such as a rainy season, were suitable.

PLANT COMMUNITIES OF THE MESOZOIC

Although large sphenopsids and lycopods were still present in the Mesozoic, they were rare. The *Lepidodendron* type of lycopod tree became extinct at the close of the Triassic, and the *Calamites* type of large sphenopsid survived only a little longer. Four groups of plants were clearly dominant in the Triassic and Jurassic landscapes. The dominant understory plants were still ferns, which include a variety of foliage types. A middle story of plants was quite diverse, including tree ferns, seed ferns, and a group called the cycadeoids (Figure 14.17). Known as **cycads**, these are extinct relatives of the living cycads, which are gymnosperms. They have a stem or trunk that looks like a large pineapple and is composed of the coalesced bases of large leaves. The leaves break off during growth, leaving a cluster of sturdy bases surrounding the stem (Figure 14.18). The leaves are large and palmlike, in a cluster at the tip of the stem. The cones of these cycadeoids tend to be embedded in the stem, whereas in living cycads the cones are produced at the top in the center of the leaf whorl. The upper story of Triassic forests was formed by a variety of conifers. These had distinctive patterns of cells in the wood that permit us to ally them with primitive living conifers called **araucaria** (Figure 14.19). Today, these kinds of conifers are restricted to small areas in the southern hemisphere in Australia, New Zealand, and South America. The Norfolk Island pine is an example that is commonly grown as a houseplant or outdoor specimen on lawns. The best known fossil occurrence of these conifers is the Petrified Forest National Park in northeastern Arizona. Here, immense logs of araucarians are exposed where they have weathered out of Upper Triassic rocks called the **Chinle Formation**. This conifer forest covered a large area of the southwestern United States, probably growing on broad, flat floodplains and in coastal areas.

Another new group of gymnosperms, the Ginkgoales, or **ginkgos**, put in an appearance in the Triassic. These gymnosperms are typically small to large, slow-growing trees. Each individual is either male or female, bearing small cones of one sex or the other. The leaves are quite distinctive, having a fan shape with parallel

FIGURE 14.17
Compound leaf of a fossil cycad from Mexico. The large, much-divided leaves issued directly from the top of a short, barrel-shaped trunk. (Courtesy of Field Museum of Natural History, Chicago.)

FIGURE 14.18
The short, stubby trunk of a fossil cycad of Cretaceous age from Maryland. The trunk is about 45 cm high and is composed of the coalesced bases of large, much-divided leaves that issued from the top of the trunk. The triangular scars where the leaves attached cover the outside of the trunk. (Courtesy of National Museum of Natural History.)

veins and the outer margin split or entire (see Figure 8.13, page 128). The group was quite common in the Mesozoic all over the world. By the Cenozoic, however, it had dwindled in abundance and had virtually disappeared from the fossil record. The group is represented by a single living species, *Ginkgo biloba* (the maidenhair tree), that survives only in domestication. This interesting tree lived in northeastern Asia into historical times when the last wild specimens were apparently cut down

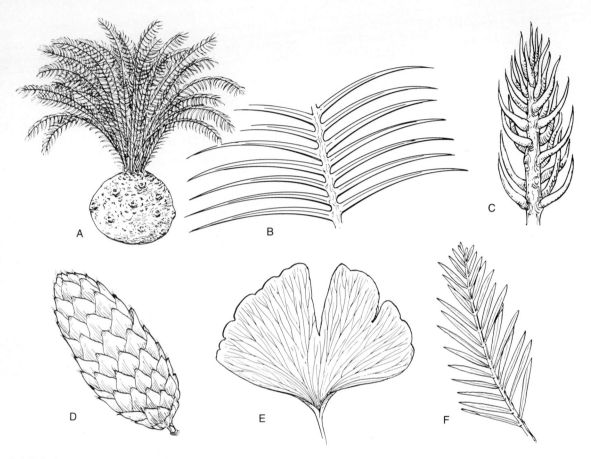

FIGURE 14.19
Mesozoic plants. **A.** Reconstruction of a fossil cycad, about 2 m tall. **B.** Enlargement of a
small part of a cycad leaf showing several leaflets. **C.** Twig of an araucarian pine. **D.**
Cone of a fir conifer. **E.** Ginkgo leaf. **F.** Twig of *Metasequoia*.

by the early Chinese for wood. The tree did survive in domestication, however,
because it was considered to have spiritual qualities and was planted in the grounds
of Buddhist monasteries and temples. Now *Ginkgo* is growing all over the world
again because it is attractive and remarkably resistant to disease and decay.

The plants that accompanied the dinosaurs and other reptiles in the Jurassic
landscape were predominantly a gymnosperm flora. Various kinds of conifers,
including araucarian pines, were the principal large trees of the Jurassic. The other
most commonly found kinds of plants are various sorts of ferns, both tree ferns
and herbaceous ferns which did not have a stem or trunk and gymnospermous
cycadeoids which formed a small tree or middle story to the vegetation.

FIGURE 14.20
A sandstone slab containing several angiosperm leaves of late Cretaceous age from Idaho. These plants all belonged to species that are now extinct. The leaf types indicate a mild climate in the northwestern United States at the close of the Cretaceous. (Courtesy of National Museum of Natural History.)

Evolution of Flowers

The major feature of plant communities during the Mesozoic was the appearance and sudden expansion of the flowering plants, the **angiosperms**. How did these most advanced plants, which make up over 90 percent of the land plants alive today, evolve? What were the traits that made them so successful? First we will look at the record.

No angiosperms are known with certainty from the Jurassic. The oldest flowering plants are found in Lower Cretaceous rocks in Greenland. They are rare and not very diverse. Beginning with the very base of Upper Cretaceous rocks, angiosperms begin to appear in abundance and on a worldwide basis (Figure 14.20). Once they got going, they rapidly came to dominate virtually all terrestrial habitats, even those with the most severe climatic contrasts of hot and cold, wet and dry. This extraordinary dominance over groups of plants that had held sway for millions of years—the conifers, cycads, ginkgos, and ferns—has several explanations. No one morphological change was responsible for the success of the flowering plants; rather, it involved a combination of several factors.

Some people erroneously consider seed plants and flowering plants to be synonymous. This is not the case at all, as we have seen. The seed habit—encasement of the developing embryo in a protective coat with a contained food supply—originated with seed ferns, as early as the Devonian. All gymnosperms, the dominant plants of the Mesozoic, are seed plants. In order to understand how a **flower** evolved, we can begin with the familiar pine cone. Many primitive land plants have the reproductive structures grouped together into a cone. This is the case in some lycopods and in seed ferns and other gymnosperms. Each segment of the cone is a modified leaf, bearing either male or female reproductive structures on the inner side. In a pine, the seeds are exposed; if you break off segments of a ripe female cone, you can see the large seeds.

FIGURE 14.21
Diagrammatic cross sections
of a pine cone, **A,** and an angiosperm flower, **B,** showing
four whorls of reproductive
structures. These are all alike
in the gymnosperm above
and highly modified in the
angiosperm flower.

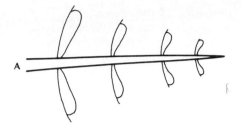

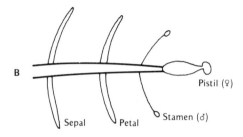

Now let us imagine the following modifications of a cone. We will specify that the cone be situated at the tip of a branch. The leaves, which are **sporophylls,** or spore-leaves, at the base of the cone become sterile, without reproductive organs. The sporophylls are arranged in whorls around the cone. Two whorls of sporophylls become sterile and serve to encase and protect the immature, budding cone. In flowering plants, these become an outer whorl of sepals and an inner whorl of petals (Figure 14.21). The next series of sporophylls along the cone bear pollen-producing male reproductive structures. These become modified into long, tubular stalks that bear pollen-producing sporangia. Each modified sporophyll is called a **stamen.** The terminal sporophylls of the cone become modified into egg-producing structures called **pistils.** Thus, a flower is a collection of highly modified leaves, some of which bear reproductive structures. Rather than having each sporophyll similar, as within a cone of a gymnosperm, the different whorls of sporophylls are highly modified and differentiated, resulting in a quite complex structure.

We can now ask why these modifications and evolution of the flower structure were so important to the success of the angiosperms. One of the most conspicuous differences between angiosperms and gymnosperms has to do with the usual mode of achieving pollination. Living gymnosperms are generally wind pollinated; there is no reason to believe that their fossil relatives were not also so pollinated. Wind pollination is a haphazard or, more properly, a random process. In order to be successful they must produce a tremendous quantity of pollen that is easily liberated and that is small and light enough to be carried considerable

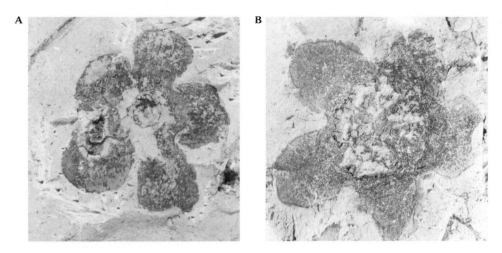

FIGURE 14.22

A Cretaceous flower. **A.** Blossom showing a five-part calyx (five sepals), slightly enlarged. **B.** The same kind of flower, about twice natural size, showing remains of fruit in the center. From Rose Creek, Dakota Formation, central Kansas. (Courtesy of David Dilcher, University of Florida.)

distances by the wind. The number of female receptors must also be great to ensure adequate fertilization.

Many, but by no means all, angiosperms have departed from this pollination method and have evolved a consortium with insects as pollinators. We can imagine that initially, during the Cretaceous, or perhaps in the Jurassic, beetles or other insects that fed on cones of gymnosperms had pollen of male cones accidentally attach to their body, so that when they then fed on female cones, some of the female gametophytes were fertilized in the process. Natural selection, over millions of years, may have favored those plants that were increasingly attractive to insects or made themselves more readily accessible to insect cropping in order to achieve this method of fertilization. The evolution of petals to attract insects, or of distinctive smells and nectar, can be viewed as part of the selective process that led to the evolution of flowers (Figure 14.22). Concurrent with these changes, there was no longer a need to produce such large quantities of pollen or of pollen receptors; thus, the number of stamens and pistils could be progressively reduced, as in more advanced kinds of flowers. The pollen could be larger, heavier, and sticky—the better to cling to insects. This important change in effecting fertilization had a tremendous impact on plant life. After all, plants are mostly stationary organisms; they cannot move together for reproduction as do most animals. Any adaptation that would increase their reproductive potential should have been, and was, exploited to the fullest. Flowering plants have evolved a bewildering variety of forms and especially with respect to their flowers.

The heightened relationship between flowering plants and insects also had a great effect on the insects themselves. Many became dependent on flowers as a source of food. The advent of flowering plants in the Cretaceous undoubtedly led directly to the tremendous increase in abundance and variety of modern groups of insects (e.g., butterflies, bees). We live in a terrestrial world that is primarily a flowering plant–insect world. Mammals, including humans, birds, worms, and other animals, play a much smaller role in shaping our natural environment than do these two groups. It is very doubtful that insects had anything like their present diversity prior to the evolution of flowering plants in the Cretaceous. Many species of flies, beetles, bees, and other insects have life cycles that are predominantly controlled by and dependent upon specific kinds of flowering plants.

A second aspect of the success of flowering plants has to do with their successful dominance of several habitats of terrestrial communities. The various kinds of gymnosperms—the conifers, the seed ferns, the cycads, and ginkgos—are mostly woody plants, in the form of either trees or woody shrubs. In Cretaceous terrestial plant communities what seems to be lacking is an understory of nonwoody, small, soft, herbaceous plants. This role was seemingly filled by ferns, and by the few vestiges of small lycopods and sphenopsids that had survived into the Mesozoic, as well as by mosses. One of the conspicuous differences between flowering plants and gymnosperms is that the former very successfully invaded the understory habitat, producing a great variety of small, low, nonwoody plants. A great part of the diversity of flowering plants is contained within this category. There is, proportionately, a much smaller number of flowering plants that are woody trees or shrubs. And, it is exactly in the understory group of angiosperms that one finds the greatest emphasis on insect pollination. Many angiosperm trees are still wind pollinated, as are willows, birches, and other trees. Some angiosperm herbaceous plants, such as ragweed, are pollinated by wind, but the great majority depend on insects. The dominance of flowering plants in the modern world, then, can be largely understood on the basis of two radical innovations that evolved in these plants: a change in the mode of achieving fertilization and a successful invasion and exploitation of many understory habitats.

PLANT COMMUNITIES OF THE CENOZOIC

We have already noted that by the close of the Cretaceous period the flowering plants had already become the dominant floral type in terrestrial communities. They were widespread and diverse. There are several aspects of Cenozoic land plant distribution that are important. These include evidence for progressive increase in diversity of angiosperm floras, especially in the early part of the Cenozoic; the gradual modernization of the floras up through the Cenozoic, so that they come to look more and more like modern floras; and the biogeographical distribution of the plants, which reflects the shifting of ancient climates.

No other region of the world has yielded as many collecting sites for Cenozoic plants as has western North America. Many floras from individual localities are

FIGURE 14.23
A large fan-palm leaf from the Eocene of Wyoming. The presence of palm trees this far north during the Eocene is a clear indication that climates in this area were much milder at that time than they are today. (Courtesy of National Museum of Natural History.)

quite diverse and may include over a hundred species. For this reason, Cenozoic paleobotany is considerably better known for North America than it is for any other part of the world. Our discussion, then, will focus primarily on North America, where conclusions concerning these fossil plants can best be documented.

Beginning with Paleocene time, several localities in California, Oregon, and Washington have yielded extensive floras that consist primarily of leaves of trees. These include magnolias, figs, a persimmon related to Asian forms, a custard apple, and other forms (Figure 14.23). The relationships of these plants point to two important conclusions. First, in Eocene time the northwestern United States had a subtropical climate (Figure 14.24), with rainfall in the range of 170 to 200 centimeters (70 to 80 inches) a year and a mean annual frost-free temperature of about 20°C (68°F). Second, these floras show a relationship to plants that are now widely scattered in distribution. Some of the fossils closely resemble trees that still live along the Pacific Coast and that have apparently lived in this area for many millions of years. Others represent plants that are now confined to Asia, while still others are related to living plants that are confined to subtropical and tropical parts of the western hemisphere, especially southern Mexico and Central America.

Proceeding farther north along the Pacific Coast, several Eocene floras have been found in Alaska. These have an aspect that is quite different from that of the Oregon and Washington fossils. The Alaskan floras include a preponderance of plants that are considered to be temperate or north temperate in present climatic distribution. Some of these are walnuts, chestnuts, elms, oaks, pines, cypresses, and *Sequoia*. There is also a less common element of warmer climate plants, including relatives of the breadfruit tree and the avocado.

Taking these two areas together, we can conclude that Eocene climates were considerably more moderate than they are today. The boundaries between tropical and subtropical zones and between subtropical and temperate zones were situated conspicuously further north in North America than is true at the present.

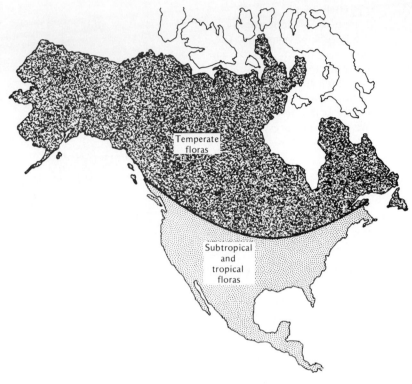

FIGURE 14.24
Distribution of early Cenozoic (Paleocene and Eocene) floras in North America. Note the
northern extension of subtropical floras.

Accompanying this benign climate, plant communities were seemingly much more
cosmopolitan and widespread than they are today. Floras very similar to those of
Alaska have also been found in Greenland and Spitzbergen, well within the Arctic
Circle. These high-latitude temperate forests had a circumpolar distribution in the
Eocene.

In Oligocene rocks of Oregon, fossil floras are known that indicate a distinct
cooling off of the climate from that of Eocene time. All of the Oligocene plants are
of temperate affinities. They include redwoods, hawthorns, beeches, alders, and
oaks.

In the Miocene, the cooling trend seen in the Oligocene continues (Figure
14.25). Not only are the fossil plants indicative of temperate climatic conditions,
but now they begin to show evidence of decreasing rainfall and somewhat more
arid conditions (Figure 14.26). The fossils continue to show a much more
cosmopolitan distribution of certain plants than they have today. For instance,
these Miocene floras include several plants that are today found as natives only in
Asia, such as the ginkgo, tree-of-heaven, and a water chestnut. A second element

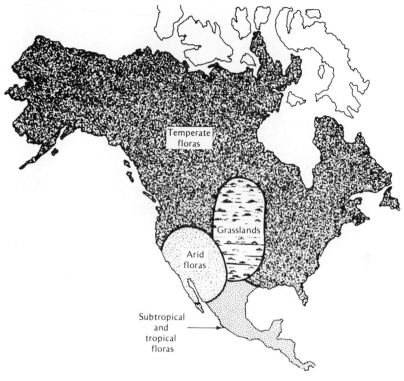

FIGURE 14.25
Distribution of Cenozoic floras during the middle of the Cenozoic (Miocene). Note that temperate floras now extend much further south than they did during the early Cenozoic and that grassland prairies and desert floras are now evident, as a result of climatic changes.

consists of trees that are now native to eastern North America, having been gradually excluded from western areas. These include elms, sweet gums, magnolias, and figs.

During the Pliocene, several floras indicate exclusion of Asian and eastern North American elements from western floras. Virtually all of the fossils found have a close relationship with plants still growing in the general western area. Several of the earlier Miocene floras indicate rainfall not in excess of 75 centimeters a year (30 inches), and the Pliocene plants record a continuation of this tendency toward aridity.

This sequence of fossil plants from the Eocene into the Pliocene records a climate that very gradually became progressively cooler and drier. These climatic trends culminated in the Pleistocene with the onset of the first continental ice sheets covering much of Canada and the north-central part of the United States (Figure 14.27). Obviously, the distribution of plants was profoundly affected by

FIGURE 14.26
Two angiosperm leaves of Miocene age. **A.** This specimen from Idaho is related to the grape family. **B.** An oak leaf from Oregon. (Courtesy of National Museum of Natural History.)

glaciation and interglacial periods when the ice sheets had melted. At the maximal limit of glaciation, the coniferous forests of the north pushed far south into the central United States. Concurrently, temperate hardwood forests shifted into the southern United States and Mexico with similar restriction of subtropical and tropical forests in southern Mexico and Central America. During interglacial periods, these various plant communities migrated northward again to positions close to those of today (Figure 14.28). This extensive migration of forests north and south occurred four times during the Pleistocene, with the waxing and waning of major ice sheets. The result was an intensification of the process of fragmentation of plant distribution, resulting in sharply different plant communities in different parts of the continent.

Accompanying these changes in climate were two other effects that had a profound influence on the nature and distribution of modern floras in North America. We have already seen that there is clear evidence for increasing aridity in the western states in the later Cenozoic. This culminated in the onset of truly desert conditions in the southwestern United States and in northwestern Mexico. A host of plants evolved that were specially adapted to these severe conditions, including many kinds of cactus, shrubs, and small trees that could survive very low rainfall.

A second major development involved the area of the Great Plains of the United States and Canada. This area stands in the rain shadow of the Rocky

FIGURE 14.27
Distribution of floras during the Pleistocene at a time of maximum glacial advance. Temperate floras are pushed far to the south, as are subtropical plants.

Mountains. As the area became progressively drier, starting in the Miocene, larger plants, especially many trees, were gradually excluded or confined to banks of major streams. A variety of grasses became increasingly prominent, leading to the establishment of the extensive grassland prairies that still persist over much of this area. These extensive prairies offered large areas of new habitats for mammals. The fossil record of these prairies has been preserved in the Great Plains where Miocene and Pliocene deposits contain the very resistant siliceous seed husks of grasses.

By far, the great majority of Cenozoic fossil plants that are known consist of leaves of trees, fossil wood, and **pollen.** Fruits, seeds, flowers, or the remains of small herbaceous plants are much less common. Tree leaves tend to be rather tough and leathery and commonly do not decay very rapidly. Furthermore, trees live for years to produce millions of leaves over their lifetime. On the other hand, herbaceous plants have soft leaves and the entire plant tends to die at once. Consequently, we have a very much poorer fossil record for small, soft angiosperms than we do for the longer lived trees. Furthermore, most of the former plants are

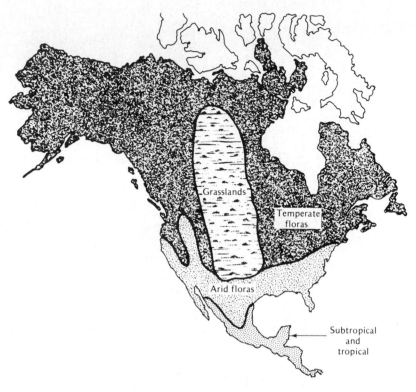

FIGURE 14.28
Distribution of floras in the Pleistocene during a time of maximum glacial retreat. Grasslands and temperate forests extend far to the north; low rainfall at such times results in
the spread of arid and semiarid plants.

either annuals or biannuals; they live only for a season or two, set their seed, and
then die. By contrast, trees and shrubs, the woody angiosperms, are all perennials;
they live for many years. This is one of the very marked differences between the
plant forms evolved among the angiosperms and those of the earlier dominant
groups of plants. Among gymnosperms there are virtually no annual or biannual
life forms known. The conifers, cycads, ginkgos, and others are all perennials. The
development of an annual habit is one strategy for avoiding adverse, harsh
conditions of either extreme cold or aridity. By germinating, growing, and setting
viable seed within a short growing season, the annual plant can ignore winters
and droughts. This way of life has been exploited to the fullest by angiosperms but
not, so far as the fossil record indicates, by other major groups of plants. It is
possible that the number and diversity of herbaceous plants increased gradually
during the Cenozoic as climatic conditions became cooler and drier. Unfortunately,

the fossil record is so poor for these groups that this hypothesis cannot yet be tested.

Another distinctive feature of Cenozoic angiosperms is the progressive development of the deciduous habit, in which all of the leaves are shed each year. This is in contrast to the evergreen habit, in which only some leaves are lost during the course of a year. There are a few deciduous gymnosperms, such as the conifer *Metasequoia* and the bald cypress, the latter deriving its common name from the fact that it loses its leaves each autumn. The ginkgo is another example. The great majority of gymnosperms retain their leaves for more than a year, losing a few and growing some new ones each year, so that the total leaf complement is replaced only over a period of years. Angiosperms, on the other hand, include a great variety of deciduous trees and shrubs. The deciduous habit, like the annual habit, is an adaptation that is especially suited for survival under adverse conditions. After it drops its leaves, a deciduous tree goes into a period of dormancy; in this state, cold weather or drought cannot affect it nearly to the extent as if it were evergreen. It seems likely that the deciduous habit evolved and spread through the angiosperms during the course of the Cenozoic as the climate gradually deteriorated. Many angiosperms in tropical and subtropical climates are evergreen; the proportion of those with a deciduous habit increases conspicuously in higher latitudes, both north and south.

KEY TERMS

angiosperms	flower	rhizome
araucaria	ginkgos	seed
catkin	gymnosperms	sphenopsids
coal ball	lycopods	sporangia
cone	phloem	stamen
conifers	pistil	wood
cycads	pollen	xylem
ferns	psilopsids	

READINGS

Beck, C.B., ed. 1988. *Origin and Evolution of Gymnosperms*. Columbia Univ. Press. 504 pages. The most up-to-date ideas on gymnosperm paleobotany written by a group of authorities. Suitable for readers with a background in botany or paleobotany.

Cronquist, A. 1988. *The Evolution and Classification of Flowering Plants*. New York Botanical Garden. 555 pages. A revision of an important book dealing with family level and higher evolution and classification.

Friis, E.M., and others. 1987. *The Origins of Angiosperms and Their Biological Consequences*. Cambridge Univ. Press. 358 pages.

Gastaldo, R.A. 1986. *Land Plants: Notes for a Short Course.* Studies in Geology 15. Univ. of Tennessee, Dept. Geological Sciences. 226 pages. This set of notes includes several papers on various groups of fossil land plants, including gymnosperms and angiosperms.

Thomas, Barry. 1981. *The Evolution of Plants and Flowers.* Peter Lowe Press. 116 pages. A short, easy-to-read, illustrated book on the evolution of plants. Emphasis is on the angiosperms.

Tidwell, W.D. 1975. *Common Fossil Plants of Western North America.* Brigham Young Univ. Press. 173 pages. An excellent handbook for the identification of fossil plants. Very well illustrated and useful for any area, although emphasis is on western fossil floras.

White, M.E. 1988. *Australia's Fossil Plants.* Reed Books. 144 pages. Lots of color and black-and-white photographs of Australian fossil plants, with an emphasis on Gondwana flora.

15
Early Consumers on Land: Reptiles

DIAGNOSTIC FEATURES

Amphibians are predominant through the Mississippian and, as far as we know, were the only tetrapods present in the coal swamps of the early Pennsylvanian. By middle Pennsylvanian time the first reptiles had appeared, representing a great advance over the amphibians. As one of their diagnostic features, reptiles have an **amniote egg**, a reproductive character that would eventually allow them to dominate many available land habitats (Figure 15.1). This is an egg that is covered by a hard but porous shell. Inside, the embryo is surrounded by tissues that nourish it and take care of wastes; respiration can take place through the shell. The amniotic egg can be laid on dry land without drying out, unlike the soft amphibian egg that must be laid in water. The reptiles also forego the intermediate larval or tadpole stage of the amphibians; in the former group, there is direct development from the embryo to the adult.

The evolution of amphibians to reptiles was seemingly gradual, with intermediate forms that combine a blend of typical amphibian and reptilian characters being fairly common (Figure 15.2). Generally speaking, the early reptiles stabilized on a particular style of construction of the backbone. Another diagnostic feature is the lack of an otic notch in reptiles, the ear being situated at the rear of the skull. Bones of the back part of the reptilian skull are reduced in number and size, a continuation of the trend seen from crossopterygian fishes to amphibians. The skull tends to be somewhat narrower and higher than that of amphibians. The

FIGURE 15.1
Fossil dinosaur eggs from the
Cretaceous of Mongolia. Al-
though fossil amniote eggs
are rare, they have been
found in rocks as old as the
Permian. (Courtesy of Field
Museum of Natural History,
Chicago.)

FIGURE 15.2
A mounted skeleton of the Permian amphibian *Seymouria* from central Texas. This small
animal, about 2 ft long, shows a unique blend of amphibian and reptilian characters but
is much too young to have been the direct ancestor of reptiles. Notice that an otic notch
is still present in the back of the skull. Other anatomical features resemble those of reptiles. (Courtesy of National Museum of Natural History.)

bones of the two girdles are enlarged and have wider areas of support with the
backbone. The limbs still spraddle at the sides, but the bones tend to be somewhat
longer and more slender. The bones of the wrist and ankle are reduced in number,
and the finger and toe bones are stabilized into a consistent pattern of 2–3–4–5–
3. In this system, each number indicates the number of bones per digit, the first
number representing the inside digit (big toe) and the last, the outside digit (little
toe).

Most important in distinguishing one kind of early reptile from another is
the structure of the bones in the temple region of the skull, behind the eye. Many

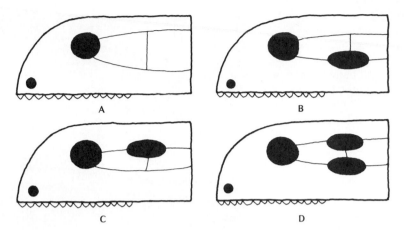

FIGURE 15.3
The basic types of reptilian skulls. **A.** Anapsid type, with solid bone and no openings in the temple region (postorbital bone behind the eye and squamosal bone at the back of the skull). **B.** Synapsid type, with the postorbital and squamosal bones meeting above the opening in the temple region. **C.** Euryapsid type (aquatic and semiaquatic reptiles), with the temple opening above the two bones. **D.** Diapsid type, with two openings and a bony bar between.

reptiles had one or two openings in the skull in this location, presumably to accommodate bulging jaw muscles. The nature and arrangement of these openings, called **temporal** or **temple openings**, provides data that is used in subdividing all major reptile groups (Figure 15.3). The group with a skull that is solidly encased in bone, with no opening at the rear of the side, is considered to be the most primitive (Figure 15.4). The reptiles of this group are referred to as **stem reptiles** because they are the ones from which the other, more advanced, reptiles are thought to have evolved. The only living reptiles of this group, which is called the **Anapsida**, are the turtles and tortoises. Another group, the **Synapsida**, or **mammal-like reptiles**, has a temporal opening low on the side of the skull. They are extinct but very important because mammals evolved from them. The third group, named **Euryapsids**, has a single opening high on the skull, a condition derived from a diapsid ancestor. An assorted lot of reptiles are included, most of which had an aquatic or semiaquatic way of life. The final main group has two openings, one above the other, separated by a bony bar (Figure 15.5). These are the **Diapsida**, or **ruling reptiles**. They include the dinosaurs of the Mesozoic Era, as well as most living reptiles—the crocodiles, alligators, snakes, and lizards.

LATE PALEOZOIC REPTILES

In the Pennsylvanian, when reptiles first appear, the fossils are relatively rare and without very much variety. By Permian time, reptiles appear in much greater

FIGURE 15.4
Side view of the skull of a stem reptile (anapsid), *Labido-saurus,* from the Permian of Texas. The solid bony temple region is behind the large orbit for the eye. (Courtesy of Field Museum of Natural History, Chicago.)

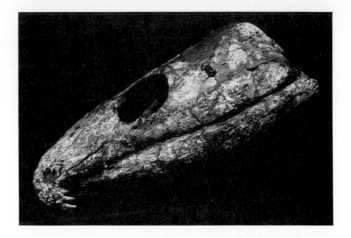

abundance and variety and are clearly predominant over the amphibians. This ascendancy of the reptiles was due, at least in part, to changes in climate. The coal swamp forests of the Pennsylvanian, denoting abundant rainfall, had dwindled by Permian time, giving way to seasonal rainfall and perhaps to more extremes in temperature. The various groups of amphibians that had typified Pennsylvanian rocks are still present in the Permian but with a somewhat different aspect.

Amphibians had evolved along two quite distinct lines with respect to habitat. Some of them had become increasingly land-dwelling animals. These included some which were of relatively large size, up to 2 meters long, with very massive short limbs and a large, flat, alligatorlike skull. The majority of amphibians had taken a different route. They had given up life on land and returned to a dominantly

FIGURE 15.5
Skull and lower jaw of a duck-billed dinosaur, *Lambeo-saurus.* Distinctive features include the flattened beak that lacks teeth, the long row of sturdy grinding teeth that denote a plant diet, a peculiar bony crest on the forehead, and the two large openings in the temple region of the skull (diapsid condition). (Courtesy of Field Museum of Natural History, Chicago.)

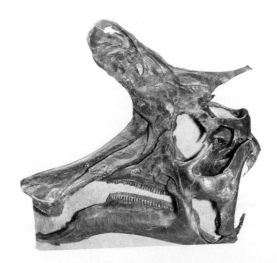

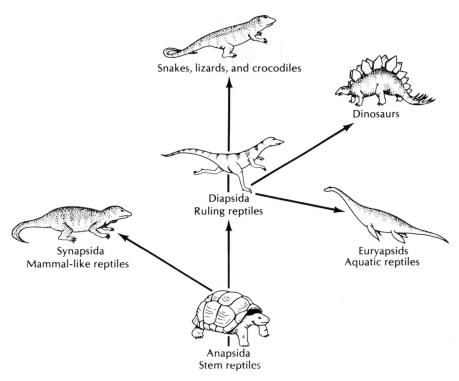

FIGURE 15.6
Evolutionary relationships among major groups of reptiles.

aquatic life, living all or most of their time in fresh water. In some of these amphibians, the limbs are so reduced in length and size that the animals could not have supported themselves on land. In these types the backbone becomes smaller and weaker. The tail is a swimming structure and the skull is typically very low, flat, and broad. At least part of this change in amphibians was surely due to change in climate. The immense area of freshwater swamps to which they could retreat for reproduction was simply no longer available. Reptiles, with somewhat larger brains, also were providing more intense competition. Due to their ability to lay their eggs on land, the reptiles could flourish under these drier conditions. Reptiles underwent a conspicuous radiation of types in the Permian and evolved a wide variety of forms that were probably herbivorous, and others that were carnivorous. How can we determine the diet of these extinct animals? One clue is body shape. Carnivores tend to be slender, whereas herbivores have a barrel-shaped trunk. Both body types are present in Permian reptiles. Some of the small, slender reptiles with sharp pointed teeth are thought to have been insect eating.

Two groups of reptiles were dominant during the Permian Period. One was the **anapsids**, or stem reptiles (Figure 15.6). These primitive reptiles are thought

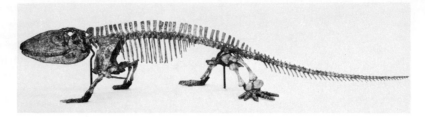

FIGURE 15.7
A mounted skeleton of a primitive pelycosaurian synapsid from the Permian of New Mexico. The obscure single large temporal opening is behind and below the orbit for the eye. (Courtesy of Field Museum of Natural History, Chicago.)

to have been the ancestors of the synapsids, the mammal-like reptiles. The **diapsids**, which were barely represented in the Permian but which dominated the Mesozoic scene, led to the euryapsids (Figure 15.7). The synapsids flourished during the Permian Period and were represented by an archaic group, the pelycosaurs. These reptiles evolved into the Triassic therapsids, or advanced mammal-like reptiles, that in turn evolved directly into mammals. The pelycosaurs included *Dimetrodon*, the sail-back reptile. There has been controversy concerning the function of the conspicuous sail (Figure 15.8). It has been suggested that the sail served primarily as a secondary sexual character; as a device for defense, making the animal look larger and fiercer than it really was; and also as a thermoregulatory device. This last idea has gained general acceptance. It is postulated that the animal, when it was cold, would turn itself broadside to the sun's rays. The sail, being richly

FIGURE 15.8
A mounted skeleton of the Permian sail-back reptile *Dimetrodon.* Notice the large single opening in the rear of the skull behind the orbit for the eye. The highly elongate bones rising from the vertebrae supported a fleshy sail in life. The animal was probably an aggressive carnivore, as witnessed by the impressive teeth in the jaws. (Courtesy of National Museum of Natural History.)

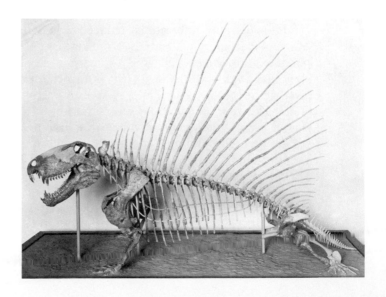

supplied with blood vessels, served to warm up this cold-blooded animal. If *Dimetrodon* became too hot, it simply turned itself 90 degrees until the sail was parallel to the rays of sunshine. Mammal-like reptiles of the Lower Permian bear little resemblance to mammals, but by late Permian time they show a series of features clearly indicative of their evolution along lines that would ultimately culminate in mammals. These late Permian forms are quite rare in North America, but they are very well known from a sequence of high Permian and Triassic terrestrial rocks in South Africa known as the **Karroo Group**.

MESOZOIC REPTILES

The Mammal-like Reptiles of the Triassic

Reptiles flourished in the Mesozoic Era, with two major groups dominating the landscape. The first of these was the mammal-like reptiles in the Triassic Period. They had evolved considerably from their Permian ancestors. A variety of features can be seen in these forms that indicate a transition toward mammals. Their limbs became longer and more slightly built than those of other reptiles. They are tucked under the body instead of spraddled at the side, making for swifter locomotion and more efficient support of the body. The braincase became progressively larger in proportion to the rest of the skull. The number of bones in the toes of the feet was reduced in number to a 2–3–3–3–3 formula.

These reptiles also developed a secondary hard, bony palate. This consisted of a bony extension at the front of the mouth from bones of the upper or primary palate. The secondary palate served to separate air taken in from food in the mouth. This feature is especially useful in maintaining body temperature, as the animal can breathe at the same time it is ingesting food. It has been suggested that the presence of a bony second palate in these reptiles is a clue that they may have already become warm-blooded—one of the important characters of mammals, but one that is nearly impossible to detect directly in the fossil record.

The teeth of these reptiles also underwent fundamental change. The number of teeth was reduced and confined to the bones of the jaw edges; they were not scattered over the roof of the mouth as in many other reptiles. The teeth came to be differentiated into nipping incisors in front; stabbing canines next; followed by shearing, cutting, and grinding premolars and molars at the back of the jaw. Most other reptiles have simple, cone-shaped teeth that are basically all alike. This differentiation of the tooth row is another feature in which the mammal-like reptiles resembled mammals.

One final feature must be mentioned. Mammals are characterized by the presence of a single bone in the lower jaw, the **dentary**. This bone articulates with a bone of the upper jaw, the **squamosal**. All reptiles have more than one bone in the lower jaw, and one of these, the **articular**, articulates with the **quadrate** bone of the skull. Advanced mammal-like reptiles approach the mammalian condition in that the dentary is much enlarged and other bones of the lower jaw are reduced in number and size. The articular and quadrate bones especially become quite small. This feature is used as an important diagnostic character to distinguish advanced mammal-like reptiles from primitive mammals. If the jaw articulation is effected by the dentary-squamosal, the specimen is a mammal; if by the articular-quadrate, it is a reptile. We can now ask what happens to the articular and quadrate bones in mammals after they lose their articular function. You will recall that reptiles have a single bone for sound transmission in their middle ear—the stapes—which is the old fish hyomandibular. This bone is joined by two others in mammals, which have three bones in the middle ear. The two new bones are the old articular and quadrate of reptiles. These bones are shifted over to the ear, which is situated close to the jaw joint in mammal-like reptiles and in early mammals. The bones, now called the **malleus** and the **incus**, are the hammer and the anvil bones of the ear.

Not all of these advanced features are found in a single fossil reptile. It is not completely clear whether mammals all evolved from a single reptilian ancestor or whether mammals evolved from different lineages of mammal-like reptiles. Although the fossil record for the reptiles is reasonably good, the early mammalian record is very poor. Recently discovered skulls from the Triassic-Jurassic boundary give us a better idea of what the earliest mammals were like; they were very primitive and quite small (mouse- to rat-sized).

The Ruling Reptiles

The other major group of land animals in the Mesozoic was the diapsid, or ruling, reptiles. One important group was the **thecodonts**, which before the close of the Triassic gave rise to the groups known as **dinosaurs**. The thecodonts first appear in the early Triassic and become extinct at the close of the period. While some thecodonts ran on all four legs, being **quadrupedal**, others exhibited a new **bipedal**, or two-legged, stance. They used the front feet for food handling, grabbing, and so on. Most thecodonts were relatively small reptiles, from the size of a chicken to about 2 meters long. Their bones were thin and lightly constructed, some having hollow leg bones, like birds. The skull was thin-boned with large vacuities, or areas where bone was not developed. The hind limbs were quite elongate and powerful for swift movement. The front legs were short and weak. They give the impression of having been fast runners for their time, in some respects resembling large, ground-dwelling birds, such as ostriches. They may have been insectivorous, or perhaps they caught other small prey, but they had feeding habits requiring speed. Their swiftness may also have been advantageous in escaping larger, heavier predators, such as some of the amphibians and mammal-like reptiles. By the close of the Triassic, the thecodonts had given rise to a group of more advanced reptiles, the Saurischia, which represents one of the two main groups of dinosaurs. The term comes from *sauria*, meaning reptile, and *ischia*, referring to the ischium bone

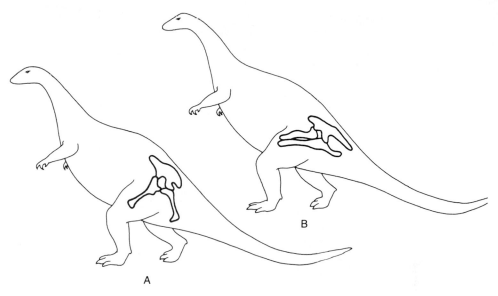

FIGURE 15.9
Pelvic girdles in bipedal dinosaurs. **A.** In saurischians the girdle of three bones (ilium, ischium, pubis) is triradiate. **B.** In ornithischians the pubis bone extends both fore and aft.

of the pelvis (Figure 15.9). These dinosaurs had a pelvis built like that of many other reptiles. From them evolved the other major group of dinosaurs, the **Ornithischia**, which appeared at the end of the Triassic. This group had a birdlike pelvis (Figure 15.10).

The saurischians were relatively common by late Triassic time and became the most conspicuous reptilian group in the Jurassic. As we have already seen, the earliest saurischians were bipedal, small, and lightly constructed, with insectivorous or carnivorous food habits. From these, in the Jurassic, evolved the major group of large and small predaceous animals of this time, the **theropods**. These were all bipedal animals. Many of them were quite small, 6 feet or less in length. Others, by the close of the Jurassic, had become very large beasts, up to 30 feet long, with heavy bodies and hind legs (Figure 15.11). The front legs became progressively smaller and weaker. The teeth evolved into long, spearlike, stabbing teeth. These animals must have been formidable predators. The theropods were the only carnivorous dinosaurs. They clearly were at the apex of the food pyramid during the Jurassic and continued so into the Cretaceous (Figure 15.12).

In addition to the carnivorous dinosaurs, the saurischians include one other major group called the **sauropods**. These were strictly herbivorous animals, and they were quadrupedal, walking on all four limbs. Again, some were small, but others became very large; sauropods include the largest land-dwelling animals known—the dinosaurs, among them *Diplodocus* (Figure 15.13), *Apatosaurus*, and *Brachiosaurus*. The heaviest ones are estimated to have weighed up to 40 tons and

FIGURE 15.10
A mounted skeleton of the duck-billed dinosaur *Anatosaurus*. This dinosaur lived during
the Late Cretaceous. Note the typically ornithischian structure of the pelvic girdle bones.
The jaw lacked teeth in the front of the mouth, the teeth having been replaced by a
horny beak; hence, the name duck-bill. (Courtesy of National Museum of Natural
History.)

the longest ones were nearly 90 feet long. The legs were massive and pillarlike,
constructed much like those of an elephant, to support the tremendous weight of
the body.

The legs of most sauropods clearly show that these large animals evolved
from a bipedal ancestor; except in Brachiosaurus the front legs were shorter and
less massive than are the hind limbs. The neck and tail were commonly elongate
and the head was quite small. The teeth were small and peglike. Because of the
small teeth and restricted size of the throat, some paleontologists have questioned
whether these huge animals could have consumed enough plant material to keep
their bodies well nourished. It has been suggested that they must have eaten large
quantities of soft, lush vegetation that was easy to crop and to digest. Exactly what
plants formed the bulk diet of the sauropods is not certain, and the known kinds
of Jurassic land plants (mostly gymnosperms such as the cycads and pines) present
somewhat of a puzzle in this regard. Sauropods possessed gizzard stones in a crop
that surely helped shred tough vegetable material. There has also been controversy
concerning whether these animals were truly land-dwelling, or whether they spent
most of their lives in freshwater pools and lakes, where the water could help buoy
up their tremendous weight. The legs of sauropods are massively built to support
their weight, and footprints have been found very deeply impressed into what was

FIGURE 15.11
A large carnivorous dinosaur of the Mesozoic Era. Note the bipedal stance, the triradiate structure of the pelvic girdle, the reduced forelimbs that were used for grasping prey, and the large teeth. (Courtesy of National Museum of Natural History.)

FIGURE 15.12
The skull and lower jaw of one of the large carnivorous dinosaurs, *Albertosaurus*. The specimen is about 1 m long. (Courtesy of Field Museum of Natural History, Chicago.)

FIGURE 15.13

The giant sauropod *Diplodocus* of Late Jurassic age from Utah. Note the arch shape of the spinal column and massive legs, designed to support many tons of weight. This dinosaur is the longest one known, close to 90 ft in length, but was not the heaviest, having a somewhat lighter build than did other large sauropods. (Courtesy of National Museum of Natural History.)

once soft sediment. For these and other reasons most paleontologists now agree that sauropods were able to walk around on dry land. Their large size was probably effective at protecting them from all but the largest Jurassic predators, such as *Allosaurus*, which was 9 meters long. Perhaps several of these large predators ganged up on one of the gigantic herbivores, in pack fashion, in order to prey on it.

The other major group of dinosaurs, the ornithischians, is much less diverse and less common in the Jurassic than the saurischians. The bird-hipped dinosaurs were all herbivorous, and they may have evolved from a bipedal saurischian ancestor. Although there are four distinct groups of ornithischians, only two of these are present in Jurassic rocks; the other two did not evolve until the Cretaceous. One Jurassic group is called the **ornithopods**, or duck-billed dinosaurs (see Figure 15.5, Figure 15.14 and Figure 15.15). These were bipedal, with reduced front legs. Some of the Jurassic forms had relatively large front legs that could have been used for walking. Many were small, from 1.5 to 4.5 meters long. A characteristic feature is that the tooth row did not reach the front of the mouth; it is probable that these animals had a horny beak. The other Jurassic group is the **stegosaurs** the best-known example of which is *Stegosaurus* (Figure 15.16). These animals were quadrupedal, but show clearly that they evolved from a bipedal ancestor. The front limbs are typically shorter and less heavily constructed than are the hind

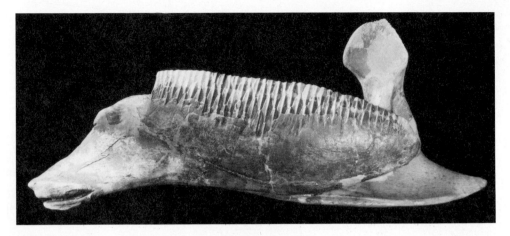

FIGURE 15.14
Inner view of the lower jaw of an ornithopod dinosaur, possibly *Kritosaurus* showing the battery of large grinding teeth for shredding plant debris. Although quite different in origin, the tooth battery somewhat resembles the single molar tooth of modern elephants. Total length is 90 cm. (Neg No. 120324 [photo by J. Kirschner] courtesy Department Library Services, American Museum of Natural History.)

FIGURE 15.15
Enlargement of the preserved skin from the rib area of a mummified duck-billed dinosaur, *Edmontosaurus annectens*, from the Kirtland Formation, San Juan County, New Mexico, of Cretaceous age. The skin surface is covered with large and small polygonal to oval bony tubercules. Actual width of specimen is 50 cm. (Neg. no. 35608 [photo by Anderson] courtesy Department Library Services, American Museum of Natural History.)

FIGURE 15.16
The ornithischian dinosaur *Stegosaurus,* of Jurassic age. Stegosaurs were about 20 ft long and characterized by two rows of large bony plates along the backbone. Notice the long bony spikes on the end of the tail. These dinosaurs were herbivores. (Courtesy of National Museum of Natural History.)

legs. The stegosaurs reached lengths of about 7 meters. They had a very small head and brain cavity for their size and an enlarged nerve center in the pelvic region of the spinal column.

The outstanding features of the stegosaurs are the distinctive skeletal devices they evolved. All along the back, from the neck to near the end of the tail, they had a double row of heavy, triangular, bony plates, or **scutes**, that were arranged in alternating, overlapping positions. These plates may have functioned for temperature regulation, much like the sail of *Dimetrodon*. The end of the tail bore long bony spikes or spines. The scutes presumably helped protect the vital spinal cord from attack from above by large predators. The spines on the end of the tail would also have been a formidable defensive weapon when the tail was lashed back and forth. The predatory pressure on these animals must have been considerable in order for them to have evolved such elaborate defensive mechanisms.

Dinosaurs continued to dominate land animals right until the close of the Cretaceous, when both saurischians and ornithischians became extinct. The giant sauropods of the Jurassic had dwindled in ferocity in the Late Cretaceous. They are exceedingly rare in North American Cretaceous rocks, but they have been found in greater numbers in India and Mongolia. The predaceous carnosaurs of the Cretaceous were the largest and probably the most ferocious of the dinosaurs. The largest and most popular of these animals is *Tyrannosaurus rex*, the "king tyrant lizard," known from Upper Cretaceous rocks. This reptile is surely one of the largest predators known. It was 13 meters long and weighed up to 8 tons. Its spearlike teeth were up to 15 centimeters long. One might think that this giant animal would have fed on the large sauropods, but the remains of the two reptiles

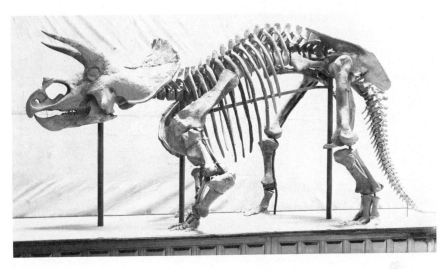

FIGURE 15.17
A mounted skeleton of *Triceratops,* the youngest and most advanced of the herbivorous ceratopsian dinosaurs. There was a horny beak at the front of the mouth and three horns on the skull. A large bony frill extended back from the skull, protecting the neck region. The specimen is late Cretaceous in age. (Courtesy of National Museum of Natural History.)

are not found together in the same rocks. The ornithischians were the dominant group of dinosaurs in the Cretaceous. The duck-billed dinosaurs were abundant and diverse. Many evolved elongate nasal passages that served as sounding tubes. Some of them were probably semiaquatic and developed bizarre bony crests on the top of the head that contained elongate nasal passages and a nasal opening at the top of the crest, presumably for breathing under water. The stegosaurs became extinct in the Jurassic. They may have been replaced ecologically by another group of ornithischians, the **ankylosaurs**. These were large armored tanks that somewhat resemble living armadillos. The back and sides of the animal were covered by columns of close-fitting thick bony scutes that completely encased it. In addition, there were long bony spikes at the shoulder region and on the tails of some of them. They had very feeble dentition, some lacking teeth altogether.

The final group of ornithischians, which, like the ankylosaurs, is confined to the Cretaceous, is the **Ceratopsia**. The most famous of the group is *Triceratops* (Figure 15.17). These dinosaurs were mostly of moderate size, 4.5 to 6 meters long. They are typified by having the back of the skull produced into a wide, flat bony frill that extended over the neck region and surely protected that vital and vulnerable area from attack. Another interesting feature of these animals is that they gradually evolved horns on the nasal region of the face above the eyes. The oldest and smallest ceratopsians from the Early Cretaceous lacked horns.

Typical Cretaceous dinosaurs were the large carnosaurs such as *Tyrannosaurus* and three groups of herbivorous ornithischians: the duck-bills, ankylosaurs, and the horned dinosaurs. All became extinct by the close of the Cretaceous. It is not clear whether all groups disappeared more or less simultaneously or whether the extinction took place gradually over several millions of years. The very youngest dinosaur remains known in the United States consist of scraps of ceratopsian bone from beds that were initially thought to be Cenozoic in age. Thus, it is possible that *Triceratops* outlived other groups of dinosaurs and was the last one of these large reptiles to become extinct.

Warm-blooded Dinosaurs

For the last several years controversy has arisen concerning whether some or all dinosaurs were warm-blooded. All living reptiles are cold-blooded but can maintain an elevated body temperature by behavioral means such as basking in the sun. We have seen that the sail of the *Dimetrodon* may have been a crude temperature-regulating device. The very large bulk of some of the sauropods may have maintained their more or less constant body temperatures because it would take a long time for them to heat up or cool off. The triangular scutes on the stegosaurs may have acted like radiator fins that controlled body temperature. The limb bones of living mammals and reptiles have a different microscopic structure that is related to warm- and cold-bloodedness. Some dinosaur bones have a structure more like that of modern mammals than of modern reptiles, indicating that dinosaurs may have been warm-blooded. Another argument that supports this theory has to do with the ratio of predators to prey. In a couplet of cold-blooded predator and prey, for instance, an alligator as predator and a fish as prey, it takes relatively few fish to support one alligator. In a warm-blooded couplet, however, such as a lion and an antelope, the ratio is much higher. Counts of herbivorous and carnivorous dinosaur remains in museum collections indicate that the predator-prey ratio is closer to that of warm-blooded mammals than it is to cold-blooded vertebrates. The possibility of errors in such an approach is considerable, and not all vertebrate paleontologists accept this evidence as strong support for warm-blooded dinosaurs. It seems likely that scientists and students will argue about the temperature of these extinct beasts for several years to come.

The Social Relationships of Dinosaurs

In the past few years there has been a great advance in our understanding of how dinosaurs lived and what kinds of social interactions they may have had. Much of this new information must be credited to Jack Horner, a dinosaur paleontologist with Montana State University and the Museum of the Rockies in Bozeman, Montana. He and his fellow workers have discovered the greatest concentration of dinosaur nests, their contained eggs, and baby dinosaur skeletons that has ever been found. These discoveries permit us to make inferences about dinosaur

FIGURE 15.18
Dinosaur egg clutch of a hypsilophodontid ornithopod, reduced in size. From the Upper Cretaceous of central Montana. (Photo courtesy of the Museum of the Rockies, Bozeman, MT)

behavior. The nests occur in two different layers in the Hell Creek Formation of late Cretaceous age, and they were built by two different duck-billed dinosaurs (ornithopods), adults of which have been found associated with the nests.

The nests are quite large and were excavated by the dinosaurs in the soil (Figure 15.18). The nests may be up to 6 feet in diameter and are shaped like large bushel baskets. The eggs are very carefully arranged in the nests and have a definite spiral pattern around a central egg. The eggs are elongate and standing on end, indicating a precise behavior pattern either in laying the eggs or manipulating them after they were laid. The nests are filled with vegetable material, now plant fossils. These may have helped to incubate the eggs or served as camouflage to help prevent egg-eaters, such as other small reptiles or perhaps mammals, from preying on the eggs. Decomposing plant material generates heat, as in a compost pile, and if heat was needed, this may be evidence that the dinosaurs were warm-blooded. The nests are in large clusters and have a definite spacing, about 25 feet apart. This spacing indicates that a large cluster of dinosaurs shared the nesting territory at the same time and that they may have returned to the same nesting site for more than one year. Thus, the nests provide clear evidence for social behavior.

The great majority of the duck-bill nests contain only broken egg fragments, indicating that most of the baby dinosaurs hatched successfully. However, a few nests contain whole eggs, and studying these by X-ray and CAT-scan techniques reveals that they contain the bones of embryonic dinosaurs (Figure 15.19). These are the most numerous and best preserved fossils of unhatched dinosaurs ever found. There is one very important difference between the contents of the nests in the two layers at the Hell Creek Formation. In the older layer a few skeletons

FIGURE 15.19
Egg of the ornithopod dinosaur *Troödon*, showing evidence of the embryo skeleton within, from Upper Cretaceous rocks of central Montana. (Photo courtesy of the Museum of the Rockies, Bozeman, MT)

of recently hatched baby dinosaurs have been found. These are only slightly larger than the eggs and indicate that the dinosaurs died shortly after hatching. In the other nests, however, juvenile dinosaurs have been found that are up to 4 feet long. These clearly inhabited the nests long after they hatched and underwent a conspicuous growth interval before leaving the nest. Thus, they must have been fed by one or both parents. Evidence of this kind of parental care by fossil reptiles is virtually unique to these sites in Montana.

Another indicator of social behavior by dinosaurs has been developed by study of fossil footprints. In Cretaceous rocks in the western United States dinosaur footprints are abundant in some layers of rock. Such rocks occur especially in western Canada, Montana, Wyoming, Utah, and New Mexico. The rocks were deposited on a coastal plain that bordered the western edge of an extensive inland sea that spread from Texas to northern Alaska during Cretaceous time. This was the last major flooding of the North American continent by shallow seaways. The dinosaur tracks indicate several things. First, exactly the same kinds of tracks have been found over a very wide latitudinal range, from New Mexico to western Canada. This seems to indicate that the dinosaurs may have migrated, perhaps on a seasonal basis. Second, the same kinds of tracks, especially those of some herbivorous dinosaurs, occur in dense concentrations, thus indicating group or herd behavior. This, coupled with the nest information, is a strong indication of the social instincts of dinosaurs. Jack Horner has called some of the duck-bills the "cows" of the Cretaceous, living in herds, migrating with the seasons, and taking care of their young.

How to Measure a Dinosaur

If you pick up any book about dinosaurs, you can find many vital statistics on these animals—how much they weighed, how long or tall they were, how fast they could run, and so on. But if you look into the matter of what actual bones have been found for a specific dinosaur, you may find that not a single complete skeleton has been discovered. Rather, a partial skull, a couple of leg bones, one or two vertebrae, and a couple of ribs may be all that has been found. How, then, do paleontologists come up with seemingly complete and accurate measurements of such incompletely known beasts? In a few cases specimens have been found that are complete enough to be used as a basis for estimating length or height. In other cases fossil footprints, identified as to the kind of dinosaur that made them, can be used to estimate weight, based on cross sections of leg bones.

Using footprints to obtain an estimate of animal length does not provide an overall length, but rather the trunk length, from the hipbone socket (known officially as the acetabulum) to the forelimb socket (the glenoid fossa). The length of the tail and neck cannot be directly measured from footprint data. First, one needs to identify the front footprints and the hind footprints. These may differ slightly in size, shape, number of toe impressions, and so on. The front footprints may occur slightly in front of the hind footprints or on either side of them, or the two kinds of prints may overlap, with the hind foot stepping completely or partly into the mark left by the front print. The pattern is determined by considering the relative length of the trunk and the legs. In an animal with a quite long trunk and short legs the footprints may be equally spaced. Next a sequence of four successive prints is identified. On a drawing of these the center of the left-side front footprint is connected with the center of the right-side front print. This will be an oblique line that crosses the center of the trackway. The same thing is done for two successive hind prints, left and right. Next, the center points of the two oblique lines along the center of the trackway are connected. This distance is called the stride and provides a direct measure of the trunk length. Estimates of tail and neck length are added to this to get the overall length of the animal.

The weight of an animal is related to the total volume of bone and the total volume of tissues that make up the animal. Actually, we are using weight here in a very sloppy way. What we are really calculating is the mass of the animal, not its weight (which is the mass multiplied by the force of gravity). First, we can determine the mass of various living animals, large and small. We express this weight, for large animals, in tonnes. A tonne is not the same as the English ton but rather a metric ton or 2,205 pounds. Masses of living animals range from the blue whale, at 90 to 100 tonnes (about 200,000 lbs.), to humans, at 0.05 to 0.07 tonnes (110 to 150 lbs.), to an amoeba, at .0000003 tonnes (.01 oz.). To calculate mass it is necessary to have some estimate of the volume of the animal and the average density of the stuff making up the animal—bone, muscles, gut, and so on. It turns out that most animals are about the same density as water. Some animals, such as crocodiles, which have heavy bone scutes, are slightly denser than water, about 8 percent more dense. The volume of large animals such as dinosaurs has been determined by building scale models of the animal, determining the volume of the model, and then multiplying that by the scaling factor. Such calculations of volume for a large sauropod such as *Brachiosaurus* yield a volume of between 45 and 50 cubic meters. If the dinosaur were the same density as water, its mass would be between 45 and 50 tonnes, about one-half that of a living blue whale and some nine times heavier than a modern elephant.

Another way to estimate mass is by measuring the circumference of the upper leg bones (the femur in front or the humerus behind). This has been done for different sizes of living animals, and a roughly straight line relationship exists. The leg bone sizes of the dinosaur are then compared to those of living animals. The result of such analysis may result in estimated masses that are somewhat less than those resulting from scale-model analysis. For instance, *Brachiosaurus* comes out to weigh 32 tonnes by this method. A different formula must be used for bipedal than for quadrupedal animals.

Once we know the weight of the animal, we can begin to estimate how fast the animals could run. Again, we study footprints. We determine the animal's stride, just as we did above in determining the length of the animal. The faster the animal is moving, the greater the stride length. Speeds of various animals have been determined and related to the animals' stride. The speed, adjusted to take account of gravity, is called dimensionless speed. Using this and leg length it is possible to calculate the speed of the animal. The greatest speeds for dinosaurs obtained this way are for ornithopods, which had running speeds up to about 25 to 27 miles per hour. This is faster than a human can run and slightly slower than a horse can gallop. The speed assumes an animal that weighed about 1300 pounds. No tracks have been found for the largest dinosaurs that suggest they moved at a running speed. The really big ones all seem to have ambled along at a much slower pace.

Scientists have also calculated how strong dinosaurs were, in terms of their bone strength and athletic ability. This is done by comparing length and diameter of leg bones with those of various modern animals.

Finally, it has been possible to measure the tones (sounds) that dinosaurs may have made. The crested duck-billed dinosaurs had long tubular nasal cavities in the head crest that connected with the roof of the mouth. These hollow cylinders may have been sound-producing chambers. If so, the pitch of the sounds that they made, assuming air passed over the opening to the cavity as in an organ pipe, would be directly related to the length of the chamber, just as in a pipe organ the long pipes produce the bass sounds and the short pipes the treble sounds. The frequency of the sound in cycles per second can be figured by dividing 170 by the length in meters of the pipe. Performing this calculation for one duck-billed dinosaur results in a pitch of F natural. In any one species of duck-bills some specimens have smaller crests than others. The crests are thought to be a secondary sexual characteristic, and the females are presumed to have had the smaller crests. If this is so, then females would have made higher pitched tones than the males. Even among dinosaurs the females sang soprano.

The Extinction of Dinosaurs

One of the fascinating problems of paleontology, about which much has been written, has to do with the cause or causes of the extinction of the dinosaurs. First we should reemphasize that many dinosaurs became extinct before the end of the Cretaceous. The stegosaurs lived only in Jurassic time, and the sauropods were already scarce in the Late Cretaceous. The most common dinosaurian remains in very young Cretaceous rocks are those of the ceratopsians, carnosaurs, and ornithopods. Thus, final dinosaur extinction mainly involves these three groups. A host of different theories have been put forward to explain the disappearance of these large animals. It has been suggested (a) that the gradually developing small mammals may have eaten their eggs; (b) that they were bombarded by cosmic rays that rendered them sterile; (c) that mountain building somehow affected them; and (d) that there were changes in climate, in vegetation, and in distribution of land and sea that were unfavorable and hastened their demise.

The most recent theory about the extinction of dinosaurs, ammonites, and other groups at the end of the Cretaceous Period is that the planet was struck by an asteroid at this time. Evidence for such an impact consists of a clay layer greatly enriched in the rare element iridium at what is commonly called the **K/T boundary**. *K* is the map symbol for Cretaceous and *T* is the symbol for the Tertiary. This iridium-enriched layer has now been found at numerous localities in Europe and North America. It is suggested that the dust from the impact restricted photosynthesis and created imbalances in both the marine and terrestrial food chain. Despite this claim, flowering plants were not dramatically affected by this extinction event. This theory is so new that it will take several years for additional data to be gathered, both for and against the hypothesis.

Older, more traditional explanations for the extinction of dinosaurs have to do with physical changes in the environment or with changes in vegetation. Most paleontologists agree that the two most likely causes may have had to do with physical changes in the environment, on the one hand, or with changes in vegetation on the other. Most dinosaur fossils are found in rocks that were laid down on vast flood plains or coastal flats, where there were river channels, lakes, and swamps in which the skeletons became entombed. Withdrawal of the seas at the end of the Cretaceous and uplift of many areas during mountain building may have decreased the available areas of suitable habitats. Changes in worldwide temperature may also have contributed to the extinction. The gymnosperm diet of Jurassic dinosaurs was reduced in abundance by the flowering plants that were flourishing within the Cretaceous. If the herbivores became rare because of lack of appropriate food supplies, the large predaceous dinosaurs that preyed on them would have become much rarer still, and extinction of the herbivores would have led swiftly to extinction of their predators. If either food supply or suitable habitats, or both, came into short supply, various dinosaurs may have been so widely and thinly dispersed as to affect their breeding, especially if they did not migrate in search of mates. Inbreeding and failure to breed may have become increasingly common, with concomitant effects on the structure and success of breeding populations.

Whatever caused extinction, the underlying factors may have been numerous, such that no simple explanation with a single cause would be adequate to explain this phenomenon. Both physical and biological factors may have been involved.

FIGURE 15.20
Archaeopteryx, the oldest known bird, from Jurassic rocks in Germany. In life this bird was about the size of a crow. Note the impressions of feathers along the wings and tail, the clawed toes at the leading edge of the wing, and the long bony extension of the backbone into the tail. (Courtesy of National Museum of Natural History.)

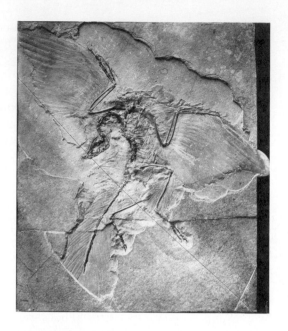

THE FOSSIL RECORD OF BIRDS

Excepting the dinosaurs, probably the most famous fossils are those of the *Archaeopteryx,* the oldest known bird (Figure 15.20). The first specimen, discovered in 1861, consisted of a single feather. The recognition of a feather was possible because of the excellent preservation in a very fine-grained lithographic limestone, so called because it was quarried commercially for making printed lithographs, including geological maps. Late that same year a complete specimen was found which was bought and sold several times before it was purchased by the British Museum of Natural History. The British Museum paid 700 pounds (about $14,000 dollars today), a very large sum of money at the time, for this specimen and an additional 1700 fossils. Since 1861 four additional specimens have been excavated, each of which was treated as a valuable commercial object and sold at least once before deposition in a scientific institution. The second specimen went to the Berlin Museum. This specimen, sold in 1881 for 20,000 German marks (about $19,000 today) was discovered in 1877 in a different lithographic stone quarry approximately 10 miles away from the first complete specimen. A third specimen was not discovered until 1956 and presently remains in private hands where it is not available to the scientific community for study. The fourth specimen, labeled as a pterosaur, a flying Mesozoic reptile, was collected in 1855 and stored in a museum drawer until 1970 when a discerning paleontologist recognized it as an *Archaeopteryx.* Finally, a fifth specimen, originally collected in 1951, was recognized in 1973, again in a museum, where it had been classified as a small dinosaur. Additional single feathers are also known.

Archaeopteryx differs from modern birds in several ways. Teeth are still present in both jaws. The tail is long and has a series of vertebrae down its midline. The

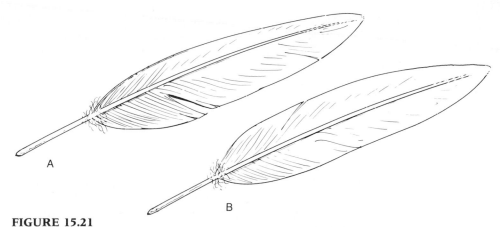

FIGURE 15.21
Primary wing feathers of birds. **A.** An asymmetrical feather of a bird that can fly. **B.** A symmetrical wing feather of a flightless bird. The oldest known bird, *Archaeopteryx*, has asymmetrical primary wing feathers and is therefore presumed to have been capable of flight.

front limbs, or wings, are short and stubby. There are still three clawed fingers at the wing tips, the wrist bones occupying the mid-length of the wing. The sternum, or breastbone, is still small and not enlarged to accommodate large flight muscles as in modern birds. Other features are more birdlike. The braincase is enlarged and solidly encased by bone. The orbits for the eyes are large, and that undoubted bird feature, feathers, are preserved. The small breastbone of *Archaeopteryx* has led to speculation that the animal may not have had large enough chest muscles to sustain flight. On the other hand, the primary wing feathres of *Archaeopteryx* are clearly asymmetrical, with the small barbs on either side of the central shaft being of unequal length. This pattern is typical of all flying birds today, whereas flightless birds have barbs of equal length on either side of the shaft (Figure 15.21). Thus, this feature would seem to indicate that the oldest known bird could indeed fly, although it may not have been as adroit at this function as many modern birds.

There has been considerable debate concerning the ancestry of birds. Although no one doubts that birds evolved from the diapsid reptiles, there is a difference of opinion as to whether they evolved from the advanced saurischian reptiles, specifically the small, bipedal, lightweight coelurosaurians, or from a more generalized group of diapsids, the pseudosuchian thecodonts. Some researchers believe that the birds and some dinosaurs are so closely related that they should be put together into a single new class of vertebrates. Whatever the ancestor may be, it certainly was a small, hollow-boned, bipedal animal with long hind limbs and toothed jaws.

There are two quite distinct theories as to the origin of flight in birds. One idea is that the bird's ability to fly originated in a climbing and gliding ancestor. In this view bird ancestors climbed trees and eventually learned to parachute to the ground, then to glide from one tree to another, and finally to fly. This trait may have originated as an escape from predation or a search for food such as leaves, seeds, and fruits. Climbing into trees and nesting in trees for safety may also have preceded gliding and flying. Several different sorts of animals can glide besides the well-known flying squirrels of North America. There are gliding marsupials in Australia as well as gliding lizards, snakes, and frogs. It has been suggested that the small clawed digits at midwing of *Archaeopteryx* allowed the animal to haul itself up into trees. An analogous feature is found in the chicks of a single living bird, the hoatzin of the Amazon Basin. This bird nests in trees overhanging water. When threatened by predator attack the chicks deliberately fall from the nest into the water. They can swim to the base of the tree and then use two small functional claws at the middle of each wing to clamber back up into the nest. They lose these claws as adults. The fingers are assuredly not a primitive feature derived from *Archaeopteryx* but rather a secondary derived character.

The second hypothesis about flight is that birds evolved from ground-dwelling ancestors. In this view feathers first evolved for temperature control, because the reptilian ancestor was becoming warm-blooded. These small reptiles ran along the ground and captured insects and other small animal prey. The feathers of the forelimbs became elongated and were used to sweep up and capture insects, much like a butterfly net. From this stage, the animals acquired the ability to jump into the air to capture prey, and the developing wings helped to stabilize the body and let it down gently back to the ground. Finally, they used the forelimbs as wings for flight. Vertebrate paleontologists continue to debate these two scenarios concerning the origin and evolution of flight in birds.

In 1990 another astonishing discovery of fossil birds was announced. A small, sparrow-sized bird of Jurassic age was found in China. This bird is about 10 million years younger than *Archaeopteryx*. It also has a considerably larger breastbone and thus was more clearly a good flyer.

After the Jurassic, a series of Cretaceous birds are now known. For many years the only Cretaceous birds were from the marine Niobrara chalk of western Kansas. These two birds, *Hesperornis* and *Ichthyornis*, were both water birds; one was small and ternlike, the other a larger diving bird with small wings. Bones of these birds were discovered in the nineteenth century, and for many years they were the only two Cretaceous birds known. In recent years at least four additional genera of birds have been found in the Niobrara chalk, and other Cretaceous birds have been found in several other areas. Cretaceous birds are now known from England, Texas, Alabama, Montana, Canada, Chile, and South Dakota.

During the Cenozoic Era birds are rare fossils. Their aerial habitat, generally small size, and thin, hollow bones all work against their preservation as fossils. Some of the most primitive and common fossil birds are the very large flightless birds commonly called ratites. These include living ostriches, emus, cassowaries, and others, as well as the large extinct dodos, elephant birds, and moas. There are twenty-seven orders of modern birds. The more advanced birds, excluding the ratites, can be divided into two main groups, land birds and water birds. Among the former, the most advanced forms are the small, perching songbirds, the passerines.

KEY TERMS

amniote egg	dinosaurs	Saurischia
anapsids	euryapsids	sauropods
Archaeopteryx	incus	scute
articular-quadrate bones	Karroo	squamosal
bipedal	K/T boundary	stegosaurs
Ceratopsia	malleus	synapsids
dentary	Ornithischia	temporal opening
diapsids	ornithopods	thecodonts
Dimetrodon	quadrupedal	theropods

READINGS

The books listed below include several recent books on dinosaurs. The literature on dinosaurs, both technical and popular, continues to grow at an astonishing rate, and is difficult to keep track of, especially much of the enormous literature aimed at children, much of which is derivative and repetitive.

Alexander, R.M. 1989. *Dynamics of Dinosaurs and Other Extinct Giants*. Columbia Univ. Press. 167 pages. This book indicates how the size, weight, running speeds, and strengths of dinosaurs can be estimated. It includes a section on the sounds that dinosaurs may have made. Other large animals including flying reptiles, marine reptiles, and giant birds and mammals are included, with simple mathematical analyses.

The Asteroid and the Dinosaur (videotape). 1981. WGBA, Ambrose Video. 60 minutes. Treats the extinction of dinosaurs due to asteroid impact.

Barthel, K.W., ed. 1990. *Solnhofen: A Study in Mesozoic Paleontology*. Cambridge Univ. Press. 236 pages. A modern study of this most important fossil area from which the oldest bird, *Archaeopteryx*, was collected. Includes many other exquisitely preserved fossils.

Benton, M.J. 1989. *On the Trail of the Dinosaurs*. Crescent Books. 143 pages. A brief, simple book on dinosaurs

Digging Dinosaurs (videotape). 1986. Centre Productions, Denver Museum of Natural History 12 minutes. A short videotape for children age 7 to 12.

Farlow, J.O. 1989. *Paleobiology of the Dinosaurs*. Geological Society of America Special Paper No. 238. 100 pages. A series of articles that focus on the biology of dinosaurs and their nesting, social behavior, diet, and temperature regulation.

Feduccia, A. 1980. *The Age of Birds*. Harvard Univ. Press. 208 pages. Excellent coverage of both fossil and modern birds, their ancestry, origin, and the evolution of flight.

Glut, D.F. 1982. *The New Dinosaur Dictionary*. Citadel Press. 286 pages. An illustrated definition of all genera of dinosaurs and the higher classification of dinosaurs.

Horner, J.R. 1988. *Digging Dinosaurs*. Workman. 210 pages. A personal narrative of important discoveries of dinosaurs and their nests and eggs in Montana.

McLoughlin, J.C. 1980. *Synapsida*. Viking. 148 pages. An advanced book on mammal-like reptiles.

Padian, K., and Chure, D.J. 1989. *The Age of Dinosaurs*. The Paleontological Society, Short Courses in Paleontology No. 2. 210 pages. A compilation of 16 articles by different authors, each an expert on some aspect of dinosaur paleontology. Especially designed as a sourcebook for college teachers.

Russell, D.A. 1989. *An Odyssey in Time: The Dinosaurs of North America*. NorthWord Press. 239 pages. An oversize book replete with many color pictures, especially of Cretaceous dinosaurs and Canadian specimens. Speculation on what dinosaur descendants, had they existed, might have been like.

Weishampel, D.B., and others. 1990. *The Dinosauria*. Univ. California Press. 730 pages. The most comprehensive and thorough modern treatment of dinosaurs. Quite technical and organized mainly by major groups, but includes essential information on evolution and paleobiology.

16
Advanced Consumers on Land: The Mammals

THE EARLIEST MAMMALS

The mammals evolved from the mammal-like reptiles that lived during the Triassic Period. Our fossil record for these earliest mammals is quite skimpy, consisting mainly of tiny teeth and jaw fragments. Only a few reasonably complete skulls or postcranial skeletons have been found that allow us to make a reasonable reconstruction of what these earliest mammals were like (Figure 16.1). The earliest mammals were all quite small, generally about the size of a mouse, and were very primitive, with a small brain and short legs compared to their advanced descendants. Mammals are clearly the predominant animals on land today, but it took them a long time after their appearance to reach this position of ascendancy. They were around for almost 100 million years before they became common, large, and diverse. During all of this time, the reptiles occupied dominant positions as the large herbivores and carnivores on land. Seemingly it was only after most of these large reptiles died out at the close of the Cretaceous that mammals were able to come into their own as a major element of terrestrial communities.

The Triassic record for mammals is very poor, consisting mostly of a few small molar teeth, except at the end of the period, when skulls and disarticulated skeletons appear. The Jurassic mammalian record is also quite fragmentary; these fossils are known from only a few sites. Mammals were still quite small, generally mouse- to rat-sized, during this time. Complete skeletons are unknown, isolated teeth and fragments of jaws being the most common fossils. Several different kinds

FIGURE 16.1
Reconstruction of an early
Mesozoic mammal, *Megazo-
strodon*. The animal was
about 3 inches long.

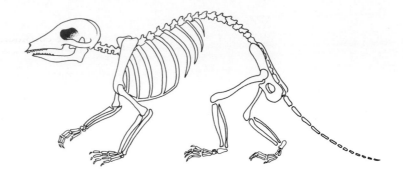

of mammals had already evolved by the Jurassic; however, all of the Jurassic
groups are now extinct, so we cannot say exactly what they looked like. Knowledge
of these animals is based mostly on their molar teeth, which are definitely
mammalian in character, but they are simpler and more primitive than are those
of most living mammals. These early mammals were clearly subordinate to reptiles.
They were probably small carnivores, insectivores, or herbivores. One group, the
multituberculates, were rat- to beaver-sized, with teeth reminiscent of those of
some living rodents. The other groups, called **pantotheres** and **triconodont
mammals**, were more likely small predators. They may have been warm-blooded
because of features we have already seen in their ancestors, the mammal-like
reptiles. They probably had hair, an insulating feature that probably first appeared
in mammal-like reptiles. Hair is a structurally new feature of vertebrates, unlike
scales in reptiles or the feathers of birds, and not derived from either.

An interesting problem concerns the mode of reproduction of these primitive
mammals. Three grades of mammals are alive today in which the method of
reproduction is quite different. The **monotremes** are quite primitive; they still lay
eggs but already have primitive mammary glands. Only two species are alive: the
spiny anteater (echidna) and the duck-billed platypus of Australia. Are these relicts
of primitive Mesozoic mammals? That is a difficult question to answer. Monotremes
have several characteristic skeletal features by which they can be identified, but
skeletons of Jurassic mammals do not show monotreme affinities. The critical
feature would be the molar teeth of living monotremes, because we know what
these structures are like in the fossils. Unfortunately, the molar teeth of monotremes
are degenerate, little more than featureless pads of enamel. Thus identification of
monotremes in the fossil record is almost impossible.

Another unusual mammalian group is the pouched mammals, or **marsupials**.
These can be recognized as fossils, but they do not appear until the Late Cretaceous.
Advanced mammals, the **placentals**, that give direct live birth, also first appear in
the Late Cretaceous. Thus, we have little evidence for their relationship to Jurassic
mammals.

Marsupials and Placental Mammals

By Cretaceous time mammals had undergone considerable evolution and diversification. Triconodonts survive into the Late Cretaceous but are rare. Pantotheres are present in the Early Cretaceous. The multituberculates continued to flourish. The most common Cretaceous mammals, however, were the marsupials. These were related to living opossums in North America and to the extensive marsupial fauna of Australia and South America—the kangaroos, wombats, koalas, and many others. Most of the Cretaceous marsupials were small; some bore an amazing resemblance to the living opossum. In addition, a few small primitive placental mammal fossils have been found. In placentals, which constitute most of the living mammals, there is considerable development of the young before birth. Placentals have well-developed mammary glands and many engage in extensive parental care of the young. In marsupials, the young are born at a very immature stage; they spend a long period attached to a teat in the pouch before they are sufficiently developed to emerge. Furthermore, the relative brain size of marsupials is smaller than that of placentals.

The oldest placentals found in Upper Cretaceous rocks are of a group called **insectivores**. As the name implies, the living representatives of this group mainly eat insects. Moles, shrews, and European hedgehogs are the best known living insectivores. These are among the most primitive living placentals. All of the Cretaceous mammals are still quite small, none being larger than a fox or beaver. Some were herbivorous, others undoubtedly insectivorous, and still others may have been small predators. The fossils of these mammals are rare. The molar teeth are the most important anatomical parts for study. By careful observation of the arrangements of cusps, ridges, and other features on the molars, it is possible to readily distinguish a marsupial molar from a primitive placental molar. In the Cenozoic, the evolutionary history of the mammals is most fully expressed in the evolution of the molars and other teeth. No other parts of the hard anatomy show such widespread and diverse changes as do these feeding structures.

CENOZOIC MAMMALS

We are now ready to examine the mammalian fossil record for the 60 million years leading up to the present day. We have seen that by the close of the Mesozoic reptiles no longer dominated terrestrial communities. They had been the main herbivores and carnivores, both large and small, for many millions of years. Just as soon as the dinosaurs and other reptiles became extinct, the mammals took over. This is one of the major examples of replacement of one large group by another shown by the fossil record. The mammals surely did not force out the reptiles or out-compete them. They played a waiting game, remaining in the background until the reptiles had had their day and many large reptiles had become extinct. Then mammals diversified very rapidly. Based on the fossil evidence, they became more abundant; there were more kinds of them living at one time than there ever were of reptiles.

It is convenient to divide the Cenozoic mammalian faunas into two major groups for discussion, so as to better understand their evolutionary history. During the early part of the Cenozoic Era, there was an **archaic mammalian fauna**. This fauna flourished during the Paleocene, Eocene, and Oligocene epochs of time. Most of the major groups of mammals that evolved during that time later became extinct, although there are still some vestiges of this old fauna alive today. Beginning with the Miocene and continuing up to today, we can think of a **modern mammalian fauna**, similar in many respects to living mammals.

The Archaic Mammal Fauna

Mammals are an extremely diverse group of animals. There are many different orders, suborders, infraorders, superfamilies, and families. In addition to their success on land, they, like the Mesozoic reptiles before them, have invaded the sea, in the form of whales and dolphins. And, like the pterosaurian reptiles, some, such as bats, acquired the ability to fly. On land we can recognize two major groups of mammals: the **carnivores**, or flesh-eating mammals, and the **ungulates**, the hoofed, herbivorous mammals. In addition, there are many smaller less diverse groups such as the primitive insectivores, the primates (lemurs, monkeys, apes), the subungulates (elephants and their relatives), the rodents, which are a very large group, the lagomorphs (rabbits and hares), and the edentates (sloths, armadillos, anteaters).

When we first get a look at Cenozoic mammals in the Paleocene, these various groups are far from being distinct. The distinction between flesh-eating and plant-eating mammals is most vividly seen in characters of the molar teeth. Many of the early mammals as yet show little specialization of these teeth. As we mentioned in considering Cretaceous mammals, the oldest and most primitive placental mammals are the insectivores, ancient relatives of shrews and moles. In the Paleocene, insectivores are still common. Most of the other mammals are also small, the largest being the size of a sheep. One of the most conspicuous groups consists of early carnivorous types called **creodonts**. The earliest ones are still very similar to their insectivore ancestors. During the Early Tertiary, these flesh eaters show a variety of specializations. Their front teeth (incisors) became bladelike for nipping and tearing flesh, and they retain their stabbing canines. Most important, a pair of upper and lower molars became increasingly larger and higher, and developed a bladelike edge. As these teeth sheared past each other, they could tear flesh and cut through sinew and bone. These specialized molars, called **carnassials**, became increasingly prominent in the jaws of carnivores during the Eocene and Oligocene. The earliest creodonts were relatively small animals with short legs. They probably were not very swift runners.

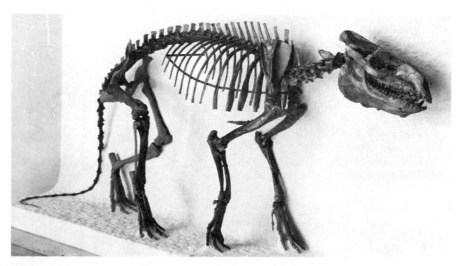

FIGURE 16.2

A mounted skeleton of an Oligocene artiodactyl, *Merychoidodon.* These primitive ungulates, called oreodonts, were relatively small and primitive in structure. The teeth were
not very specialized. Notice that there are four functional toes on each foot, although the
feet are partially raised off the ground. (Courtesy of National Museum of Natural
History.)

Although the creodonts were undoubtedly the dominant carnivores during
the Early Cenozoic, more advanced types of carnivores appear in the Paleocene
that eventually would replace them. These are the **fissipeds**, or split-footed
carnivores. The earliest creodonts were flat-footed; they walked around on the
entire soles of their feet. The fissipeds, on the other hand, developed somewhat
longer legs, and gradually the bearing surface of the feet was reduced to the
surfaces of the toes. By the Oligocene epoch, the creodonts had dwindled to a few
surviving stocks. Their place as the major predators on land was taken over by the
fissipeds. Two main groups of these advanced carnivores appeared in the Oligocene
and continue to the present day. One group includes the dogs, bears, raccoons,
weasels, and their relatives; the other consists of the cats, hyaenas, and the Old
World civet cats.

By far the most abundant and diversified mammals during the early part of
the Cenozoic were the **ungulates**, the hoofed, herbivorous mammals. Like the
carnivores, the earliest of these animals show little departure from their presumed
insectivore ancestors. They were mostly small in size, the body was relatively long
and slender, and the legs were short and little specialized (Figure 16.2). Ungulates
do show a somewhat more advanced foot structure than do the earliest carnivores
in that the former animals did not walk on the flats of their feet. Instead the feet

FIGURE 16.3
The largest mammals of the Eocene were the uintatheres, one of which is shown here as a mounted skeleton. These ungulates were heavily built, with short massive legs, peculiar bony knobs on the skull, and long defensive canine teeth. (Courtesy of National Museum of Natural History.)

were already raised up, so that only the lower surfaces of the toes were in contact with the ground. Their teeth are quite generalized initially, but soon begin to show adaptation to plant eating. The molar teeth become squared up, with a flat grinding surface for shredding leaves and other parts of plants.

The most common group of primitive herbivores is called the **condylarths**. They mostly fit the general picture of early herbivores as described in the previous paragraph. These may have been the dominant plant eaters during the Paleocene and Eocene, but they died out at the close of the latter epoch.

Other archaic ungulates included several groups that attained a rather large size quite early. One such group was the **pantodonts**; they were about the size of a sheep in the Paleocene but became cow-sized, about 8 feet long, in the Eocene. Some of these had strong claws instead of hooves, and long, heavy upper canine teeth. These are thought to have been specializations for a root-grubbing style of food gathering. Another group was the **uintatheres**, known mainly from Eocene rocks of the western United States (Figure 16.3). The name comes from the Uinta Mountains of Utah. Some of these animals were as large as a rhinoceros and are the largest archaic mammals known, apart from early toothed whales that were 25 meters long. They had peculiar bony swellings on the top and face of the skull, presumably for defense. The males had long, stabbing upper canine teeth. The animals had short, stubby legs to support their considerable weight.

A final group of primitive ungulates goes by the name **Notoungulata** (Figure 16.4). These animals were common very early in the Cenozoic. Their remains have been found in Paleocene rocks in North America and Asia. They soon became extinct in North America but persisted in South America. Apparently South America was in land communication with the other continents during the Paleocene, when the notoungulates moved into this area. Then the land bridge, the present-day Isthmus of Panama, was submerged. The notoungulates and other primitive mammals (especially marsupials) survived in South America in splendid

FIGURE 16.4
Side view of the skull and lower jar of a South American notoungulate, *Homalodotherium*, of Miocene age. The long battery of large grinding teeth indicates that the animal was a herbivore. (Courtesy of Field Museum of Natural History, Chicago.)

isolation up until fairly recent times. These primitive ungulates radiated into a host of different types. Some of them became quite large and were cattlelike in appearance. Others resembled horses and camels. Still others became rodentlike; some of these were very large, among them the bear-sized beavers. This is a clear example of parallel evolution, the notoungulates radiating into the great variety of habitats offered in South America. In the process they evolved various features of the skeleton (especially of the skull, legs, and teeth) that resemble, but are not related to, features that evolved in other groups of ungulates elsewhere.

In addition to these various groups of archaic ungulates that did not survive the Cenozoic, there were two other groups of hoofed mammals that originated quite early in the era and went on to become the dominant hoofed mammals alive today. These are the **Perrisodactyla** (odd-toed) and the **Artiodactyla** (even-toed) hoofed mammals. In the first group, the axis of the foot runs down the center of the foot, through the third digit (Figure 16.5). In the second group, the axis is between the third and fourth digits (your ring and middle fingers). Living representatives of the perissodactyls include the horse, tapir, and rhinoceros. The horse has a single (third digit) functional toe, the others have three toes. After the Eocene artiodactyls are considerably more diverse than are the odd-toed ungulates. Many have the number of functional toes reduced to two, the third and fourth, resulting in the split- or cloven-hoofed ungulates. Pigs, camels, sheep, goats, antelope, deer, and cattle are all examples of artiodactyls.

In the Early Cenozoic, the various groups of odd- and even-toed ungulates are already well differentiated and do not share a direct ancestor. One of the best known of these early mammals is *Hyracotherium*. This animal was about the size of a fox, and of slender build. It had three functional toes on each hind foot and four toes on each of the front feet. The teeth were low crowned and of a generalized plant-eating character. Although *Hyracotherium* is generally taken as the ancestral stock for all later horses, it is sufficiently generalized that it could also be considered

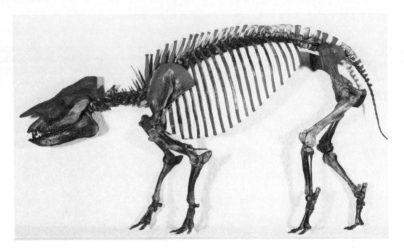

FIGURE 16.5

Skeleton of *Palaeosyops,* a primitive Eocene perissodactyl that was quite large compared to most early Tertiary mammals. This ungulate belongs to an extinct group of odd-toed herbivores called titanotheres, some of which attained much larger size than the specimen shown here. (Courtesy of National Museum of Natural History.)

the ancestor for all of the perissodactyls. It was surely a browser, living in the forest or on forest edges, nibbling low leaves from trees and shrubs.

Although they are mainly a Late Cenozoic assemblage, one final group of archaic mammals should be mentioned—the elephants and their relatives. These animals originated in Africa. The oldest and most primitive **proboscideans** (from *proboscis*, or trunk) are found in Eocene rocks in Egypt. As adults, these were about the size of a baby elephant. They had four short tusks, each a few inches long, two each in the upper and lower jaws.

In summary, we can now describe the general character of early Cenozoic mammalian communities. The primary consumers, or herbivores, included a wide variety of hoofed ungulates. Many of them were rather small by modern standards, generally smaller than a sheep, but a few, the pantodonts and uintatheres, reached the size of a cow, and brontotheres were rhino-sized. Most conspicuous of these were the condylarths. The notoungulates were largely confined to South America. Secondary and tertiary consumers were represented first by the creodonts and later by early members of the fissipeds. All of these carnivorous mammals gradually developed adaptations of the skeleton and teeth that made them progressively more effective predators during the Early Cenozoic.

In addition to these main groups, there were many other kinds of mammals that are generally represented by a somewhat poorer fossil record than are the principal groups of herbivores and carnivores. These include insectivores, bats, primates, a variety of rodents, and edentates (sloths and their relatives). The oldest known whale fossils are also found in marine rocks of Eocene age. Thus, there was a tremendous burst of radiation and diversification of mammals during the early part of the Cenozoic. Some of the habitats they occupied probably had been left vacant by reptilian extinctions in the Mesozoic; but other ways in which these early mammals made a living were undoubtedly new. The tree-dwelling habit, the fruit and flower diet of some primates, and the nut and hard seed diet of many rodents were adaptations that were not pursued by the reptiles. At least some of these new ways of life were directly related to the characteristics and expansion of the flowering plants during this same time interval.

Modern Mammal Fauna

In later Cenozoic communities, beginning with the Miocene Epoch, many of the mammal groups that had been conspicuous during the Paleocene, Eocene, and Oligocene had either become extinct or had dwindled to an insignificant role. In the northern hemisphere, almost 75 percent of the mammal families known from the Miocene still persist today. On a worldwide basis, this percentage is lower, about 50 percent, because of the larger number of endemic families, confined to isolated South America, which later became extinct.

The fissipeds were now the dominant carnivorous mammals. They underwent a conspicuous radiation during the latter part of the Cenozoic, which saw all of the modern kinds of carnivores appear. There was an abundance of dog or wolflike forms. Some of these were of large size, approaching small bears in length and weight. In addition to a variety of true cats, there were several different kinds of saber-toothed cats that were a conspicuous element of many faunas (Figure 16.6). They originated in the Eocene and continued into the Pleistocene Epoch when they became extinct, not very many thousands of years ago. They evolved long, stabbing, and slicing upper canines. The lower jaw was hinged so that it could be opened widely. The molars were virtually reduced to very large, shearing carnassial teeth. The limbs of advanced saber-tooths were massively constructed. These carnivores presumably preyed on huge mammals, perhaps mastodons and elephants. In addition to a variety of weasels, otters, and their relatives, bears became a conspicuous predator element in the later Cenozoic. A final group of carnivores to first appear was the marine predators—seals and walruses.

Among the hoofed mammals, the two dominant groups were the perissodactyls and the artiodactyls, the latter of which has tended to be more diverse than the former. In both groups, a number of common evolutionary trends can be seen (Figure 16.7). There is an overall tendency for the size of the animals to increase. This is accompanied by an increase in the length of the legs and a larger brain size and skull. The feet change their relationship to the ground so that only the lower surfaces of the digits have ground contact. The "palm" or "sole" of the foot is raised, serving to further increase leg length. In the odd-toed forms, this trend culminates in the advanced horses. Beginning in the Pliocene, only the third digit is functional in these animals, and the feet are raised so that the animal runs on

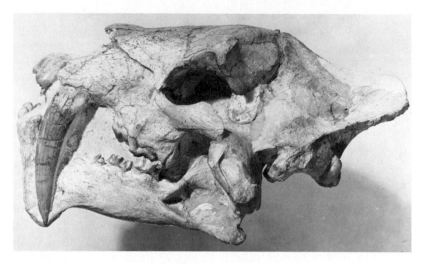

FIGURE 16.6
The skull and lower jaw of an Oligocene sabertooth cat, *Hoplophoneus*. Note the long, stabbing canine in the upper jaw, the bony projection on the lower jaw that helped protect the stabbing teeth, and the large, shearing carnassial teeth at the back of the jaw. (Courtesy of National Museum of Natural History.)

the tip of a single toe on each foot. In artiodactyls, this tendency results in only two functional toes on each foot (the third and fourth digits) and, again, only the tips of the digits touch the ground. These evolutionary features can be viewed as adaptations for greater speed to escape predators, which are partly a result of a major change in habitat from forest-dwelling browsers to plains-dwelling grazers.

Another trend in the hoofed mammals is for the jaws to become longer and the facial region of the skull to elongate. Accompanying this trend is progressive modification of the teeth. These become larger and more nearly square in outline. The premolars come to look much like the molars. There is an increasingly complex pattern of enamel on the grinding surface of the molars and premolars. Instead of being low, the teeth become high crowned, growing throughout the life of the animal. These changes result in a long battery of grinding teeth in both jaws that are very resistant to wear, especially from grasses high in silica content which comprised much of the plant life of expanding prairies (Figure 16.8).

Not all of the ungulates show such advanced features. Many forms remained browsers and did not develop all or any of the advanced features we have just listed. Among the perissodactyls, both the tapirs and rhinos tended to be much more conservative than the horses in evolutionary changes (Figure 16.9). Even among the horses, not all of them became herd-dwelling grazers; some still remained browsers and three toed.

Among the even-toed ungulates, deer, cattle, and camels showed the greatest increase in diversity in the later Tertiary. These are the ruminants, or cud-chewing

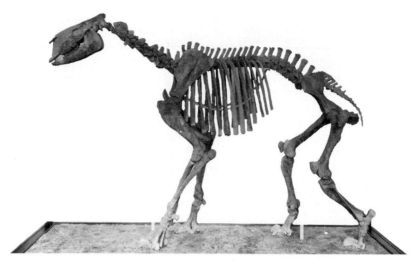

FIGURE 16.7

A Miocene perissodactyl belonging to an extinct group called chalicotheres. The animal, *Moropus*, was about the size of a small horse, but was heavily built and probably slow moving. This extinct form is unusual in that there are claws on the feet, rather than hooves. These are probably related to a root-digging habit. (Courtesy of National Museum of Natural History.)

FIGURE 16.8

Bottom view of the skull of a Pliocene horse that was slightly smaller than a modern horse. Notice the long battery of molars and premolars in the jaws; these teeth are high crowned and have complex patterns of enamel. There is a conspicuous gap in the tooth row between these grinding teeth and the nipping incisors at the front of the jaw. (Courtesy of National Museum of Natural History.)

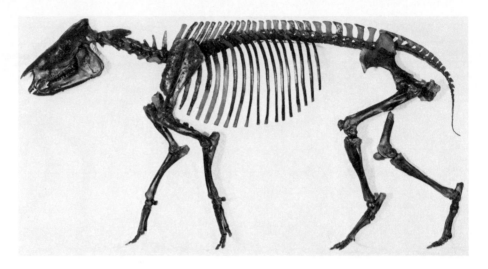

FIGURE 16.9
A primitive rhinoceros from the Eocene, *Hyrachyus*. The skull is about a foot long. Note the three-toed feet, typical of primitive perissodactyls, or odd-toed ungulates. (Courtesy of National Museum of Natural History.)

ungulates. The second stomach in these animals is an adaptation for eating the harsh grasses of extensive prairies that came into existence during this time. Swine and their relatives, the hippos, were important nonruminant groups of artiodactyls.

Elephants underwent a spectacular evolutionary history during the later Cenozoic. They increased tremendously in size, and their tusks became very long. Some had two lower tusks, some two upper tusks, and others had four tusks. The skull became high and the neck short to support the weight of the massive skull, trunk, and tusks (Figure 16.10). The legs became pillarlike to support the great weight of the animals. Two main groups can be recognized: the **mastodons** and the **mammoths** and elephants. The mastodons had a conservative tooth structure with several low-crowned molars in each jaw half. The elephants and their hairy relatives, the mammoths, had only four molars, but they were very large teeth and crossed by many ridges of enamel.

SOME ASPECTS OF SPECIAL INTEREST

Several aspects of mammalian paleontology during the Cenozoic are of special interest. These are (a) the existence of centers of evolution and migration from those centers of specific groups of mammals, (b) the isolation and eventual unification of South America with North America, and (c) the extinction within the last few thousand years of many of the larger mammals.

FIGURE 16.10
Skull and lower jaw of an advanced elephant, the mammoth, of Pleistocene age. Notice the high back part of the skull that supported massive neck muscles. A single very large molar tooth is present in each half of the upper and lower jaws. The snout contains a large cavity where a tusk was emplaced in life. (Courtesy of National Museum of Natural History.)

Centers of Evolution

Each of the major lineages of mammals initially evolved in a restricted area and then migrated away from that center of evolution to appear as fossils in widely separated parts of the world. For certain groups of mammals, we have good documentation for the place of origin and times of migration. For other groups, the fossil record is too poor for us to be certain about these biogeographical aspects.

Horses.   One of the best known of the Cenozoic mammalian records is that for horses. The oldest known horse, *Hyracotherium*, is found in both North America and Europe. These occurrences provide evidence for ease of migration between these two continents during the Eocene. During the Oligocene, there is a succession of horse genera found in North America. These horses show progressively more modern features, such as larger size and longer legs (Figure 16.11). This sequence is confined to North America. There were no horses in Europe during Oligocene time. From the Eocene up into the Pleistocene, there is an unbroken record of horse evolution in North America (Figure 16.12). Occasionally, one or more of these types of horses migrated from North America to Eurasia, probably via the Bering Strait, which was the major land bridge between the Americas and Eurasia during Cenozoic time. European horses can be viewed as a series of migrants with intermittent, local evolutionary lineages. During most of the Cenozoic, North America was clearly the center of horse evolution until the close of the Pleistocene, only a few thousand years ago. Horses were still in evidence when humans first

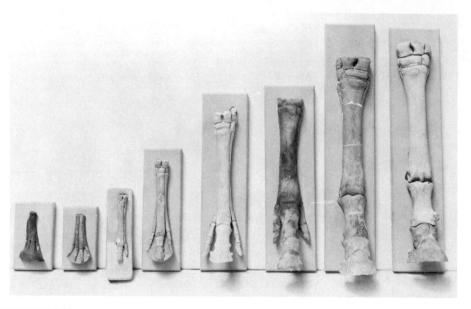

FIGURE 16.11
Evolution of the forefoot in horses. The two small, four-toed examples on the left are Eocene in age. The furthest left is *Hyracotherium*. The third from the left is Oligocene in age, the next two Miocene, and the right-hand examples Pliocene, Pleistocene, and Recent (from the modern horse, *Equus*). Notice the increase in size and length of the foot and the change from four to three to one functional toe. (Courtesy of National Museum of Natural History.)

invaded North America, again across the Bering Strait. Artifacts and fossil horse bones have been found together. Then, for some reason, horses became extinct in North America, although they continued to survive in Europe and Asia; in Africa they were represented by zebras. The first wild horses of the western states in modern times are descended from horses that escaped from the Spaniards during their early exploration of the Americas.

Camels.   A group of mammals with a pattern of evolution similar to that of the horses is the camels (Figure 16.13). They seemingly originated in North America where we find them in the Late Eocene. Again, there is a continuous sequence of fossil camels through the remainder of the Cenozoic in North America, until we come to the end of the Pleistocene. By this time, some of the North American camels had become very large, much larger than those living today. One group invaded South America during the Pliocene and survive today as the llama, but in North America camels became extinct at the close of the Pleistocene. They continue to survive in Eurasia in the modern Bactrian camel and dromedary

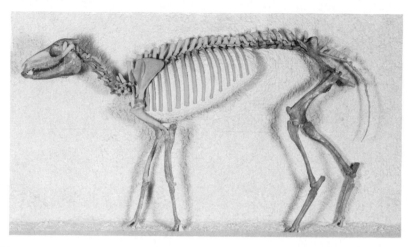

FIGURE 16.12
A mounted skeleton of the Oligocene horse *Mesohippus*. This horse was smaller than modern horses, with a shorter skull and less specialized teeth. There were still three functional toes on each foot, and the legs were not yet highly elongate. (Courtesy of National Museum of Natural History.)

FIGURE 16.13
A Miocene camel, *Stenomylus,* from Nebraska. Note the elongated vertebrae in the neck and the elongate legs. Camels are artiodactyls, and this specimen shows two functional toes on each foot. (Courtesy of National Museum of Natural History.)

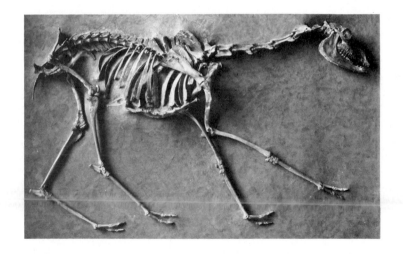

Elephants and Other Mammals.

Several major groups of mammals seem to have had a center of origin in Africa, from where they migrated north into Europe, east into Asia, and finally into North America. One such group is the elephants and their relatives. Mastodons are found in Africa in the Oligocene; by the close of the Miocene they are found in Europe and North America, where they persisted until their extinction at the close of the Pleistocene. Other groups that may have evolved in Africa, or at least in the Old World, and that never did migrate to North

FIGURE 16.14
Comparison of the molar
teeth of a mastodon, **A,** and
a mammoth, **B.** The mammoth tooth is considerably
larger than the mastodon
tooth, has many parallel
ridges of enamel, and occurs
singly in the jaw. The mastodon tooth, on the other
hand, occurs with one or two
other molars, has only a few
rounded cusps, and is
smaller. (Modified from Os-
born, 1940.)

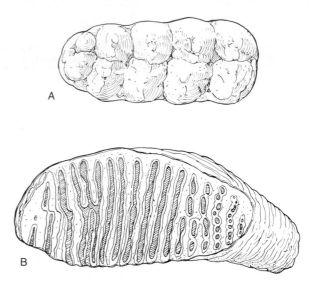

America include the giraffes, true hippos, and many of the African and Asian
antelopes, which are not related to the North American pronghorn.
    The distribution of large elephantlike animals during the Pleistocene in North
America forms an interesting pattern. First, there are two quite different kinds of
such large animals, the mammoth and the mastodon. Many people not familiar
with the paleontology of these animals tend to confuse these two early mammals.
The mastodon belongs to a different family from the mammoth, which is in the
Elephantidae and is closely related to the modern living elephants, which are
placed in two distinct genera; the African elephant is *Loxodon*, the Indian elephant
is *Elephas*. Perhaps one reason for the confusion is the similarity of the scientific
generic names of the two fossil animals; the mammoth name is *Mammuthus*, the
mastodon name is *Mammut*. The mastodon has relatively short, straight tusks, and
each jaw contains two or three rather small molar teeth at any given time. These
have a few rounded cone-shaped crests suitable for browsing. On the other hand,
the mammoth has highly elongate and strongly curved tusks and a single, very
large molar tooth in each half of each jaw (Figure 16.14). These molars are slowly
replaced during life, but there is only a single functional molar at any one time.
Each tooth has numerous cross ridges of enamel that provide a broad, hard surface
for shredding harsh vegetable material, especially grasses.
    Neither of these animals was present in North America until the very close
of the Pliocene Epoch, when they presumably crossed over the Bering land bridge
from Asia. The mastodon is represented by a single species, *M. americanus*. This
large beast was a forest dweller and lived primarily in the enormous forested areas
of eastern North America. Most of the fossil animals found in peat bogs, old river
sandbars, and other ice age deposits in the eastern United States are of mastodons.
Mammoth fossils are present but relatively rare in this area

In contrast the mammoths underwent considerable species evolution after they arrived in North America, although the precise taxonomy of these animals is in considerable disarray. A conservative view of their evolution would count three distinct species during the Pleistocene, rather than the sixteen species that were proposed by some earlier workers. The imperial mammoth is a very large animal found primarily in the southwestern and southeastern states. It does not range very far north. The Columbian mammoth, on the other hand, is a temperate animal and is found throughout the central United States. Finally, there is the wooly mammoth, the northernmost species that ranged from Alaska and Canada down into the Great Lakes region. These species are differentiated in minor details of the molar teeth, tusks, and skull.

Mastodons and mammoths, as well as a variety of other animals, have been found in the **Rancho La Brea** tar pits in downtown Los Angeles, one of the most famous fossil localities in the world. These pools of tar form by upward seeping of oil from buried sands. The oil loses light hydrocarbons that could be refined into gasoline, leaving behind heavy tarlike deposits that do not evaporate easily. Animals that stray onto the surface of the pools are caught in the sticky tar which cannot support their weight, so they are trapped and eventually engulfed in the tar. The tar is constantly churning as new material wells up from the depths, so there is no internal stratigraphy to the deposits. During the Ice Age thousands of animals were trapped and buried in these pools. Their bones are beautifully preserved. Mammals are the best known fossils from this locality, although birds, insects, reptiles, and other animals are also found. A single human (Amerindian) fossil is known. Interestingly, over 90 percent of the animals recovered, counted as individuals, are carnivores and only 10 percent are herbivores. The meat eaters were presumably lured onto the surface of the pool by other animals trapped in the tar. The predators in turn became trapped. The carnivores include sabertooth cats and a large extinct wolf, the dire wolf. Other flesh eaters found in fewer numbers include an extinct American lion, the mountain lion or cougar, bobcat, fox, coyote, weasel, badger, skunk, and short-faced bear. The herbivorous animals include both the mastodon and mammoth and the camel, bison, peccary, antelope, horse, tapir, and two kinds of sloth.

South American Isolation

We have already alluded to the isolation of the South American land mass very early in the Cenozoic. When this separation of North and South America occurred in the Paleocene, the principal mammals on the latter continent included a variety of marsupials and very primitive ungulate groups. None of the early carnivorous creodonts are known from South America. The endemic groups evolved and radiated into a variety of habitats. The hoofed mammals became horselike, rhinolike, rodentlike, and bearlike in many of their skeletal adaptations. Yet they also retained many very primitive features from their notoungulate ancestors. The marsupials likewise diversified, and they included most of the larger carnivorous

FIGURE 16.15
A mounted skull of a marsupial sabertooth cat, *Thylacos-
milus,* from Pliocene rocks of
South America. (Courtesy of
Field Museum of Natural History, Chicago.)

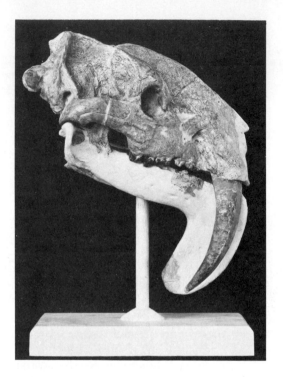

types in South America. One marsupial developed large, stabbing canines, like
those of the sabertooth cats of the rest of the world (Figure 16.15). Others came
to resemble dogs and wolves in structure.

All of this changed abruptly toward the close of the Pliocene Epoch, when
the present-day Isthmus of Panama was established. Now there was a two-way
highway for migration southward of advanced placentals and for northward
movement of endemic South American forms. The invasion of South America by
advanced carnivores, especially large cats, such as the puma, and dogs and wolves,
spelled the end for the various primitive hoofed mammals that had successfully
persisted in South America for so long. They all became extinct. Other groups,
such as the tapirs, the llama, and other advanced artiodactyls, probably competed
indirectly with the notoungulates for the same habitats. The marsupials, relatively
slow animals, with a smaller brain than that of advanced placentals, also became
extinct, except for a variety of "opossum." It was able to survive and become a
successful northerly migrant. Other endemic forms that were able to move
northward and compete successfully against their more advanced brethren included
the extant armadillo, the porcupine, and giant ground sloths, which survived in
North America until very recently.

A similar sort of situation is taking place today in Australia. In this case, the
continent was isolated from mainland Asia but shared land connections with South
America and Antarctica. The fauna included monotremes and marsupials. Only a

TABLE 16.1
Extinct Pleistocene mammals worldwide. (Compiled from Anderson, 1984.)

Groups	Total Known Genera	Extinct Genera	Extinct Families	Percentage of Extinct Genera
Marsupials	70	18	4	26
Insectivores	21	4	1	19
Edentates	40	32	4	80
Primates	27	11	0	41
Carnivores	63	12	0	19
Rodents	127	35	1	28
Lagomorphs	9	3	0	28
Litopterns*	2	2	1	100
Notoungulates*	3	3	2	100
Perissodactyls	12	6	1	50
Artiodactyls	103	45	0	44
Tubulidentates	2	1	0	50
Hyracoids	2	1	0	50
Sirenians	2	1	0	50
Proboscideans	8	6	3	75
Deinotherioids	1	1	1	100

* Endemic South American herbivore groups

few placental rodents (mice and rats) were apparently rafted across the straits that separated Australia from the mainland of Asia. Other placentals included bats, which were not as seriously affected by water barriers. With the coming of humans, the dog (dingo) was introduced by the aborigines, and Europeans brought dogs, cats, rabbits, and additional rodents. These have been competing very successfully with the ancient marsupial fauna, and some of the native species have already become extinct or are very rare. Without proper conservation and care, Australia may eventually become a second South America, with almost total loss of its primitive fauna.

Mammalian Extinctions

During the entire Cenozoic, mammals tended to evolve quite rapidly (Table 16.1). There are many genera and families of mammals which became extinct during the course of the last 60 million years. Yet one of the most significant waves of extinction occurred almost yesterday, within the past few thousand years. These extinctions are puzzling for two reasons. First, mostly large mammals became extinct. Smaller relatives of these animals survive today. Second, all of the mammals that became extinct at this time had survived through four major advances and retreats of continental ice sheets during the Pleistocene Epoch. The last million years or so of earth history has probably been as variable with respect to world temperature and climate as has any time in earth history. Immense ice sheets covered the northern part of North America, Europe, and parts of Asia. The Greenland and Antarctic ice caps are the dwindled remnants of these icy periods. The final glaciers, called the Wisconsin glaciation in North America, retreated about 10,000 years ago. It was after that retreat that the extinctions of large mammals, especially in North America and Europe, began to occur. On the North American continent, from 30,000 to 5,000 years ago, the mastodon and wooly mammoth became extinct, as did the imperial elephant, a camel, the horse, a giant beaver, large ground sloths, giant bison, the dire wolf, sabertooth cats, large deer, and moose. The comparable situation in Europe saw the extinction of large cave bears, wooly rhinos, and the Irish elk. In many respects it is fair to say that the living North American native mammalian fauna has low species diversity, especially of larger forms. These extinctions did not seem to affect the African faunas nearly to the extent they did those of the northern hemisphere. The currently rich mammalian faunas of Africa may be indicative of what the North American fauna looked like before this wave of extinction took place.

It is difficult to ascribe these extinctions to climatic change. These animals lived through four major glaciations and three interglacial periods when the climate was at least as mild as it was 30,000 years ago. Some scientists have argued that early humans hastened the demise of these large animals by selectively preying on large forms. It is not yet clear whether the timing of the extinctions and of the arrival and spread of humans in North America is sufficiently close to allow this as a viable hypothesis. Certainly humans lived in Eurasia long before the extinctions began to take place there. It is also difficult to imagine that the relatively small populations and crude hunting methods of the human beings who initially colonized North America would have placed sufficient hunting pressure on these animals to directly result in their extinction. Prior to the invasion of North America by the Europeans, the American Indians seemingly lived in balance with the mammalian fauna, without causing undue predatory pressures. Thus, this time of extinction, like that at the close of the Mesozoic and the Paleozoic, raises many questions that we cannot yet answer satisfactorily. Both biological and physical factors may have been involved, including the crucial dependence of secondary consumers on specific primary consumers. The loss or depletion of just a few large herbivores may have triggered waves of loss throughout the larger carnivorous mammals.

COMPARISONS AND CONTRASTS

In summing up the record of mammals during the Cenozoic, it is worthwhile to make some comparisons and contrasts between the record of land animals and that of the reptiles in the Mesozoic. In the first place, we can see a similar proportion of primary and secondary consumers in the two faunas. Among the mammals there are many more kinds of ungulates and other herbivorous types than there are carnivores. The same statement can be made about the dinosaurs. Only one group of dinosaurs was carnivorous—the carnosaurs. All the other saurischians and all of the ornithischians were herbivores. It is quite clear that among Mesozoic reptiles and Cenozoic mammals, herbivorous types were much more diverse than carnivores in terms of total diversity, and very probably they also were more diverse at any one moment in geological time. This is in accord with our ideas about food pyramids. Just as it takes a large amount of plant material to sustain one herbivore, it takes a number of herbivores to keep one carnivore going. The basic aspects of community structure among Mesozoic reptilian communities were probably comparable to those of Cenozoic mammalian communities.

Mammals cannot be considered in isolation from other dominant groups of land life in the Cenozoic. If we could take a stroll through the woods at this time, it is clear that three groups would have made up the great majority of animal life that we would see and hear. Except for an odd snake, turtle, worm, or frog, we would see or hear mammals, birds, and insects. We have already discussed the impact of the flowering plants on the diversification of mammals, especially on the various groups of forest- and plains-dwelling ungulates. It seems very unlikely that without the evolutionary diversity of flowering plants there would be anything like the great number of different kinds of mammals that are alive today or that thrived in the past. The increase in the number of mammalian primary consumers has probably also had a strong influence on the variety of mammalian secondary consumers or carnivores.

Among the insects, many are especially adapted to feed from the leaves, flowers, seeds, wood, or other parts of flowering plants. Even though insects evolved back in the Paleozoic and have always been successful animals, they undoubtedly underwent an almost explosive radiation as they began to exploit the increasing number of plants that evolved during the Cenozoic. Even the few insects that do not relate directly to plants, such as mosquitoes and fleas, are intimately associated with warm-blooded mammals, which, in turn, are primarily or secondarily dependent upon angiosperms.

The tremendous variety and abundance of birds is also closely related to the success of the angiosperms. Early birds of the Mesozoic, of which we have a very scanty record, were probably all animal eaters. They fed on insects and other arthropods; fish; and small reptiles, amphibians, or mammals. They led a life comparable to that of shore and water birds, hawks, and vultures today. The great increase in variety of so-called song birds, most of which have plant-related diets, is probably a direct consequence of the dominance of the flowering plants. The greatly varied seed-eating birds all depend mainly on flowering plants for their food, although a few do feed on conifer seeds, and others, such as hummingbirds, feed directly from flower nectar. Even woodpeckers, which are primarily insect eaters, will eat seeds during adverse conditions.

Thus, the successful adaptation and proliferation of the flowering plants during the gradually changing conditions of the Cenozoic provided the prime impetus for a whole series of complex changes that ultimately led to the natural world we live in today.

KEY TERMS

artiodactyls	mammoth	pantotheres
carnassials	marsupials	perissodactyls
carnivore	mastodon	placentals
condylarths	monotremes	proboscideans
creodonts	multituberculate	Rancho La Brea
fissipeds	notoungulates	triconodonts
insectivores	pantodonts	ungulates

READINGS

Kurtén, B. 1988. *On Evolution and Fossil Mammals*. Columbia Univ. Press. 301 pages. A series of essays on fossil mammals, especially those from the Cenozoic in Europe.

Kurtén, B. 1986. *How to Deep-Freeze a Mammoth*. Columbia Univ. Press. 121 pages. A collection of short essays on various aspects of fossil mammals and the Ice Age.

Lillegraven, J.A., Kielan-Jaworowska, Z., and Clemens, W.A., eds. 1979. *Mesozoic Mammals*. Univ. of California Press. 311 pages. A technical book with chapters by various authors.

Savage, R.J.G. 1986. *Mammal Evolution: An Illlustrated Guide*. British Museum of Natural History, Facts on File. 259 pages. An oversize book with many illustrations of bones and reconstructions, many in color. Little coverage of human evolution.

Sutcliffe, A.J. 1985. *On the Track of Ice Age Mammals*. British Museum of Natural History. 224 pages. An excellent, lavishly illustrated book on Pleistocene mammals worldwide.

17

Primate and Human Evolution

Of all the kinds of fossils that are found, from tiny one-celled algae to giant dinosaurs, none are as fascinating as the fossils of human and human-related animals. Human or near-human fossils are exceptionally rare and are found in remote parts of Africa. The amount of money, time, and effort that is needed to find these remains far exceeds the expenditures for any other kinds of fossils. How do these ancient human and humanlike remains fit into the fossil record of life?

Our species name is *Homo sapiens*. The generic name *Homo* means man, and the trivial name *sapiens* means wise. We belong to an order of mammals named **Primates**, meaning first. How do we decide whether a specific fossil should be included in the genus *Homo*? If the fossil is included in *Homo*, then should it be included in our species, *sapiens*, or in another species? How does the paleontologist go about answering these questions? The following criteria are now generally agreed upon to be important in making distinctions at the family, genus, and species levels in *Homo*like fossils. The family Hominidae, which includes humans, is characterized by upright posture. The forelimbs are not used for locomotion and, hence, are usually shorter and less robust than the hind limbs. In addition, the teeth are characteristically small and are all about the same size, in contrast to the larger teeth of apes and monkeys which have larger incisors and canines than their other teeth (Figure 17.1). The teeth of apes and monkeys, when viewed from above or below, make a U-shaped pattern, rather than a rounded, semicircular outline as in humans. In lower primates the teeth of males, especially the canines, are commonly larger than in females. These sexual differences cease to exist in

FIGURE 17.1
Comparison of the upper
jaws of a human and an ape.
Note the rounded, semicircular outline and the same
small teeth of the human jaw
compared to the *U*-shaped
jaw with large canine and
front incisor teeth of the ape.

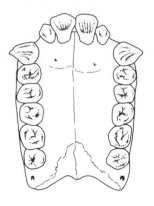
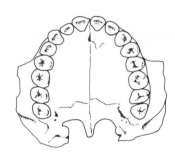

humans. Two genera are recognized in the family Hominidae: *Australopithecus*, which is a genus generally recognized as the ancestor of humans, and *Homo*. These two genera are differentiated by brain size. The several species of australopithecines characteristically have a brain volume of between 600 and 800 cubic centimeters. Within the genus *Homo* there are three species, *H. sapiens* with a brain size of 1200 to 1800 cubic centimeters and two extinct species, the older *Homo habilis* with a brain 700 cubic centimeters in volume and the *H. erectus* with a cranial capacity of 800 to 1200 cubic centimeters. Thus, we see a progressive increase in size of the brain within this group (Figure 17.2). The enlargement especially affects the forebrain or cerebral cortex that is concerned with observations, memory, and comparisons.

Recent computer-assisted analyses of the shape of the palate of fossilized skulls indicate that *Australopithecus* could not pronounce all of the sounds that we can and had limited speech. In some European caves the bones of **Neanderthal** man, now recognized as our own species, are found near splendid color cave drawings called pictographs. We assume that this artwork was associated with the skill of oral communication.

The most common skeletal parts of fossil **hominoids** are the teeth, just as in other fossil mammals. Primate teeth, especially the molar teeth, are useful in classification and study of these fossils, including humans. Primates generally have quite generalized teeth that can reveal whether animals are insectivorous or fruit- and leaf-eating herbivores. Insectivorous primates are generally quite small and weigh a pound or less. The herbivorous primates generally weigh 5 or more pounds. In the most evolved primates, chimpanzees and humans, the molars reflect an omnivorous diet, including both plants and flesh.

One of the important characteristics that we always want to determine from hominoid fossils is whether the animal could walk upright or whether the animal normally used the forelimbs for locomotion. Freeing the forelimbs from a role in walking, running, or swinging through trees was an important initial step in use of the arms to manipulate tools. It is now known that at least one living ape, the chimpanzee, uses twigs or grass stems as simple tools to accomplish specific tasks.

FIGURE 17.2
Side views of skulls of various primates discussed in the text. **A.** *Notharcts*, a primitive Eocene primate, skull 3 in. long. **B.** *Aegyptopithecus*, the oldest known ape, Oligocene in age. **C.** *Proconsul*, the Miocene stem of hominoid apes. **D.** *Australopithecus*, the oldest hominoid. **E.** *Homo*, a modern human.

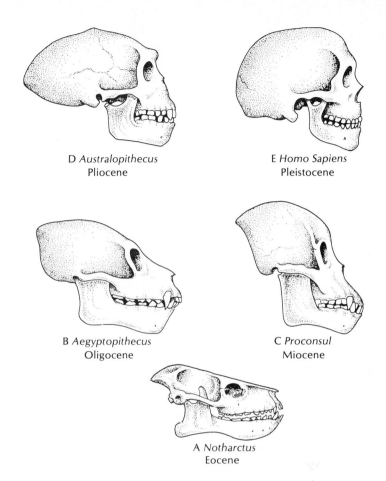

D *Australopithecus*
Pliocene

E *Homo Sapiens*
Pleistocene

B *Aegyptopithecus*
Oligocene

C *Proconsul*
Miocene

A *Notharctus*
Eocene

At one time, the ancestors of humans surely went through comparable stages of tool development. Even with fragmentary fossil bones of the limbs and the shoulder and hipbones, it is possible to tell whether a specific animal had an upright stance. For many years, it was assumed that development of an upright stance for walking and use of simple tools would have come after conspicuous enlargement of the brain. We now know that this is not the case. The remains of small, upright-walking *Australopithecus* suggest that this primate had a remarkably small brain case, no larger than that of a chimpanzee. The brain enlarged presumably in response to the changes in the way of life that accompanied development of walking.

The first worked tools that primates used, insofar as we can tell from the fossil record, were smooth rounded stream cobbles. These were used by hunters as hammers to break bones and extract the marrow of animals that had been killed for food. Such cobbles are called **eoliths**, or dawn stones. They can be recognized by typical wear patterns on them that are the result of repeated pounding. These eoliths have been found in sediments that have been dated to be 2.1 million years old and were used by *Australopithecus* long before *Homo* evolved. They greatly antedate any other evidence of humans, for instance, evidence of fire.

THE PRIMATES

The order of mammals to which humans belong, the Primates, is one of the longest surviving and most generalized groups of mammals. The order first appeared in the Eocene and continues to the present day, represented by about 200 living species. The order is divided into two main groups or suborders.

The first group of primates is the suborder Prosimii; these are the **prosimians** or premonkeys. This suborder consists of three groups: the living lemurs and the tarsiers and lorises, as well as the extinct **adapids** in the Eocene. The living prosimians are all small, tree-dwelling animals of India, Madagascar, and southeastern Asia. The lorises and **tarsiers** are insectivorous and the considerably larger lemurs are fruit- and leaf-eating. Most of these have a partly opposable thumb and, hence, have grasping hands. The muzzle or snout is greatly shortened, resulting in a flattened face. This indicates a decrease in the sense of smell and increased emphasis on sight. The eyes face forward rather than to the side, and thus, stereoscopic or three-dimensional vision is possible. Unlike many other mammals, the primates can also see in color. All of the primates mentioned so far have either thirty-eight or thirty-six teeth, differing from humans and apes with thirty-two teeth in having three rather than two premolars in each half of the upper and lower jaw. Lemurs and tarsiers are first known from the Early Eocene.

The second group of primates is the suborder Anthropoidea which includes the monkeys, apes, and humans. This group apparently evolved from a prosimian stock in the Late Eocene or Early Oligocene Epoch. Two major groups of monkeys are the Old World monkeys of Africa and Asia, which have thirty-two teeth and lack a prehensile tail, and the New World monkeys of South and Central America, which have thirty-six teeth and have such a tail. Other important differences between these two groups indicate that the New World monkeys diverged early from the main line of anthropoid evolution and have existed in isolation for a considerable span of geologic time. Fossil records for both groups begin in the Oligocene Epoch.

The Superfamily Hominoidea includes five living genera in addition to humans. These are the chimpanzee, gorilla, orangutan, and two genera of gibbons. In addition, the extinct genus *Australopithecus* is judged to be the ancestor of humans. These are all placed in three families in the superfamily. In addition, many other exclusively fossil genera, sixteen genera in total, are considered to be ape or near-ape hominoids. These range back into the Oligocene Epoch when they are represented by *Aegyptopithecus* (Figure 17.3).

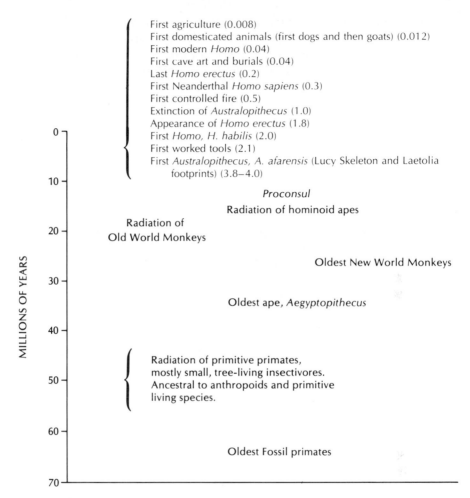

First agriculture (0.008)
First domesticated animals (first dogs and then goats) (0.012)
First modern *Homo* (0.04)
First cave art and burials (0.04)
Last *Homo erectus* (0.2)
First Neanderthal *Homo sapiens* (0.3)
First controlled fire (0.5)
Extinction of *Australopithecus* (1.0)
Appearance of *Homo erectus* (1.8)
First *Homo, H. habilis* (2.0)
First worked tools (2.1)
First *Australopithecus, A. afarensis* (Lucy Skeleton and Laetolia footprints) (3.8–4.0)

Proconsul
Radiation of hominoid apes
Radiation of Old World Monkeys

Oldest New World Monkeys

Oldest ape, *Aegyptopithecus*

Radiation of primitive primates, mostly small, tree-living insectivores. Ancestral to anthropoids and primitive living species.

Oldest Fossil primates

MILLIONS OF YEARS

FIGURE 17.3
Major elements of the evolution of primates during the Cenozoic Era. The time scale at the top of the figure has been expanded to accommodate very recent events within the past 4 million years.

The following characteristics are now generally agreed to differentiate humans, *Homo*, from other genera in the family. The family Hominidae is characterized especially by upright, bipedal (two-footed) locomotion. Within the family, *Homo* is characterized by a brain cavity of 800 cubic centimeters or larger. The only other genus currently placed in the family, *Australopithecus*, has a brain capacity of 600 to 800 cubic centimeters or less. Our species *Homo sapiens* is separated from other species in the genus by having a brain size of at least 1200 cubic centimeters. Other species, *Homo habilis* and *H. erectus*, have brain sizes of 700 and between 800 and 1200 cubic centimeters, respectively. Thus, the size of the brain cavity in the skull of fossilized hominoids is especially crucial in determining relationships. If at least fragments of the skull are not found, then doubt remains as to the proper familial, generic, and species relationships.

Modern studies of the biochemistry of living anthropoids, including comparisons of their proteins, indicate clearly that the two African apes, especially the chimpanzee and to a lesser extent the gorilla, are much more closely related to humans than are the Asian apes, the orangutan, and gibbons.

THE FOSSIL RECORD OF PRIMATES

The oldest fossils of primates are found in Eocene rocks. The fairly common adapids are aboreal, insectivorous, and fruit-eating and are about the size of a mouse or woodchuck. The adapids include the ancestors of later primates as well as many rodentlike forms, eleven of which have been named. These include the ancestors of lemurs, tarsiers, and lorises. Some of these were larger plant eaters and arboreal tree dwellers, weighing up to 5 pounds. This group of primitive primates became extinct by the end of the Eocene Epoch. Little is known about the fossil record linking these early primates to their living descendants. Rare fossil lorises are known from the Miocene to the Recent.

We turn now to the primate group of central interest to us, the hominoids. These can be divided into two distinct groups at the beginning of the Oligocene Epoch, the New World monkeys on the one hand and all other hominoids on the other. The American monkeys are judged to have had Old World ancestors but are not thought to be on the main line of hominoid evolution. Their tooth arrangement was primitive, but they did have prehensile tails that are lacking in old world monkeys.

At the base of the hominoid radiation is the Egyptian Oligocene genus *Aegyptopithecus*. This animal weighed up to 10 pounds and was a fruit-eating tree dweller that exhibited distinct hominoid features. It clearly was an ancestor of early apes of the Miocene such as the *Proconsul* (Figure 17.3) and, at the same time, was structurally similar to Old World monkeys. The features that separate monkeys and apes developed progressively with time as these two groups evolved during the later Tertiary. A variety of primitive apes evolved during the Miocene and Pliocene, both in Africa and in Eurasia. From this group, in the late Cenozoic, the direct ancestor of humans, *Australopithecus*, ultimately evolved.

The fossil record of primitive apes extends from 23 to 14 million years ago in eastern Africa and from 15 to 10 million years ago in Asia. Between 7 and 4 million years ago, there are almost no hominoid fossils known. Filling in this gap is a challenge for anthropologists and paleontologists. The fossil record picks up again in East Africa with *Australopithecus afarensis*, known from 4.0- and 3.8-million-year-old sites. This species was fully upright, as is demonstrated by the remarkable bipedal trackways found at Laetoli, Tanzania by Mary Leakey, and by the nearly complete skeleton of Lucy.

At least four species of *Australopithecus*, which disappeared about 1 million years ago, are known. These species all have larger brains than apes, reduced canine teeth, and show little evidence of sexual dimorphism. *Australopithecus* (Figure 17.3) soon evolved into two distinct lineages. One of these was a heavy-bodied large type that had especially large molar teeth. This kind of australopithecine is considered to have been a herbivore, using the heavy molars to shred coarse vegetal material. There are two species of this robust kind of australopithecine, both of which became extinct without descendant species; these are *Australopithecus boisei* and *A. robustus*. The other kind of australopithecine was smaller, with thinner bones, a more slender body, and smaller molar teeth. This is called the gracile type of *Australopithecus* and is represented by *A. africanus*, which is thought to be the direct ancestor of our own genus, *Homo*. Because the oldest remains of humans are found in rocks that are 2.0 million years old, it is clear that *Homo* and australopithecines coexisted for about 800,000 years in eastern Africa. This oldest species of humans, *Homo habilis*, was soon replaced by *H. erectus*, the remains of which were found in rocks 1.8 million years old. This species had a brain size of up to 1000 cubic cm, had humanlike dentition, and was a tool user. All of these australopithecine and hominoid fossils are found so far only in Africa. About 1 million years ago *Homo erectus* migrated out of Africa into Asia where fossils have been found in China and Java. This species clearly evolved directly into our present species *Homo sapiens*. The oldest identified fossil of our species is about 300,000 years old and consists of archaic Neanderthal types with stocky builds, heavy brow ridges, and receding chins. More modern individuals of our species date back about 40,000 years ago.

Despite general agreement on the overall nature of human evolution, many details are still unknown or in dispute. Gaps in the fossil record remain to be filled just as in any other field of science. These gaps should not be viewed as negating the enormous amount of detailed information that has been gathered concerning the fossil record of humans, human ancestors, and human relatives. It is clear that hominoids originated in Africa and evolved there in isolation for a considerable period of time. This evolution was not a simple, straightline evolution but included many side branches that ultimately became extinct without leaving descendant forms. Humans spread from Africa into Asia and Europe about a million years ago and then evolved into our present species. Two waves of modern humans colonized the Americas, first from Asia and then more recently from western Europe.

THE HUMAN CONDITION

We have inherited a complex mixture of physical traits from primate ancestors as well as newly acquired unique features that characterize us as a species. In some aspects the human skeleton is quite primitive and generalized. The presence of five digits on each limb, the number of bones per digit (two in thumb and big toe, three in other digits), and the large number of teeth are all characteristics that we share with primitive mammals. Most advanced herbivorous and carnivorous mammals have evolved in a reductionist way from this state, by decreasing the number of teeth and digits.

Some features of humans are related to an arboreal habit and an early insect diet. The opposable thumb and related manual dexterity, flat nails instead of claws, stereoscopic vision, and reduction in reliance on the sense of smell are probably all related to an early insect diet. An upright posture, freeing the forelimbs first for catching insects and climbing, then for grasping tools, and finally for caring for young is again related to a tree-dwelling, insect-eating lifestyle. Thus, we carry with us abundant indications of our primate ancestry and relationship. The increasingly large forebrain of primates that controls the ability to remember, observe, and compare was surely important in insect catching and tree climbing. It led to intelligence, language, art, culture, and social bonds that are unique to our species.

Many of these important human traits cannot be studied through the fossil record because they are not preservable in the ordinary sense that teeth, bones, or stones used for tools are. The oldest tools predate *Homo sapiens* by about 2 million years. *Australopithecus* clearly could control fire. Primitive Neanderthal man certainly had some kind of language, practiced a high level of art, had elaborate burials, and surely had religious customs. Modern humans first controlled the growth of grasses for grains and domesticated sheep and goats.

The unique aspects of humans relate largely to their elaborate social behavior. The underdevelopment of human young at birth is compensated for by long periods of postnatal parental care. Families, clans, and societies are more complex than the social structures of any other social animals, such as apes, birds, or insects.

Some scientists argue that humans have evolved so extensively and rapidly in a social sense that ordinary physical evolution, as we know it for all other animals, has ceased to be meaningful to us. There are at least hints that this is not the case and that we do continue to evolve physically, given that we have only a very short time scale on which to observe such evolution. Two examples of possible physical evolution are cited here. An increasingly large number of people never develop their third molar or wisdom teeth. We may ultimately have only twenty-eight teeth rather than the thirty-two characteristic of all other hominoids. A second example involves human birth. The underdevelopment of human infants at birth may be at least partly related to the relationship between the large brain of adult humans and the size of the pelvic birth canal in human females. The fact that newborn humans' skull bones are soft, flexible, and unfused indicates that the head and brain size is at the limit for passage through the female pelvis. With increasing numbers of cesarean births we may be selecting in favor of young with larger heads, so that ultimately natural childbirth will be possible only in a few instances. This selective process may ultimately lead to increasingly large-headed, larger-brained individuals and loss of the pelvis as a viable passageway for birth.

Such physical evolutionary changes in humans are at least conceivable in the future.

KEY TERMS

Aegyptopithecus

Australopithecus

eolith

Homo

hominoids

lemurs

Neanderthal

primates

prosimians

tarsiers

READINGS

Badgley, Catherine. 1984. "Human Evolution." In R.A. Gastaldo, ed., *Mammals: Notes for a Short Course*. Studies in Geology 8. Univ. of Tennessee Dept. of Geological Sciences. Pages 182–199. A summary of the evolution of humans and their near relatives.

Campbell, B.G. 1988. *Humankind Emerging*. Scott, Foresman. 522 pages. A popular book, based in part on the Time-Life series on physical anthropology, prehistoric man, and human evolution.

De Rousseau, C.J. 1990. *Primate Life History and Evolution*. Monographs on Primatology No. 14. Wiley-Liss. 366 pages. A compilation of articles on primates based on a conference held in 1987.

Feder, K.L., and Park, M.A. 1989. *Human Antiquity: An Introduction to Physical Anthropology and Archaeology*. Mayfield Publishing Company. 412 pages. An up-to-date college-level textbook in physical anthropology.

Johanson, D.C. 1989. *Lucy's Child*. Morrow. 318 pages. A contemporary account of recent developments in human paleontology in East Africa by one of the principal participants.

Lewin, R. 1987. *Bones of Contention: Controversies in the Search for Human Origins*. Simon and Schuster. 348 pages. A well-written survey of modern research on the origin of humans and human evolution including a discussion of the divergent theories of active researchers, written by a well-known science writer.

Martin, R.D. 1990. *Primate Origins and Evolution*. Princeton Univ. Press. 804 pages. A very detailed and technical book on all of the primates, with emphasis on living primates.

Mysteries of Mankind (videotape). 1988. National Geographic Society. 60 min. Anthropological research on ancient man in Africa.

Poirier, F.E. 1990. *Understanding Human Evolution*. Prentice-Hall. 356 pages. A college level text in anthropology.

18
Conclusions

We have now looked, in a broad and elementary way, at the history of life on earth. We have seen that life has been present here for more than 3 billion years. During this enormous span, there has been a whole series of startling changes in the nature and diversity of life. We can contrast a Precambrian vista, where only photosynthetic cyanobacteria and bacteria existed in the oceans, with the modern scene dominated by flowering plants, insects, and mammals. The strategies for making or catching food have undergone conspicuous change, as have the dominant groups that engaged in one form or another of food procurement.

The reader should now understand that paleontology is almost a perfect blend of geology and biology. Both the physical and organic aspects of earth history must be taken into account if we are to arrive at a rational view of life on earth. If geological occurrences of fossils are given disproportionate emphasis, exciting biological concepts that could be applied to fossils may be overlooked. If fossils are considered strictly from a modern biological viewpoint, inaccurate judgments about past physical conditions, or even the relative ages of the fossils, may result. Thus, paleontology is interdisciplinary in the broadest possible sense.

It should be clear that there is much more to paleontology than simply describing and documenting new species of fossils that someone happens to find. Synthesis of a wealth of existing data is required in order to reduce the information we have to manageable form, suitable for testing important hypotheses. The literature on fossils is so vast that no one today can possibly master all of it. Most paleontologists compromise by becoming specialists in one or two groups or ages of fossils. The important thing is not which group of fossils is studied, or how large or complicated it is, but rather how the study of fossils is viewed. The right approach in considering any group of fossils can shed new light on important problems if the paleontologist is aware of such problems, asks the right questions about them, and tries to find viable solutions. Paleontology is an exciting and challenging field. There are still many important questions, as yet without satisfactory answers, that will undoubtedly be answered in time.

What will the future bring for paleontology? Of course, we cannot predict with absolute certainty. But current research always points to the way in which future research will proceed. Many of the important recent developments in paleontology have come from several sources. First is the application of current biological concepts to the fossil record. For instance, ideas developed in the study of modern populations and communities of plants and animals have been successfully applied to fossils. Such new ideas that derive from studies of living organisms ordinarily cannot first come from the paleontologist. He or she should not be slow, however, in exploring the applicability of new ideas to fossils. Thus, to a certain extent, we are dependent on other scientists for ideas that hold promise in the paleontologic field. Equally important are new techniques. Various kinds of equipment that are developed may often have important applications in paleontology, leading to the creation of new areas of study. The use of the computer to handle large amounts of data or to make many computations, of the scanning electron microscope to elucidate fine structural details of fossils, of new methods of dating rocks, and of isotopic methods of determining ancient ocean temperatures—all of these have significant application in paleontology. The trick is, of course, to recognize the important problems that new instruments or techniques may help solve. If such new devices are used to attack mundane problems or simply to gather reams of new data, they are not going to be of very much consequence.

To some extent, paleontologists are limited by the vagaries of fossil localities. A new locality yielding unexpected fossils may shed considerable light on an important problem. Many of the Precambrian fossils now known were not known earlier because we did not know how to look, where to look, or what to look for. So, paleontology in part depends on the availability of opportunities to collect fossils and on the imagination of the collector.

With new emphasis on plate tectonics the geographic and climatic distribution of fossils has become increasingly important and will continue to be an important area of research for years to come. For instance, for a North American paleontologist the Paleozoic fossils contained in rocks of northwestern Africa, which was adjacent to our Atlantic seaboard for millions of years, take on heightened importance in the light of drift theory.

One of the most encouraging aspects of paleontology today concerns the outlook and training of future paleontologists. It was traditional for many years for paleobotanists and vertebrate paleontologists to be trained primarily as botanists and zoologists, respectively, and for invertebrate paleontologists to be trained mainly as geologists. This dichotomy is gradually disappearing, so that all paleontologists are receiving more thorough grounding in important aspects of both geology and biology. We are already seeing the results of such broad training, and it will be even more important in the future. The outlook for paleontology is stimulating and exciting. The next ten to twenty years will undoubtedly produce new ideas, new theories, and new data that will greatly enhance our understanding of the life of the past.

Now that we have reviewed the history of life on earth, does that history teach us any lessons about our present-day situation, or what the future may bring? A famous quotation by George Santayana is that those who do not remember the past are doomed to repeat it. That statement may also hold true for our understanding of past life.

The most obvious lesson surely has to do with extinction, a facet of paleontology that fascinates many people. Is our own species doomed to become extinct? is a common question. As is so often the case, there is not a single, simple answer to this question. On one hand, we can say that extinction is the ultimate fate of every species that we know anything about. There is no evidence at all that any species survives forever. That being the case, *Homo sapiens* surely will eventually become extinct. On the other hand, we are the first species on earth to understand what extinction is all about, even though we may only dimly know many underlying factors, and we may therefore come up with some ways to prevent extinction of our own kind. We are also the first species to be able to control and alter our environment deliberately and in very significant ways. At the present we are altering our environment in ways that are seriously deleterious to our continued success as a species. As erosion of fertile topsoil, destruction of forests, acid rain, increased carbon dioxide emissions from burning of fossil fuels, burning off of the ozone layer, and pollution of soil, air, and water by pesticides all continue to pervasively alter the face of the earth, we clearly are putting ourselves in an increasingly precarious ecological situation. An unexpected disadvantageous event or series of events could tip the balance against us, as it has so many other times in the history of life. Whatever such an event might be it could send our species, and others, cascading into oblivion. Nuclear war, an unexpected huge meteorite impact, or series of volcanic paroxysms induced by crustal plate movement could provide the doomsday trigger.

The lesson to be learned seems clear. We have already altered the physical, chemical, and biologic nature of this planet to the extent that it will never ever be the same as it was before the existence of modern humans. Instead of continuing to degrade our environment we should instead be actively improving the environment, in order to maximize our chances for continued survival. We are not doing that today, nor is there any comprehensive plan available for environmental enhancement. Perhaps we will be forced eventually to adopt such a policy on a worldwide basis. Hopefully this will happen before deterioration becomes so severe that it cannot be reversed. The earth will clearly be here as a body of rock for many more millions of years to come. It remains to be seen whether any forms of life as we know them today will persist for such long periods of time.

Glossary

abiotic synthesis (4)   the process by which life evolved from nonliving material; thought to have occurred in the Precambrian, but not since then.

absolute age (1)   the age of a rock in years, usually determined by radioactive dating. (See *relative age*.)

acritarchs (10)   microscopic, organic-walled cysts thought to be formed by marine planktonic algae, perhaps related to dinoflagellates; especially common in Cambrian through Devonian rocks.

adapids (17)   primitive primates of early Cenozoic age that include the ancestors of all later primates.

adaptive radiation (6)   an evolutionary mechanism by which one group of organisms breaks through into a new adaptive zone and then branches out into a series of new habitats; reflected by increased diversity of the group.

Aegyptopithecus (17)   the oldest known fossil ape, found in Egypt, of Oligocene age.

Agnatha (12)   the most primitive fishes; those without jaws.

ahermatypic corals (12)   a type of coral that does not contain photosynthetic zooxanthellae and thus is not a reef builder.

algae (2)   a general term for all water-dwelling, nonvascular (marine or freshwater) plants.

allopatric speciation (6)   the evolutionary process of the formation or splitting of a species by geographic isolation.

amino acid (4)   simple organic molecules that form the basis for proteins in plant and animal tissues. Four amino acids are important constituents in DNA and RNA.

ammonite septum (12)   a type of septum of ammonoid cephalopods; characterized by secondary wrinkles or sutures on every lobe.

ammonoids (12)   a major group of cephalopod mollusks that is extinct; characterized by an external shell and wavy partitions (septa) dividing the shell into chambers.

*Numbers in parentheses after each word or phrase refer to the chapter(s) in which that word or phrase is important.

amniote egg (15)   the shelled egg of reptiles that can be laid on land. The shell protects the embryo from drying out, nourishes it, and takes care of wastes. Respiration can take place through the shell.

amphibian (13)   a class of vertebrates that typically has the initial life stages in water and later development on land; characterized by an egg that is essentially like a fish egg. Amphibians are the oldest and most primitive of the tetrapods, which include living toads, frogs, and salamanders.

anaerobic (4)   a term applied to an atmosphere or environment that is reducing or devoid of oxygen.

analogous structures (6)   parts of different organisms that have similar function but not necessarily similar origin; for example, bird wings and insect wings are analogous.

anapsids (15)   the stem reptiles, characterized by lack of an opening in the temple region of the skull; includes living turtles.

angiosperms (14)   the flowering plants.

animal (2)   a multicellular, heterotrophic organism at the tissue or higher level of organization.

araucaria (14)   a primitive group of pines (conifers) now restricted to the Southern Hemisphere; common in the Triassic petrified forest of Arizona.

archaeocyathids (12)   an extinct phylum of spongelike organisms, confined to Cambrian rocks, that built small patch reefs.

Archaeopteryx (15)   the oldest known bird; found in Jurassic limestones of Germany.

Archean (1,5)   the oldest interval of geologic time, from the origin of the earth about 4 to 2.5 billion years ago.

Archimedes (8)   an extinct genus of late Paleozoic bryozoan, characterized by a central spiral axis that supported a lacy frond.

arthropods (9)   a phylum of animals that includes trilobites, insects, and crabs; characterized by jointed limbs.

articular-quadrate bones (15)   the two bones that join the upper and lower jaws in reptiles; these become two bones of the middle ear in mammals.

artiodactyls (16)   even-toed herbivorous animals with an axis down the foot between the third and fourth digits; include many four-toed and two-toed advanced mammals, such as bison, camels, hippos, and pigs.

atmosphere (4)   the gaseous envelope surrounding the solid earth.

Australopithecus (17)   the first hominoid, the ancestor of humans, that first appeared about 3.8 to 4 million years ago.

autotroph (1,4,9)   an organism capable of manufacturing its own food; such as cyanobacteria, some protists, and plants.

Azoic (1)   an old name for the Precambrian interval of time. The word means "without life," a reference to the fact that it was once thought that fossils did not occur in these rocks.

banded iron formations (5)   thick sequences of finely banded, silica-rich rocks of Precambrian age that contain high proportions of oxidized iron materials, indicating the presence of free oxygen.

Baragwanathia (13)   the oldest known land flora of vascular plants; found in Upper Silurian rocks of Australia.

barrier (8)   any impediment to the migration of plants or animals, marine or terrestrial. An ocean is a barrier to land life; land is a barrier to marine life.

belemnites (8)   an extinct group of cephalopod mollusks that had an internal, cigar-shaped skeleton.

benthos (9)   a general name for all organisms, whether plants or animals, that live on the bottom sediment in an aquatic environment. The habitat may be marine or freshwater.

bilateral symmetry (2)   a form of symmetry in which one plane and only one plane divides an organism into two equal halves that are mirror images of each other.

biofacies (3)   an assemblage of fossils that occur together and characterize a particular environmental setting.

biostratigraphy (3)   the science of determining the relative ages of rocks by the use of fossils.

bipedal (15)   a term applied to an animal that uses its hind legs only for locomotion; contrasts with a four-legged (quadrupedal) gait.

Bitter Springs (5)   a locality in central Australia that has yielded many Precambrian microfossils about 1 billion years old.

bivalves (11)   a class of mollusks characterized by a shell of two valves hinged at the top. Also called pelecypods, they include clams, mussels, and oysters.

black chert (5)   chert rock that contains a lot of carbon, and is hence black in color. Almost all microscopic Precambrian fossils are found in black chert associated with stromatolites.

blastoids (11)   an extinct group of stalked echinoderms that are found in Ordovician through Permian rocks.

bony fishes (12)   a general name for the advanced fishes, excluding primitive groups and sharks; includes the ray-finned and lobe-finned types. The scientific name is class Osteichthyes.

brachiopods (9,11)   a phylum of invertebrate animals characterized by two valves composing the shell that are situated on the top and bottom of the animal.

branch reduction (10)   an evolutionary trend in graptolites in which the number of branches in the colony is gradually reduced, normally following a pattern of 16 to 8 to 4 to 2 to 1 branch.

branch rotation (10)   an evolutionary trend in graptolite colonies in which the branches gradually change from a downward stance to an upright position.

bryozoans (11)   a phylum of exclusively colonial animals with a stony, cylindrical, or lacy skeleton; mostly marine and most varied in the Paleozoic. The individuals are very small individuals, each with a lophophore

Burgess Shale (9)   a Middle Cambrian rock unit in British Columbia, Canada, that has yielded carbon films of the soft parts of many primitive invertebrates.

calcite (3)   a mineral composed of calcium (Ca) and carbonate (CO_3); the most common shell material of many invertebrates.

Cambrian (1)   the oldest period of time (570–515 million years ago) in the Paleozoic Era.

carbon 13 (5)   an isotope of the element carbon that is stable and depleted in organic material. Ratios of C_{12} and C_{13} are used to determine the organic or inorganic nature of carbon in Precambrian rocks.

Carboniferous (1)   the period between the Devonian and the Permian. The European classification; called the Mississippian and Pennsylvanian periods in North America. Carboniferous rock is divided into lower and upper portions. The period runs from 360 to 286 million years ago.

carbonization (3)   a mode of fossil preservation in which most of the organic material is destroyed, leaving behind a carbon-film impression of the organism.

carnassials (16)   specialized premolar or molar teeth in advanced mammalian carnivores characterized by a high, bladelike shearing edge.

Carnivora (16)   an order of mammals, including cats and dogs, that consists of most of the flesh-eating mammals.

carnivore (9)   any flesh-eating animal.

carnosaurs (15)   bipedal, carnivorous dinosaurs of the Mesozoic Era.

cast (3)   the natural filling of a mold left behind after a fossil has been removed from surrounding rock by solution.

catkin (14)   a drooping set of either male or female reproductive organs in land plants.

Cenozoic (1)   the youngest era of geologic time, from 66 million years ago to the present. The term means "young life" or "recent life."

cephalopods (12)   a class of mollusks, including living squid, octopus, and Pearly Nautilus, as well as many fossil nautiloids, ammonoids, and belemnites.

ceratite septum (12)   the septum in ammonoid cephalopods with a suture pattern of intermediate complexity; every other lobe has secondary crinkles on it.

Ceratopsia (15)   a group of ornithischian dinosaurs, including *Triceratops*, characterized by a frill and parrotlike beak; all but the most primitive also have bony horns on the head. The group is confined to the Cretaceous Period.

chalk (10)   a type of sedimentary rock composed largely of the skeletons of microscopic coccoliths and foraminifera; especially widespread during the Cretaceous Period.

chemical fossil (3) chemical molecules, usually organic, that indicate life and that are preserved in rocks. Precambrian occurrences of the degradation products of chlorophyll; pristane and phytane are outstanding examples.

chert (5) a type of sedimentary rock formed by the chemical precipitation of silicon dioxide (SiO_2); chemically the same as flint and quartz.

chitinophosphatic (3) a term applied to the shell material of some invertebrates that is composed of alternate layers of an organic compound, chitin, and calcium phosphate ($CaPO_4$).

class (2) a category in the classification of life that ranks below the phylum level and above the order level.

clastic (3) a term applied to any sedimentary rock composed of particles that have been transported to a site of deposition.

coal ball (14) a concretion formed in coal of calcium carbonate or pyrite in which the remains of coal swamp plants may be preserved in exquisite detail.

coccolith (10) the skeleton of tiny marine algae, composed of calcium carbonate, one of the principal constituents of chalk.

coelacanth (12) a group of marine lobe-finned fishes represented today by a single living genus, *Latimeria*.

coelom (2) the fluid-filled internal cavity that surrounds the vital organs in advanced groups of animals.

condylarths (16) primitive mammalian herbivores, especially common in the early part of the Cenozoic Era.

cone (14) a cluster of modified leaves and their associated reproductive structures in vascular plants.

conglomerate (3) a clastic sedimentary rock composed of rounded particles larger than those of sand; contains pebbles or cobbles.

conifers (14) a group of mostly evergreen gymnosperms that were especially widespread and diverse during the Mesozoic Era.

conodonts (10) a group of extinct animals that persisted from the Cambrian to the Triassic; known principally from microscopic toothlike parts of the body. They were very small, marine, and nektonic.

continental drift (7) the movement of continents on the earth's surface; also the theory that the continents were once one giant supercontinent that later broke up and drifted apart.

convergent evolution (6) a pattern of evolution in which two or more unrelated groups develop forms that come to look very much alike (homeomorphs).

core (4,7) the central part of the earth, partly molten, very dense, and composed of metallic iron and nickel.

correlation (3) in stratigraphy, the process by which units of rocks are determined to be equivalent from area to area, either in terms of their physical character or their relative age.

corridor (8) a broad pathway of migration and dispersal without barriers that allows many kinds of animals to move along it.

cosmopolitan (8) a term applied to a species that is very widespread.

creationism (6) the idea that all observed records of ancient life on earth can be explained by the story of creation in the Old Testament.

creodonts (16) a group of archaic carnivorous mammals that were especially common during the early part of the Cenozoic Era.

Cretaceous (1) a period of geologic time that closes the Mesozoic Era, occurring between the Jurassic Period and the Tertiary and lasting from 144 to 66 million years ago. The name is derived from the Latin word for chalk and named for the chalk beds of France and England.

crinoids (11) a class of echinoderms, usually with a stem or stalk atop which was a body with a few to many feeding arms; the only living stalked echinoderms.

crossopterygians (13) a group of lobe-finned fishes from which the amphibians evolved during the Devonian.

crust (4,7) the uppermost layer of the earth; includes the continents and rocks underlying the ocean.

cyanobacteria (2,5) photosynthetic prokaryotes,

also known as blue-green algae; the most common fossils in Precambrian rocks.

cycads (14)   a group of gymnosperms, now mostly extinct, that had a short, barrel-shaped trunk and many long leaves issuing from the top; especially common during the Mesozoic Era.

cycle (6)   any phenomenon that is repetitive on a regular basis. In this book cycles relate mainly to astronomical rotational features that affect climate and hence life.

cyst (10)   an organic-walled microscopic resting or reproductive structure found in the life cycle of many algae. Most fossil acritarchs are thought to be cysts.

cystoids (11)   a group of primitive stalked echinoderms that had short, armlike appendages; common in Ordovician through Devonian rocks but superceded by the blastoids and crinoids.

dentary (15)   a major bone in the lower jaw of vertebrates; the only such bone in mammals.

dermal layer (2)   a skin layer; one of the three basic cell layers of advanced animals—the epiderm, mesoderm, and endoderm.

detrital (3)   a term applied to any sedimentary particle that has been transported by water, wind, or ice.

detritus feeder (11)   any animal that ingests organic particles from atop or within the sea floor.

Devonian (1)   one of the periods of the Paleozoic Era, occurring after the Silurian and before the Mississippian; named after Devonshire in southwestern England. The absolute age is 408 to 360 million years ago

diagenesis (3)   the process by which loose, unconsolidated sediments are transformed into sedimentary rocks; involves burial, heat, and pressure.

diapsids (15)   the ruling reptiles, including dinosaurs, snakes, lizards, and crocodiles, characterized by having two openings in the temple region of the skull.

diatomite (10)   a rock composed largely of the siliceous skeletons of diatoms; especially common in some marine rocks of Miocene age in California.

diatoms (10)   a group of microscopic freshwater and marine algae with a skeleton composed of silica.

dichotomous branching (13)   a branching pattern in which each branch is equal, resulting in a tuning fork or *Y* pattern; typical of many primitive vascular plants.

Dimetrodon (15)   the sail-back reptile of Permian age; a member of the mammal-like reptiles, the Synapsida.

dinoflagellates (10)   a group of microscopic algae that are especially important as primary producers in the oceans; common as fossils since the Mesozoic.

dinosaur (15)   a popular name for two groups of diapsid reptiles of Mesozoic age that include many of the very large predators and herbivores of the time.

diploid (13)   having twice the number of chromosomes that normally occur in a sex cell. All nonreproductive cells of most animals and land plants are diploid. (See *haploid*.)

DNA (6)   abbreviation for deoxyribonucleic acid, an organic molecule important for the storage and replication of genetic information within the nucleus of cells.

echinoids (11)   a class of nonstalked echinoderms characterized by a globular or flattened test and movable spines; includes sand dollars and sea urchins.

Ediacaran fauna (5)   a late Precambrian fauna of soft-bodied metazoans from South Australia and elsewhere by correlation. The oldest metazoan fossils known.

element (1)   a basic substance of which matter is composed.

endemic (8)   native to and/or restricted to a specific area or region, for example, kangaroos and many other marsupials are endemic to Australia.

enterocoels (2)   animals in which the internal cavity, the coelom, is formed from outpocketings of the larval gut; includes the echinoderms and the chordates.

Eocene (1)   an epoch of the Tertiary Period, occurring after the Paleocene Epoch and before the Oligocene and lasting from 58 to 37 million years ago; based initially on rocks in western Europe.

eoliths (17)   the earliest tools, such as cobble stones, used by hominoids; also called dawn stones.

epifauna (9,11)   animals that live on a sea bottom.

epoch (1)   an interval of geologic time, intermediate in duration between longer periods and shorter ages; most commonly used to describe the subdivisions of the Tertiary Period (e.g., Eocene Epoch).

era (1)   one of the longest subdivisions of the geologic time scale, consisting of one or more periods of geologic time. Commonly used era names are Paleozoic, Mesozoic, and Cenozoic.

eukaryotes (2,5)   organisms having a membrane around the cell nucleus and highly organized chromosomes.

euryapsids (15)   a group of aquatic and semi-aquatic reptiles with a single opening high on the temple region of the skull.

Eurydesma (7)   a large, thick-shelled marine clam restricted to Gondwana during the Permian and closely associated with glacial deposits.

evolution (2)   the origin of new kinds of life from preexisting forms of life; the fundamental genetic and morphologic change in biological populations.

extinction (6)   the demise of any group of plants or animals.

facies (3)   a term used to distinguish a portion of a rock unit with a distinctive aspect (rock type or fossils) from other portions of the same general rock unit. Facies changes are changes in aspect.

facies fossil (3)   a fossil that occurs in a specific kind of rock and, by inference, in a specific kind of environment.

family (2)   a formal category in the classification of organisms that is more inclusive than a genus but less inclusive than an order (which may contain one or more families).

faunal province (8)   a large region characterized by a distinctive assemblage of animals.

ferns (14)   a group of primitive vascular plants, characterized by spore reproduction and large, much-divided leaves.

Fig Tree (5)   a Precambrian formation in Africa that has yielded some of the oldest known fossils.

filter bridge (8)   a migration pathway through which only some animals can pass; some migrate successfully, others are filtered out.

filter feeder (9,11)   an aquatic animal that obtains food consisting of small organic particles by straining them out of the water in a variety of ways.

fissipeds (16)   advanced carnivorous mammals that have the sole of the foot raised off the ground; includes cats, dogs, and others.

flower (14)   the reproductive structure of angiosperms, consisting of several series of specially modified leaves and their associated sex parts.

foraminifera (10)   a group of heterotrophic protists with pseudopodia. They commonly have a shell, are mostly microscopic, and are important as fossils.

formation (3)   a mappable rock unit; the basic division of stratigraphy that is given a geographic name.

fossil (1)   any indication preserved in rocks of the former existence of life.

fossilization (3)   the process by which the hard parts or (much less frequently) the soft parts of an organism become buried in sediments and preserved.

fossil zone (3)   an interval of geologic time characterized by rocks that contain a distinctive fossil or suite of fossils.

gametophyte (13)   the haploid phase, produced by spores in the life cycle of a vascular plant, that itself produces sex cells.

Gangamopteris (7)   see *Glossopteris*.

gastropods (11)   one of the classes of the phylum Mollusca, characterized by an unchambered, helically spiral shell; also called snails.

gene (6)   a unit of heredity in eukaryotes consisting of about 1500 nucleotide pairs on a chromosome.

genotype (6)   the genetic makeup of an individual, as contrasted with the physical appearance (phenotype) of that organism.

genus (2)   one or more closely related species that share common descent.

geologic time scale (1)   the divisions of time relative to earth history, consisting of several long eras of time and smaller subdivisions.

gill arch (12)   the tissues and bones that separate and support adjacent gills in a fish.

ginkgos (8,14)   a group of tree-sized gymnosperm plants with fan-shaped, parallel-veined leaves; especially common in the Mesozoic Era.

Glossopteris (7)   a seed fern with large, tongue-shaped leaves restricted to Gondwana during the Late Paleozoic; differs from *Gangamopteris* in having a distinct midvein down the center of the leaf.

Gondwana (7,8)   the southern supercontinent consisting of South America, Africa, Australia, Antarctica, and peninsular India before these drifted apart; the southern half of Pangaea.

goniatite septum (12)   the simplest type of septum in ammonoid cephalopods, consisting of a wavy pattern without any secondary wrinkles.

grade (2)   a level of organization of life, usually associated with the complex arrangement of cells into tissues, organs, and organ systems.

graptolites (10)   a group of extinct colonial animals, some benthonic, some planktonic, related to the chordates; important Paleozoic fossils.

Gunflint (15)   a Precambrian formation along the north shore of Lake Superior that has yielded black cherts and stromatolites containing microscopic algae.

gymnosperms (14)   primitive seed plants in which the seeds are exposed; important land plants in the late Paleozoic and Mesozoic, consisting of conifers, cycads, ginkgos, and others.

half-life (1)   the amount of time necessary for a naturally occurring radioactive element to decay so that only one half of the atoms remain.

haploid (13)   the number of chromosomes in a sex cell, created by the process of meiosis which reduces the number of chromosomes to half that in a non–sex cell of the same species. (See *diploid*.)

herbivore (9)   an animal that feeds on plants.

hermatypic corals (12)   corals that have symbiotic photosynthetic zooxanthellae in their soft tissues; responsible for building modern coral reefs.

heteromorphic (12)   means "different form"; applied specifically to ammonoid cephalopods that do not have the usual form of coiled shell.

heterotroph (2,4,9)   an organism that cannot manufacture its own food; applies to the animals and some protists.

hierarchy (2)   the classification of life into categories from most inclusive (kingdom) to least inclusive (species).

homeomorphic (6)   means "same form"; applies to two animals that have very similar appearances but are not closely related.

hominoid (17)   a group of primates that includes the apes and humans.

Homo (17)   the genus that includes our own species, *H. sapiens; homo* means "man."

homologous structures (6)   structures or parts of organisms that have the same origin but may or may not have the same function.

hyomandibular arch (13)   the third gill arch of primitive fishes that ultimately becomes the stapes earbone in tetrapod vertebrates.

ichthyosaurs (12)   a group of extinct, carnivorous marine reptiles, dolphinlike or porpoiselike in aspect, that flourished in the Mesozoic.

Ichthyostega (13)   one of the oldest and most primitive known tetrapods, or land vertebrates; an amphibian from Upper Devonian rocks of Greenland.

igneous rock (3)   a rock that was originally molten and then cooled and crystallized; includes basalt, granite, and others.

inarticulate (11)   a term applied to a group of brachiopods having a chitinophosphatic or calcareous shell, the two valves of which are not hinged together.

incus (15)   the anvil bone in the middle ear of mammals, derived from the reptilian quadrate bone.

infauna (9)   aquatic animals that live within sediments rather than on top of them.

insectivores (16)   a group of primitive placental mammals that are first found as fossils in the Cretaceous; includes moles and shrews.

iridium (15)   a rare trace element that occurs in greater abundance in meteorites than it does on earth; used as evidence for ancient asteroid impacts.

isolation (6,8)   a condition in which a group of plants or animals is separated for a considerable length of time from other groups; caused by geographic or ecologic factors.

isotope (1)   one of two or more forms of an element having the same atomic number (same number of protons and electrons) but differing in atomic weight (having different numbers of neutrons).

Isthmus of Panama (8)   the land connection between North and South America that formed near the close of the Cenozoic; ended isolation of endemic South American mammalian fauna.

jellyfish (10)   a member of the phylum Coelenterata related to corals and sea anemones. Although they lack a skeleton, jellyfish are sometimes found as fossils.

Jurassic (1)   a period of geologic time in the Mesozoic Era, occurring after the Triassic and before the Cretaceous. The absolute age is from 213 to 144 million years ago.

Karroo (15)   an area in southern Africa characterized by Permian and Triassic rocks that yield abundant specimens of mammal-like reptiles.

kingdom (2)   the most comprehensive division of life. In paleontology the four most important kingdoms are the Monera, Protista, Animalia, and Plantae.

K/T boundary (15)   scientific shorthand for the Cretaceous (K)-Tertiary (T) boundary.

Latimeria (8,12)   the only living genus of coelocanth fish; hence, the only living lobe-finned fish.

Laurasia (7)   the northern supercontinent consisting of North America, Europe, and Asia; the northern half of Pangaea.

lead-uranium (1)   the ratio of naturally occurring uranium and its decay product, lead; provides the basis for radiometric dating.

limestone (3)   a sedimentary rock composed primarily of calcite, which may be in the form of fossil shells.

lobe-finned fish (12)   one of the two main groups of bony fishes; important as the ancestors of land vertebrates, the amphibians.

lophophorates (2)   a group of invertebrates including two phyla, the brachiopods and bryozoans, that have a conspicuous feeding structure called the lophophore.

lophophore (11)   the loop-shaped ciliated feeding structure of brachiopods and bryozoans.

lycopods (14)   a group of primitive land plants characterized by a spiral arrangement of long, slender leaves on the stem; conspicuous tree-sized plants in coal-swamp forests.

macroevolution (6)   the process by which a small group of organisms invades a new habitat and undergoes radiation and increase in diversity; also called *adaptive radiation.*

magnetic reversal (7)   a change in the relative position of the earth's north and south magnetic poles. Because reversal has happened many times in the past, it provides a means for dating the sea floor and proving oceanfloor spreading.

malleus (15)   a bone in the middle ear of mammals commonly called the hammer bone; derived from the articular bone of reptiles.

mammoth (16)   an extinct elephant group that included wooly forms.

mantle (4,7)   the middle layer of the earth between the core and crust; very thick and composed of dense silicate minerals.

marsupials (16)   the pouched mammals, a group that first evolved in the Cretaceous; includes the opossum and kangaroo.

mass number (1)   the sum of the numbers of protons and neutrons in the nucleus of an atom. Different isotopes of the same element have different mass numbers.

mastodon (16)   a large-tusked extinct mammal related to elephants and typified by less specialized teeth than in elephants and mammoths.

megaevolution (6)   a major shift in habitat that may occur rather suddenly and in a small group; especially the change from aquatic to terrestrial habitat.

meiosis (13)   the process in which cell division takes place such that the resultant cells have half the number of chromosomes as the parent cell; the process by which sex cells (haploid) are produced from non–sex cells (diploid).

Mesosaurus (7,8)   a small, freshwater extinct reptile confined to South America and Africa; important in early debates on continental drift.

Mesozoic (1)   an era of geological time from 245 to 66 million years ago; the age of dinosaurs and ammonoids. The word means "middle life."

metamorphic rock (3)   either igneous or sedimentary rock that has been altered by heat and pressure; includes slate, marble, and quartzite.

metaphyte (13,14)   a macroscopic larger plant composed of many cells; the group includes many land plants but also macroscopic seaweeds.

Metasequoia (8)   a genus of conifer that was widespread during the Cenozoic but survived only in a refuge in China.

metazoa (2,5)   the multicellular animals; includes all phyla of animals except the sponges.

meteorite (4)   a fragment of a planetary body that has fallen to the earth; may yield information about the composition of the earth's interior.

Milankovitch cycles (6)   periodic variations in various aspects of the earth's rotation that are thought to affect glacial and interglacial intervals; named after the Yugoslavian meteorologist who first worked out the theory.

Miocene (1)   one of the epochs in the middle part of the Tertiary Period, older than the Pliocene and younger than the Oligocene; the absolute age is 24 to 5 million years ago.

Mississippian (1)   one of the periods of time in the Paleozoic Era, occurring before the Pennsylvanian and after the Devonian; named for outcrops along the upper part of the Mississippi River in Illinois, Iowa, and Missouri.

mitosis (13)   the normal process of cell division in which no change in chromosome number takes place, as during the growth of an individual.

mold (3)   the impression left in the surrounding rock by the decay of organic material or dissolution of a shell.

molt (11)   the shed skeleton of an arthropod that is discarded during a growth interval.

Monera (2)   a kingdom of life made up of the prokaryotes; includes bacteria and cyanobacteria. (See *prokaryotes.*)

monopodial branching (13)   a pattern of branching in land plants in which there is a single large central branch with smaller side branches issuing from it at intervals along the stem.

monotremes (16)   the egg-laying mammals for which there is virtually no fossil record, includes the living duck-billed platypus and the spiny echidna of Australia.

mosasaurs (12)   large marine lizards of the Cretaceous.

mudstone (3)   a sedimentary rock composed of a mixture of clay- and silt-sized particles.

multituberculate (16)   one of the extinct groups of early mammals that inhabited the Mesozoic and Early Cenozoic; larger than most contemporary mammals, from mouse- to woodchuck-sized, and with specialized rodentlike teeth.

mummy (3)   a preservation in which soft tissues are preserved by a process of either freezing or dehydration.

mutation (6)   any change in the structure or arrangement of chromosomes in a cell nucleus.

natural selection (6)   the process by which certain members of a population, because of their genetic makeup, are reproductively successful or unsuccessful.

nautiloids (12)   a group of cephalopods characterized by very simple sutures, represented today only by the Pearly Nautilus; important Paleozoic predators.

Neanderthal (17)   the name for an early group of humans in Europe, Africa, and western Asia; sometimes regarded as a separate species, a subspecies of Homo sapiens, or simply an early variety of our species.

nekton (9,10)   aquatic animals that actively swim, as opposed to those that passively float (plankton).

Neopilina (8) a genus of mollusks that is the only survivor of a Paleozoic group, the monoplacophorans, that was long thought to be extinct.

neutron (1) one of the building blocks of atomic nuclei that lacks an electrical charge. Changes in the number of neutrons in atoms of an element result in different isotopes of that element.

notoungulates (16) a group of primitive placental mammals common in the Early Cenozoic; they were isolated and survived until the Pleistocene only in South America.

Olenellus (8) a genus of Lower Cambrian trilobites characteristic of most of North America; provides the name of a faunal province.

Oligocene (1) one of the epochs of the Tertiary Period, younger than the Eocene and older than the Miocene, from 37 to 24 million years ago.

order (2) one of the categories in the hierarchial classification of life consisting of groups larger than family and smaller than class.

Ordovician (1) a period of the Paleozoic Era, younger than the Cambrian and older than the Silurian; named after an early tribe in ancient Britain that lived in the central area where the rocks occur.

Ornithischia (15) the bird-hipped dinosaurs, all of which were herbivores, such as duck-bills, stegosaurs, ankylosaurs, and ceratopsians.

ornithopods (15) the duck-billed dinosaurs and others, bipedal, herbivorous, and most common in the Cretaceous.

ostracoderms (12) jawless fishes with a heavy external bony armor; the most primitive vertebrates.

otic notch (13) the ear notch, across which stretched the ear drum, present in the rear of the skull in amphibians; absent in reptiles.

oxygen isotope (8) the ratio of two isotopes of oxygen, O_{16} and O_{18}, that is used to calculate the water temperature at which ancient shells were constructed; a paleothermometer.

paleobiogeography (7) the study of the geographic distribution of prehistoric life.

Paleocene (1) the oldest epoch of the Tertiary Period, from 66 to 58 million years ago.

paleoecology (3) the science of interpreting the ancient environments of the earth.

paleogeography (3) the study of ancient distribution of land and sea, and of mountain ranges and other geographic features. The study of the ancient geographic distribution of life is called paleobiogeography.

paleontology (1) the study of fossils.

Paleozoic (1) an era of geologic time about 570 to 245 million years ago; between the Precambrian and the Mesozoic; means "ancient life."

Pangaea (7) a supercontinent in which all of the present continents were grouped into a single land mass.

pantodont (16) a primitive group of large herbivorous mammals that flourished in the early part of the Cenozoic Era and became extinct in the Oligocene.

pantothere (16) one of the extinct groups of small Mesozoic mammals.

parallel evolution (6) the process by which two distinct and unrelated groups of organisms undergo a series of similar changes through time; can be seen in fossil graptolites.

pedicle (11) the horny and muscular stalk by which some brachiopods are attached to the sea floor.

pelagic (9) a term applied to organisms that live in the water column, floating or swimming.

Pennsylvanian (1) a period of the Paleozoic Era that occurs after the Mississippian Period and before the Permian Period; named after the coal-producing areas of Pennsylvania.

period (1) an interval of geologic time represented by a system of rock; the principal subdivisions of an era that are further subdivided into epochs.

peripheral isolates (6) a population of a species that is separated from neighboring populations of that species and where speciation is most likely to occur.

perissodactyls (16) the odd-toed hoofed mammals, generally with one or three functional toes; includes the horse, tapir, rhinoceros, and others.

Permian (1) the final period of the Paleozoic Era,

named after the town of Perm in the Soviet Union.

permineralization (3)   the process by which shell or skeletal material is infiltrated by mineral matter making the hard part denser and heavier; also called petrifaction, as in petrified wood.

phenotype (6)   the outward physical aspect of an organism.

phloem (14)   vascular conductive tissue in a land plant.

photic zone (9)   the upper layer of oceans and lakes that is penetrated by sunlight and within which photosynthetic organisms can live. The depth of the photic zone depends on water clarity and the amount of sediment in the water.

phyletic evolution (6)   the gradual change through time of one species population to another, phenotypically distinct, species population.

phylum (2)   a major group of organisms, one or more of which make up a kingdom and which may be divided into classes.

phytane (5)   an organic compound that is a degradation product of chlorophyll; found in some Precambrian rocks as an early indicator of photosynthesis.

phytoplankton (10)   floating photosynthetic organisms (protists and algae) in aquatic environments.

pistil (14)   one or several female reproductive structures in the center of a flower within which the seeds develop.

placentals (16)   advanced mammals whose young develop within the mother's body. The young are born in an immature state and require nurturing.

placoderms (12)   the earliest jawed fishes, with heavy bony armor and more than two pairs of lateral fins.

planet (4)   one of the larger bodies circling the sun, including the earth.

plankton (9)   any organism that floats in the water.

plant (2)   any of numerous organisms of the kingdom Plantae, typically photosynthetic and multicellular; may be either aquatic or terrestrial.

plate tectonics (7)   the mechanisms by which large parts of the earth's crust are formed, moved, and destroyed.

Pleistocene (1)   one of the two epochs of the Quaternary Period, lasting about 1.6 million years; the time of the ice ages.

plesiosaurs (12)   a group of marine aquatic reptiles of the Mesozoic Era with long or short necks, a large turtlelike body, and large flippers.

Pliocene (1)   the final epoch of the Tertiary Period, occurring after the Miocene and before the Pleistocene, from 5 to 1.6 million years ago.

pollen (14)   the fine, dustlike grains of seed plants that contain the male gametophyte and produce the sperm that fertilizes the plant egg.

potassium-argon (1)   the end members of a radioactive series that is used to date rocks in years. Radioactive potassium is the isotope with which the decay series starts, and the gas argon is the stable end product.

preadaptation (6)   the evolutionary process by which an organism acquires certain characteristics useful to it in its present state of existence and also advantageous in a new habitat.

Precambrian (1)   the rocks, time, and earth history prior to the beginning of the Paleozoic Era, from about 4 billion to 600 million years ago; encompassing the Archean and Proterozoic eras or the corresponding system of rocks.

primary consumers (9)   heterotrophs that feed directly on primary producers (plants or protists); herbivores.

primary producers (9)   the autotrophs; organisms that can manufacture complex organic materials themselves, through photosynthesis or otherwise.

primates (17)   an order of mammals that includes humans and their relatives.

pristane (5)   see *phytane*.

proboscideans (16)   an order of mammals that includes elephants, mammoths, and mastodons.

prokaryotes (2,5)   simple organisms, generally microscopic, that lack a highly organized nucleus with a surrounding membrane; includes the bacteria and cyanobacteria.

prosimians (17)   the less advanced and earliest appearing of the primates; includes the living lemurs and tarsiers.

Proterozoic (1,5)   an interval of geologic time between the Archean and the Paleozoic eras, from about 2.5 to 0.6 billion years ago.

protists (2)   members of the kingdom Protista; mostly one-celled organisms, some of which are autotrophs and some heterotrophs.

psilopsids (14)   the simplest and most primitive vascular land plants, lacking true leaves or roots; common in the Devonian.

punctuated equilibrium (6)   an evolutionary pattern in which morphological changes occur rapidly, separated by much longer intervals where little change occurs (periods of stasis).

pyrite (3)   a mineral composed of iron and sulfur; may be a replacement of fossil shell or bone.

quadrupedal (15)   any animal that walks on all four legs, as opposed to a bipedal animal which walks on the hind legs only.

Quaternary (1)   the youngest period of geologic time, including the last 1.6 million years of earth history.

radial symmetry (2)   a form of symmetry in which more than one plane divides an organism into two equal halves.

radioactivity (1)   a property of certain isotopes that change into other elements by the discharge of particles from their nuclei.

radiolarians (10)   a group of heterotrophic protists that have a radially symmetrical silica skeleton; microscopic and unicellular.

Rancho La Brea (16)   famous tar pits in Los Angeles that have yielded thousands of well-preserved bones of Pleistocene mammals and other animals.

ray-finned fish (12)   a group of advanced bony fishes with many fine parallel bones supporting the fins; includes the most common freshwater and saltwater fishes today.

Recent (1)   modern or historic time, since the end of the Pleistocene; essentially the past 8,000 years.

recrystallization (3)   the process by which the original microstructure of fossil hard parts is destroyed by growth of new crystals.

reducing atmosphere (4)   an atmosphere lacking in oxygen. (See *anaerobic*.)

reef (12)   an organically built structure raised off the sea floor with a rigid framework of skeletal material.

refuge (8)   a restricted area in which a species that originally had a much wider geographic range can survive; a haven or asylum.

relative age (1)   the geologic age of a fossil or rock defined relative to the age of another fossil or rock (as older or younger) rather than in years.

remnant magnetism (7)   the direction and polarity of the earth's ancient magnetic field, preserved and recorded by small magnetic grains in rocks.

replacement (3)   a process of fossilization in which the original mineral material of a hard part is replaced by another kind of mineral.

rhizome (14)   an underground stem that serves to anchor a land plant; the original anchorage before true roots had evolved.

RNA (6)   abbreviation for ribose nucleic acid, an organic molecule responsible for transferring information from DNA molecules to build proteins.

rubidium-strontium (1)   one of the radioactive decay series that is used to date rocks. Radioactive rubidium is the initial unstable isotope that eventually decays to stable strontium.

rudistid (8)   a group of large, thick-shelled bivalves that flourished in tropical waters during the Cretaceous Period, built reefs, and became extinct at the K/T boundary.

rugose corals (12)   a group of extinct Paleozoic corals, solitary or colonial, that had prominent septa; also known as horn corals.

sandstone (3)   a sedimentary rock composed primarily of sand-sized particles regardless of their mineral composition; usually composed of quartz particles.

Saurischia (15)   the reptile-hipped dinosaurs, including bipedal carnivores and quadrupedal herbivores; confined to the Mesozoic Era.

sauropods (15)   the giant, quadrupedal dinosaurs that perhaps evolved from bipedal carnivorous types; one group of the Saurischia.

scavenger (9)   an animal that feeds primarily on the dead bodies of other animals.

schizocoels (2)   a major group of metazoan animals in which the coelom forms by the splitting of mesodermal cells; includes the mollusks, arthropods, and annelids.

scleractinian corals (12)   modern corals that began in the Triassic; includes reef-building hermatypic as well as ahermatypic corals.

scute (15)   a bony plate on the exterior of amphibians and reptiles that provides protection.

sea-floor spreading (7)   the process by which new sea-floor crust is formed along oceanic ridges and pushed progressively further from the ridge as younger and younger crust is formed; thus, any given segment of the sea floor gradually spreads away from the ridge.

secondary consumers (9)   those heterotrophs that feed on primary consumers; predators and carnivores.

sedimentary rock (3)   any rock formed of sedimentary particles that are either preexisting (detrital grains) or that may be chemically precipitated, such as sandstone, shale, and limestone.

seed (14)   the reproductive product of advanced land plants that contains an embryo sporophyte and a food supply within a resistant seed coat.

septum (12)   a vertical partition of skeleton in a coral; a partition separating chambers in a cephalopod.

sessile (9)   fixed or immobile because attached or cemented at the base.

shale (3)   a fine-grained sedimentary rock composed mostly of clay-sized particles; commonly splits into thin beds or layers.

shark (12)   one of the cartilaginous fishes, the Chondrichthyes, which generally lack bones except in their teeth and dermal denticles.

silica (3)   silicon dioxide (SiO_2), a common mineral constituent of various organic hard parts among protists, algae, and sponges.

silicoflagellates (10)   a group of unicellular protists that have two flagellae for locomotion and a siliceous skeleton.

Silurian (1)   one of the periods of the Paleozoic Era, occurring after the Ordovician and before the Devonian; named after an ancient British tribe that inhabited an area in central England where rocks of this age were first found.

skeleton (3)   the hard parts of any organism but used especially in describing protists and animals.

solar wind (4)   energy in the form of subatomic particles released by the sun; swept away much of the inert gases when the earth was first forming.

species (2)   one or more actually (or potentially) interbreeding natural populations that produce fertile offspring and are reproductively isolated from other such groups.

sphenopsids (14)   a group of primitive vascular plants characterized by a jointed stem and whorls of leaves or reproductive structures at the intersections of the stem segments; common in Pennsylvanian coal swamps.

sponge (11)   a phylum of animals that consist of simple assemblages of relatively undifferentiated cells, without tissues and organs.

sporangia (14)   a small, organic sac within which spores are produced and held until release by a plant; a spore case.

spore (13)   a reproductive body produced by plants; the haploid *(N)* phase of the plant life cycle, produced by the sporophyte by meiosis and growing into the gametophyte by mitosis.

sporophyte (13)   the conspicuous plant body of vascular plants; the diploid *(2N)* phase of the plant life cycle, produced by sex cells from the gametophyte and, in turn, producing spores.

spreading center (7)   an edge of one of the plates of the earth's crust where new crust is formed and spreads to either side; commonly recorded as a high ridge or rise on the ocean floor.

squamosal (15)   a bone in the skull of mammals that articulates with the dentary bone that forms the lower jaw.

stamen (14)   one of the male, pollen-producing structures in the flower of angiosperms.

stapes (13)   a bone that is the single earbone of amphibians and reptiles and one of three earbones in mammals; derived from the fish hyomandibular bone which was used for jaw support.

stasis (6)   in evolutionary terms, the idea that species populations persist for long periods of time without discernable genetic or morphologic change.

stegosaurs (15)   a suborder of dinosaurs consisting of animals with two rows of alternating large bony scutes along the back and long, bony spines on the end of the tail; confined to the Jurassic Period.

stratigraphy (3)   the study of layered sedimentary rocks, especially their formation and relationships.

stromatolite (5)   an organical structure in rocks that consists of concentric laminations (cabbage-head structure) built by algae, although not part of their skeleton or hard parts; especially common in Precambrian and early Paleozoic rocks.

stromatoporoids (12)   a group of calcareous sponges especially important as reef builders during the Paleozoic.

subduction zone (7)   an edge of one of the plates of the earth's crust where the crust is turned down into the mantle; commonly marked by a deep trench on the ocean floor.

superposition (1)   a principle of geology specifying that in an undisturbed state younger rocks are laid down on top of older rocks.

sweepstakes route (8)   a dispersal pathway characterized by a formidable barrier that prevents most migrating organisms from crossing; dispersal to an island from a mainland (island hopping) is an example.

sympatry (6)   condition in which biologic populations occupy the same area but maintain their identity without interbreeding.

synapsids (15)   the mammal-like reptiles, characterized by a single opening low on the temple region of the skull; common in the Permian and Triassic, in which latter period they gave rise to mammals and became extinct.

tabulate corals (12)   a group of corals that were all colonial and generally lacked septa; conspicuous reef builders in the Paleozoic Era, but extinct since the Paleozoic.

tarsiers (17)   a group of small-bodied, primitive, tree-dwelling primates that includes a single living genus from the Philippines and Indonesia.

temporal opening (15)   one or more holes in the side or temple region of the skull, behind the eye, that are present in many reptiles; the number and arrangement are important in reptile classification.

Tertiary (1)   the principal period of time in the Cenozoic Era, occurring after the Mesozoic Era and before the Quaternary Period, and lasting from 66 to 1.6 million years ago.

Tethys (8)   an ancient seaway stretching from between Europe and Africa eastward across Asia; occupied the present site of the Alps, Himalayas, and Mediterranean Sea.

thallus (5)   the body of an alga.

thecodonts (15)   a group of diapsid reptiles that gave rise to the dinosaurs, specifically the Saurischia; some were bipedal and lightly built.

theropod (15)   a large, bipedal carnivorous dinosaur.

tillite (8)   any sedimentary rock deposited underneath a glacier.

Tommotian (1,5)   the lowest age in the Cambrian Period characterized by the first skeletonized fossils prior to the appearance of trilobites. Tommotian fossils are small and phosphatic, and many are of uncertain affinity.

trace fossil (3,9)   any indirect evidence of the former existence of life, commonly indicative of animal activity. Common types include burrows, trackways, and fossil excrement.

tracheophyte (2)   vascular land plants.

Triassic (1)   the initial period of time in the Mesozoic Era, occurring before the Jurassic and after the Permian, and lasting from 245 to 208 million years ago.

triconodonts (16)   the most primitive group of mammals in the Mesozoic Era, small with simple molar teeth; now extinct.

trilobites (9,11)   a group of extinct arthropods, characteristic of the Paleozoic Era, with a body divisible into a head, thorax, and tail.

trophic level (9)   a level of food production or consumption of an organism within a food pyramid. Primary producers are at the bottom, followed by primary consumers, secondary consumers, and so on.

ungulates (16)   the hoofed herbivorous mammals; included are both odd- and even-toed types.

uniformitarianism (3)   the principle that general laws of physics and chemistry applied in the past just as they do today, and that geologic processes we see today also shaped the earth in the past.

variation (6)   departure from a normal or average condition. In biology the term is used to describe variability in species populations.

vascular system (13)   the conductive and supportive tissues characteristic of land plants; includes both xylem and phloem.

wood (14)   secondary xylem that serves to strengthen and support the stems of large land plants, that is, trees and shrubs.

xylem (14)   one of the two important kinds of tissue in vascular plants that provides for conduction of water and food and support of the stem in air; secondary xylem is the woody tissue of trees and shrubs.

zooplankton (10)   nonphotosynthetic organisms that float in the water; two major groups are the microscopic protists (radiolarians and foraminifera) and the extinct graptolites.

zooxanthellae (12)   unicellular, photosynthetic, microscopic protists that reside in the tissues of hermatypic corals; important as a factor in coral reef growth.

zygote (13)   the first cell formed by fertilization in the development of any new plant or animal.

Index

Boldface page numbers refer to illustrations.

Abiotic synthesis, 62–63
Absolute age, 7
Acetylene, 62–63
Acoelomata, 22, 24
Acritarchs, 151–53, **152**
Adapids, 298
Adaptive radiation, 85–87, **86**
Adenine, 78
"Advanced," as a qualifier, 92
Aegyptopithecus, **297**, 298, **299**
Africa
  and continental drift, 100–105
  South, 67
Age of periods and epochs, 10–11
Agglutinated tests, 157
Agnatha, 198
Agricola, 4
Ahermatypic corals, 189
Algae
  ancestors of land plants, 205–8
  brown. *See* Phaeophyta
  definition, 28, 29–30
  early life, 61
  green, 69, 210. *See also* Chlorophyta
  main groups, **28**
  phylum, 28
  red. *See* Rhodophyta
  yellow-green. *See* Chrysophyta
Allopatric speciation, 82
Allopatry, 83
Allosaurus, 258
Alps and continental drift, 111
Amber, fossils in, 44, **45**
Amblypods, 272
Amino acids, 61, **62–63**
Ammonia in early atmosphere, 61
Ammonite septum, 195, **195–98**
Ammonoids, 89, **90**, 194–98
Amniote egg, 247
Amphibians, 28, 212–16, 247
Anaerobic environment, 63
Analogous structures, 91, **91**
Anapsid reptiles, 249, **250**
Anatosaurus, **256**
Anchorage of land plants, 207
Angiosperms, 235–45, **235**
Angular momentum of planets, 58

Animalia, 22–28
  kingdom (defined), 22
  levels of organization, 22
  major groups, 23
  phyla, 23–24
Ankylosaurs, 261
Annelida
  phylum, 25, 26
  schizocoels, 26
Annual plants, 244
Annularia, **229**
Anthropoids, 298
Apatite. *See* Calcium phosphate
Apatosaurus, 255
Apes, 295–98
Appearances, first, **141**
Araucaria, 127, 232
Archaeocyathids, 138, **190**
Archaeopteryx, 97, 268–70
Archaic mammalian faunas, 276–81
Archean, 11, 69
Archimedes, 126, **126**, 176
Arenaceous, 157
Armadillo, 122, 276
Arthropoda
  phylum, 26, **27**
  schizocoels, 26
  trilobites, 136–37
Articular bone, 253
Artiodactyls, **277**, 279, 281–84
Asylum. *See* Refuge
Asymmetry, 23
Atlantic Ocean and continental drift, 100, **101**, 110
Atmosphere
  origin, 61
  oxidizing, 65, 71
  reducing, 61
Atoms, 15
Attachment, brachiopod, 171–74
Australia, 67, 69, 71, **110**, 120
Australopithecus, **297**, 298–301
Autotrophs, 28, 64, 144
Azoic, 12

Bacon, Francis, 100
Bacteria, 20, 28, 64, 67–69
Banded iron formations, **68**, 71

Baragwanathia flora, 208, **219**, 219, 225
Barriers to migration, 121–23
Belemnites, 119, **120**
Benthonic organisms, **143**, 157
Bentonite, 50
Bequerel, Henri, 14
Bering Strait, 121
Big Lick, Ky., 5
Bilateral symmetry, 23
Bilateria, 23–24
Biofacies, 53
Biostratigraphy, 51, **51**
Bipedal gait in dinosaurs, 254
Birds
  as chordates, 25, 27
  fossil, 268
  origin, 268
Bitter Springs chert, 69
Bivalvia, 25, 180–83, **181–82**. *See also* Pelecypods
Black chert, 66
Blastoids, 127, **127**, 177, **177**
Bony fishes, 200–202
Brachiopods, 24, **43**, 137, 170–74, **171**
Brachiosaurus, 255
Brain, 296, 299
Branches, graptolite, 159, **159**
Bryophyta, 28
Bryozoa, 24, 174–76
  cryptostome, 176
  frond, 176
  treptostome, 176
Burgess Shale, 139, **140**
Burial, 34
Burrowing bivalves, 180–83

C12/C13, 71
CaCO$_3$, 33
Calamites, **229**, 231
Calamostachys, **229**
Calcareous exoskeleton, 169
Calcite skeleton, 33
Calcium carbonate, 33, 138
Calcium phosphate, 33, 137
Cambrian
  communities, 146, **146**
  definition, 9, 11
  faunal provinces, 117

Cambrian, *continued*
  fossils, 134–40
  seaweeds, 71
Camels, 286, **287**
CaPO₄, 33, 137
Carbon dioxide in atmosphere, 61
Carbon isotopes, 16, 71
Carboniferous, 8, 12
Carbonization of fossils, 45, **45**
Carnassial teeth, 276
Carnivores
  mammals, 276
  reptiles, 251
Cartilaginous fishes, 200
Cast, fossil, **41**, 42, **44**
Catkin, 228
Cellular grade of organization, 22
Cellulose, 207
Cenozoic
  communities, **148**
  definition, 9, 10
  faunal provinces, 121
  mammals, 275–94
  plant communities, 238–45
Centers of evolution, mammal, 285–89
Centers of spreading. *See* Spreading center
Cephalopods, 25, 26, 193–98
Ceratite septum, 195, **196**
Ceratopsians, 261
Chalk, 155, **155**, 159
Chemical fossils, 46, 70–71
Chemical sedimentary rocks, 35
Chemoautotroph, 28
Chert, 66
China, early paleontology, 2, **3**
China, as a refuge, 128
Chinle Formation, 232
Chitin, 33, 165
Chitinophosphatic shells, 33, 137, 170
Chitinozoa, 133–34, **135**
Chiton, **185**
Chlorophyll, 46, 64, 70, 219
Chlorophyta, 28, 30
Chonetid brachiopods, **173**, 174
Chordata, 25, 27
Chromosome, 68, 78
Chrysophyta, 28, 30
Class, as a category, 19

Classification, 18
Clastic sedimentary rocks, 35
Claystone, 35
Climate
  ancient, 131
  continental drift, 111
  Cenozoic plants, 238–45
Clocks, geologic, **13**
Coal balls, 229
Coal swamp plants, 228–31, **229**
Coccoliths, 154–55, **155**
Coelacanth fish, 125, **125**, 200
Coelenterata, 24
Coelom, 23
Communities
  fossil, 55
  land plant, 224–45
  marine, early evolution of, 133–42
  marine, structure of, 142–49
  terrestrial, origin of, 205–16
Condylarths, 278
Cones, fossil plant, 223, **223**, 228, 235–36
Conglomerate, 35
Conifer, 228, 232
Conodont assemblage, **165**
Conodonts, 165–67, **164–66**
Continental drift, 99–113
  effect on life, 111–13
  fossil evidence, 101–4
  history, 100
  mechanism, 105–11
  sequence of events, 112
Continental seas, 113, 115–17
Continents
  on crustal plates, **110**
  matching across Atlantic, 100
Conularids, 165
Convergent evolution. *See* Evolution
Cooksonia, 225, **225**
Coprolite, **46**
Coral reefs, 189
Corals, 189–93
  rugose, 189, **191–92**
  scleractinian, 193
  tabulate, 189, **190–91**
Cordaites, 228–31, **230**
Core, 58, 108
Correlation, 50, **53**

Corridors, migration, 121
Cosmopolitan, 112, 117
Creationism, 96–97
Creodonts, 276
Cretaceous
    chalk, 155, **155**
    definition, 8, 11–12
    faunal provinces, **120**
    flower, **237**
Crinoids, 124, **124** 177, **178**
Crossopterygian fishes, 200, 212, **215**
Crust, earth's, 58, 99, 105–11, **108**
Cryptostome, bryozoan, 176
Cyanobacteria, **20**, 67–70, **72**
Cycadeoids, 232
Cycads, 232, **233**
Cycle, 93
Cyst, acritarch, 152
Cystoids, 177
Cytosine, 78

Darwin, Charles, 77
Daughter element, 15
Deciduous habit in plants, 245
Density of planets, 58
Dentary bone, 253
Deoxyribonucleic acid. *See* DNA
Dermal denticles of sharks, 200
Dermal layer, 23
Descartes, René, 58
Deserts, 242
Detrital rocks, 35
Detritus feeders, 145, 168, 183–87
Devonian, 11, 12
Diagenesis, 35
Diapsid reptiles, 249, 254–68
Diatomite, 155
Diatoms, 28, 30, 154, **154**
Dichotomous branching in land plants, 208, 218
Dictyonema, **160**
Dimetrodon, 252, **252**
Dinoflagellates, 153
Dinosaurs, 254–68
    duck-billed, **250**
    egg, 262–64, **264**
    length, weight, 264
    nests, 262–64, **263**

    social relationships, 262–64
    warm blooded, 262
Diplodocus, 255, **258**
Diploidy in plants, 210
Dispersal, 121–23
DNA, 68, 78, **79**
Dominance in marine communities, 145–49
Duck-billed platypus, 128
Dunkleostus, **201**
Duration of periods and epochs, 10–11
Dust cloud. *See* Nebular cloud
DuToit, Alexander, 100

Earth
    age, 13–16
    core, 58
    interior, 58
    origin, 57–60
Echidna, 128
Echinoderms, 23, 24, 137, 176–79
Echinoids, 185–87
Ectoderm, 23
Ediacaran fauna, 72–74, **73**, 133, **134**
Electrons, 15
Element, daughter, 15
Elephants, 276, 280, **285**, 287, 292
Endemic life, 117
Endoderm, 23
Enterocoels, 25, **26**
Eocene
    epoch, 10, 12
    plants and climate, 239–40, **239**
Eoliths, 297
Epifauna, 143
    filter feeders, 169–83
Epiflora, 143
Epoch of geologic time, 12
Equator
    in continental drift, 116
    in Mississippian time, **116**
Equisetum, 227
Equus, **286**
Era in geologic time, 9
Eryops, **215**
Eucoelomates, 24
Eukaryotes, 20–21, 68, **70**
Eumetazoa, 24
Euryapsid reptiles, 249

Eurydesma, 102, **102–3**, 118
Eurypterid, **45**
Eusthenopteron, **213**
Evolution, 77–92
    amphibians, 212–16
    Cenozoic mammals, 275–84
    centers of, 285–89
    and classification, 18
    convergent, 89–91, **89**
    divergent, **89**
    early land animals, 212–13
    genetic basis, 78–80
    graptolites, 159–63
    jaws in fishes, 198–99
    land plants, 218–24
    leaves, 219–21
    mechanism, 80–85
    parallel, 89–90, **89**
    patterns of, 85–87
    phyletic, 82–83
    and primitive features, 92
    programmed, 162
    terrestrial vertebrates, 212–16
Extinction, 93
    cephalopods, 195–97
    dinosaurs, 266–68
    mammalian, 291, **291**
    Pleistocene, 291–92

Facies, 52–55, **53–54**
    fossil, 54
Falls of the Ohio, 192
Family, as a category, 19
Faunal provinces, 117–21
Feet of reptiles and mammals, 253
Fermentation, 61
Ferns, 227
Fertilization in angiosperms, 221
Fig Tree fossils, 67, **67**
Filter bridges, 121, **122**
Filter feeders, 145, 168–83
Fishes, 25, 27, 198–202
Fissipeds, 277, 281
Flagellata, 20
Flight
    evolution, 269
Floral provinces, 101, 117
Floras, Cenozoic, 121

Flowering plants, 31, 235–45, 293–94
Flowers of angiosperms, 235–38
Food pyramid, 144
Food web, 144
Footprint, fossil, **46**
Foraminifera, 156–59, **158**
Formaldehyde, 62–64
Formation, definition, 49, **49**
Fossil
    chemical, 70–71
    definition, 1
    facies, 52–55
    in evolution, 77–78
    zone, 51–52, **51**
Fossilization, 33–35
Fox, Sidney, 62
Frond, bryozoan, 176
Fumarole, 60
Fungi. *See* Mycophyta
Fusulinids, **51**

Gamete, 210, 222
Gametophyte, 210, 221
Gangamopteris, 101
Gastropoda, 25, 183, **184**
Gene, 79
"Generalized," as a qualifier, 92
Genetics in evolution, 78–80
Genotype, 80
Genus, as a category, 19
Geographic isolation. *See* Allopatry
Geologic map. *See* Map, geologic
Geologic time scale, 10–11
Gesner, Conrad, **3**
Gill arches, 199
Ginkgo, 120, **120**, 232–34
Glaciation
    Permian, 100, **101**, 104
    Pleistocene, 241–43, **243**
Glossopteris, 101, **102–3**
Gondwana, **102**, 104, 118
Goniatite septum, 195, **195**
Grade, 23
Graptolites, 27, 89, 159–63, **159–62**
Grasses, 243
"Great Chain of Being", 77
Green algae, 69, 210
Guanine, 78

Gunflint chert, 69, **69**
Gymnosperms, 101, 227, 232–35

Half-life, radioactive, 15
Haploidy in plants, 210
Hard parts of fossils, 33, **34**, 139–42
Harding Sandstone, 198
Haven. *See* Refuge
Helicoplacoids, 137, **138**
Herbivores
    dinosaurs, 255–60
    reptiles, 251
Hermatypic corals, 189
Hesperornis, 270
Heteromorph cephalopods, 197
Heterotrophs, 28, 63, 73, 142–45
Hierarchy, 19
Himalayas, 109
Holmia, 118
Holothurian, **44**
Homeomorphs, **89**, 91
Hominidae, 295
Homo, 295, 298
    erectus, 301
    habilis, 301
    sapiens, 295
Homologous structures, 91, **91**
Hoplophoneus, **282**
Horses, 279, **283**, 285–86, **287**
Hydrogen, 59, 61
Hydrogen cyanide, 62
Hydrozoa, 162
Hyolithes, 146
Hyomandibular arch, 215
Hyracotherium, 279, 285

Ichthyornis, 270
Ichthyosaurs, **202**
Ichthyostega, 97, 214
Igneous rocks, 36, 105
Inarticulate, 170
Incus bone, 254
India, 109
Infauna, 143, **143**
Inoceramus, **181**
Insectivores, 275
Insects, **45**, 211–12, 293

Interior of earth, 58
Invertebrates
    skeletons of, 33
    terrestrial, 211–12
Iridium, 93
Iron, banded Precambrian, **68**, 71
Irregular echinoids, 185, **186**
Isolation
    in evolution, 83–85, **84**
    of South America, 289–91
Isotope, 15
    carbon, 70
    oxygen, 130
Isthmus of Panama, 278, 290

Jaws
    in early vertebrates, 199
    reptile and mammal, 253–54
    primate, 295–96
Jefferson, Thomas, 5
Jellyfish, fossil, 73, **73**, 163–65, **164**
JOIDES, 47
Joly, John, 14
Jurassic, definition, 10, 12

Karroo Group, 253
Kelvin, Lord, 13
Kingdom
    animal, 24, 25
    as a category, 19
    plant, 20, 21, 28–31
Kozlowski, Roman, 163
K/T boundary, 267

Lambeosaurus, **250**
Land bridge, 105, 121
Land plants, evolution, 206–11, 219–45
Latimeria, 125, **125**, 200
Laurasia, 104
Lead, element, 15
Leaves, evolution of, 219–20, **220**
Lemurs, 298
Lepidodendron, **226**, 228, **228–29**
Life
    classification, 18
    origin, 61–65

Lignin, 207
Limestone, 35
Lingula, 87, **87**
Liverwort. *See* Bryophyta
Living fossils, 123–29
Lobe-finned fishes, 200
Lophophorata, 24
Lophophore, 24, 172
Loris, 298
Lungfish, 200
Lycopods, **44**, 220–31, **223**, **226**, **229**

Macroevolution. *See* Adaptive radiation
Macrophagous feeders, 144
Magnetism, earth, 105, **106–7**
Malleus bone, 254
Mammals
    archaic Cenozoic faunas, 276–81
    Cenozoic, 275–84
    as chordates, 27
    extinction, 291, **291**
    and flowering plants, 282, 293
    marine, 203
    modern Cenozoic faunas, 281–84
    origin, 253–54
    Pleistocene extinction, 291–92
    teeth, 273–75
    Triassic, 254, 273
Mammal-like reptiles, 249–54
Mammoth, 280, 284, 287
Mammoth molar tooth, **288**
Mantle
    of bivalves, 183
    of earth, 58, **111**
Map, geologic, 49, **49**
Marine communities, structure of, 142–44
Marsupials, 274–75, 289, **290**
Mass number, 15
Mass, solar system, 58
Mastodon molar tooth, **288**
Mastodons, 284, 292
Mediterranean, 111
Megaevolution, 87–89
Meiosis, 210
Merychoidodon, **277**
Mesoderm, 23

Mesohippus, **287**
Mesosaurus, **103–4**, 104
Mesozoic
    definition, 9, 10
    communities, **148**
    plants, **234**
    phytoplankton, 154
    reptiles, 253–68
Mesozoic-Cenozoic extinction, 93
Metamorphic rocks, 36
Metaphytes, 71–72
Metasequoia, 128, **129**, 245
Metazoa, origin, 71–72
Meteorites, 60
Methane, 61
Microphagous feeders, 144
Mid-Atlantic Ridge, 100, **101**, **110**
Migration, 121–23
Milankovitch cycles, 95
Miller, Stanley, 61
Millipede, 211
Miocene
    climate, 240–41
    epoch, 10, 12
Missing links, 77
Mississippian, 11, 12
Mitosis, 210
Modern mammalian faunas, 281–84
Mold, fossil, 42, **43**
Mollusca, 25–26
    minor groups, 183–85, **185**
    provinces, 123
Molt, 169
Monera, 19–21, 68
Monograptus, 160, **162**
Monoplacophorans, 25, 123, **124**, **185**
Monopodial branching, 208
Monotremes, 128, 274
Moropus, **283**
Mosasaurs, 203
Mosses. *See* Bryophyta
Motile, 143
Mudstone, 35
Multituberculates, 274
Mummy, 36–39, **37–40**
Mutation, 79
Mycophyta, 28

Natural selection, 81–83
Nautiloid cephalopods, 193, **194**
Nautilus, pearly, 193
Neanderthal, 300–301
Nebular cloud, 58
Nekton
    definition, 142, **143**
    marine, 163–67
Neopilina, 123, **124**
Neuropteris, **230**
Neutron, 15
Nitrogen in early atmosphere, 61
Noble gases, 58
Norfolk Island pine, 127, 232
North America, separation from Europe, 112
Notoungulata, 278, **279**
Nucleus
    atoms, 15–16
    cells, 68

Ocean, origin, 60–61
Olenellus, **117**, 118, **136**
Olenoides, **137**
Oligocene
    epoch, 10, 12
    plants and climate, 240
Ooze, sea-floor, 108, **108**
Opaline silica, 33
Order, as a category, 19
Ordovician
    communities, 146, **147**
    definition, 9, 11
Organ level of organization, 22
Organelle, cell, 68, **70**
Organic soup. *See* Primordial soup
Organization of life, 18–31
Origin
    earth, 57–60
    land animals, 211–16
    life, 61–65
    species, 77
Ornithischia, 255–62
Ornithopods, 258, **259**
Orthid brachiopods, 173, **173**
Ostracoderms, 198, **198–99**
Otic notch, 215, **248**
Owen, Richard, 100
Oxygen

in atmosphere, 61
    isotopes, 130
Ozone in atmosphere, 63

Paleobiogeography, 99, 115–32
Paleocene epoch, 10, 12
Paleoclimate, 131
Paleoecology, definition, 48
Paleogeographic map, 55, **55**
Paleontology, geology, biology, 304–6
Paleontology, history of, 2–6
Paleozoic
    era, 9, 11
    Late, communities, **147**
    phytoplankton, 151–53
    reptiles, 249–53
Paleozoic-Mesozoic extinction, 93
Panama, Isthmus of, 120–22
Pangaea, 104, 111–13, 116
Pantodonts, 278
Pantotheres, 274
Parallel evolution. *See* Evolution
Parazoa, 24
Parent element, 15
Pearly nautilus, 193
Pedicle, brachiopod, 171, **171**
Pelagic, 142, **143**
Pelecypods, 25
Pelvic girdle of dinosaurs, 255
Pennsylvanian period, 11, 12
Pentamerid brachiopods, 173, **173**
Perennial plants, 244
Periods, geologic time, 9–11
Peripheral isolates, **84**, 85
Perissodactyls, 279, 281–84, **283**
Permian, definition, 11, 12
Permineralization, 41
Pentremites, **127**
Petal, flower, **236**
Petrified National Forest, 128, 232
Phaeophyta, 28, 30
Phenotype, 80
Phloem, 31, 208, 221
Photic zone, 143
Photoassimilator, 64
Photoautotroph, 28, 64
Photosynthesis, 28, 64–65, 67, 207

Phyla, first appearances, 141
Phyletic evolution, 82
Phylogeny, definition, 82
Phylum, category, 19
Physalia, 162
Phytane, 46, 70
Phytoplankton, 150–56
Pines, 232
Pistil, flower, 236
Placental mammals, 274–75
Placoderms, 199–200, **201**
Planet Earth, 57
Plankton
    definition, 142, **143**
    marine phytoplankton, 150–56
    prominence, **151**
Planktonic foraminifera, 156–59
Plants
    coal-swamp, 228–31
    Devonian, 225, **225**
    kingdom, 20–21, 28–31
    land, 218–45
    origin, land, 205–11
    Silurian, 224
Plate tectonics, 99, 109–11
Plates of crust, 99, 109–11
Pleistocene
    epoch, 10, 12
    mammals, 291–92
    plants, 241–43, **243–44**
Plesiosaurs, 203, **203**
Pliocene
    epoch, 10, 12
    plants and climate, 241
Polarity, magnetic, 105–6
Pollen, definition, 222
Pollination in angiosperms, 210–11, 236–38
Polymorphism in graptolites, 159
Populations in evolution, 80–85
Porifera, phylum, 24
Potassium 40, 16
Prairies, 243
Preadaptation, 88
Prebiotic soup. *See* Primordial soup
Precambrian
    age, 11, 12
    fossils, 66–75
    life, 66–75, 133–34

Predators
    gastropod, 183
    marine, 145, 188–204
Preservation
    altered, 36
    of fossils, 35–47, **41**
    origin of hard parts, 139–42
    original, 36, **44, 45**
Primary consumers, 144–45
Primary producers, 150
Primary xylem, 221
Primates, 295, 298–300
    fossil, 300–301
"Primitive," as a qualifier, 92
Primordial soup, 64
Pristane, 46, 70
Proboscideans, 280
Proconsul, **299**
Productoid brachiopods, **173**, 174
Progymnosperms, 227
Prokaryotes, 20–21, 68
Prosimii, 298
Proterozoic, 11, 69
Protists
    kingdom, 21
    phyla, 21
    radiolarians, **21**
Proto-Atlantic, 112
Proton, 15
Provinces, floral and faunal, 117–21, **120**
Pseudocoelomata, 23–24
Pseudopodia, 156
Psilophyton flora, 208
Psilopsids, 218–24
Punctuated equilibria, 83
Pyrite, 12

Quadrupedal, 254
Quadrate bone, 253
Quaternary, 10, 12

Radial symmetry, 23
Radiata, 23
Radioactive age dating, 7
Radioactivity, 14, 59–60
Radiolaria, **21**, 156, **157**
Rancho La Brea, 289
Ray-finned fishes, 200, **201**

Recent, epoch, 10, 12
Recrystallization, 41, **41**
Redlichia province, 118
Reducing atmosphere, 61
Reefs
    ancient, 188–93
    modern, 189
    rudistid, 119
Refuge, 123, 128
Regressions, 94
Regular echinoid, 185, **186**
Relative time scale, 9
Remnant magnetism, 105, **106–7**
Replacement of fossils, **41**, 42
Reproduction in land plants, 210–11, 221–24
Reptiles, 247–70
    as chordates, 27
    marine, 202–3
    origin, 247
Reversals, magnetic, 105, **107**
Rhabdopleura, 163
Rhinoceros, **284**
Rhizome, 208, 218
Rhodophyta, 28, 30
Rhynchocephalians, 128
Rhynia, 208, **209**
Ridges, sea-floor, 105–11, **110–11**
RNA, 79
Roots of land plants, 208–9
Rubidium 87, 16
Rudistids, 119, **119**
Rugose corals, 189, **191–92**
Ruling reptiles, 254

Saber-tooth cat, 281, **282**
Sandstone, 35
Sarcodina, 20–21
Saurischia, 254–62
Sauropods, 255
Scaphopod, **185**
Scavenger, 144
Schizocoels, 25, **26**
Scorpion, 211
Scute, 260
Sea-floor magnetism, 105–9, **107**
Sea-floor spreading, 100, 105–9
Seaways, shallow. *See* Continental seas

Seaweed, 71
Secondary consumer, 144
Secondary xylem, 221
Sedimentary rocks, various types, 35–36
Sediments, burial in, 34
Sediment-water interface, 142
Seed ferns, **223**, 227–28, **230**
Seeds, 222
Segmentation, 24–26
Selection, natural, 81–83
Sepals, flower, **236**
Septa
    cephalopods, 89, 195
    corals, 192
Sequoia, 239
Sessile organisms, 143
Shale, 35
Sharks, 200
Silica, 33, 154, 179
    black chert, 66
    opaline, 30
    in replacement, 42
    and skeletons, 33
Silicoflagellates, 155, **156**
Siltstone, 35
Silurian, 9, 11
Simpson, G. G., 121
Siphon, bivalve, 183
Siphonophores, 162–63
Site of deposition, 34
Skeleton
    origin of preservable, 139–42
    preservable, 33
    and silica, 155
Skull, of reptiles, 247–49, **249**
Sloths, 292
Smith, William, 5
Snider, Antonio, 100
Solar system, 57
Solar wind, 59
Solenhofen beds, 165, 268
Soup, primordial, 64
South America, 112, 289
Specialized features in evolution, 92
Speciation
    allopatric, 82
    in evolutionary patterns, 85

Species
    definition, 19
    number of living, 18
Species problem, 82
Spenopsids, 220, **227**
Spherical symmetry, 23
Spiriferid brachiopods, 172, **173**
Spondes, 24, **25**, 179, **180**
Sporangia, 208, 218, 221–24
Spores, 208, 221
Sporophyll, 236
Sporophyte, 210, 221
Spreading center, 108, **108**, 109, **110–11**
Squamosal bone, 253
Stamen, flower, 236
Stapes bone, 215
Starfish, **43**
Stasis, 83
Stegosaurs, 258, **260**
Stem of land plants, 207–8
Stem reptiles, 249, 251, **251**
Stratification of marine communities,
    170, 178–79
Stratigraphy, 48–52
Stromatolites, 67
Stromatoporoids, 189, **190**
Strophomenid brachiopods, **171**, 174
Structure of marine communities, 142–45
Subduction zone, 109, **110–11**
Subtropical climate, 239
Suess, Edward, 104
Suez Canal, 123
Sun, 58, **59**
Superposition, 8
Support of land plants, 221
Suspension feeder. *See* Filter feeder
Sweepstakes migration route, **122**, 122–23
Symmetry of animals, 23
Sympatry, 83
Synapsid reptiles, 249

Tabulate corals, 189
Tarsiers, 298
Teeth
    carnassial, 276
    mammal, 275–76
    of reptiles and mammals, 253

Temperature, past, 129–31, **131**
Temporal opening, 249
Terebratulid brachiopods, 172, **173**
Terrestrial communities, 205–16
Tertiary, definition, 10, 12
Tethys, 119–20
Thallus, seaweed, 71
Theca, graptolite, 89, 159–60
Thecodonts, 254
Theophrastus, 2
Theropod, 255
Thin section, 174
Thorium 232, 16
Thymine, 78
Tillite, 100, 132
Time
    definition, 1
    geologic, 6–7
    in paleontology, 52–55
    relative, 9
    scale, 10–11
Tissue level of organization, 22
Tommotian, 11, 74, **74**
Tongue ferns, 101, **102**
Trace fossil, 46, **46**, 138
Tracheophyta, 28–29, **30**, 31
Transgressions, 94
Tree ferns, 228
Trench, oceanic, 109, **111**
Trepostome bryozoans, 176
Triassic
    definition, 10, 12
    mammals, 253, 273–74
Triceratops, 261, **261**
Triconodonts, 274
Trigonia, **115**, **126**
Trilobites, 169, **170**
    in communities, 136–39
    provinces, **117**, 117–18
Trophic levels, 144
Turitella, **184**
Turtles, marine, 202
Tyrannosaurus rex, 260

Uintatheres, 278, **278**
Unconformity, **9**
Understory of plant communities, 238

Ungulates, 276, 281–84
Uniformitarianism, 52
Uranium, 15
Ussher, Bishop, 4

Vagrant, 143
Variation, genetic, 80
Vascular plants
  definition, 28, 31
  evolution, 206–11
  structure, 31
Vascular system, 207
Vascular tissue, evolution, 208, 221
Vertebrae of amphibians, 215
Vertebrates, as chordates, 27
Volcanos, 62, 108

Walchia, 232
Water-air interface, 142
Water vapor in the early atmosphere, 61
Wegener, Alfred, 100, 104
Wood in land plants, 221
Wood, petrified, **42**

Xenophanes, 2
Xylem, 31, 208, 221

Zone fossil, **51**, 51–52
Zooplankton, marine, 156–63
Zooxanthellae, 189
Zosterophyllum, 225
Zygote, 210, 222